Preface

Rapid and continuous industrialization and the increasing consumption of fuels by people are major concerns associated with the existing energy crises and need to be resolved by adopting any potential alternative sources. Apart from energy insecurity, harmful impacts of fossil fuel contribute to pollution, and global warming decreases environmental sustainability. To overcome these challenges, several biofuel options are emerging as potential clean forms of energy and are developed from green technologies. Among different existing biofuels, algal biofuels are known as a very efficient mode of biofuels production which can be easily adopted in the near future as a low-cost commercial fuel. Though the third-generation biofuels, algal biofuels, have tremendous advantages, there are parameters that need to be considered for practical applications of these biofuels. Therefore, this book provides information about advancements in algal biofuels production to be implemented as low-cost advanced and sustainable biofuels technology in future. Chapters 1 and 2 present latest trends and advancements of biotechnological contributions and OMICS technology application for technical improvement in algal biofuels production. Chapters 3 and 4 explore algal butanol as a biofuel and gold nanoparticle synthesis from algal biomass for bioenergy applications, and both approaches are being used as new advanced and potential tools for biofuels production. Further, Chaps. 5 and 6 discuss critical challenges in algal biofuels production for technological advancement and engineering-based process parameter strategies to improve algal biofuels production. Additionally, Chapters 7–10 provide new insights on algal biofuels and diversity based on technical advancement ground for practical improvement as well as applications. A detailed understanding of the technological and latest research concepts may help to set goals and help to plan innovative strategies to develop sustainable as well as low-cost algal biofuels production.

Varanasi, India

Neha Srivastava
P. K. Mishra

Clean Energy Production Technologies

Series Editors

Neha Srivastava, Department of Chemical Engineering and Technology, IIT (BHU) Varanasi, Varanasi, Uttar Pradesh, India

P. K. Mishra, Department of Chemical Engineering and Technology, IIT (BHU) Varanasi, Varanasi, Uttar Pradesh, India

The consumption of fossil fuels has been continuously increasing around the globe and simultaneously becoming the primary cause of global warming as well as environmental pollution. Due to limited life span of fossil fuels and limited alternate energy options, energy crises is important concern faced by the world. Amidst these complex environmental and economic scenarios, renewable energy alternates such as biodiesel, hydrogen, wind, solar and bioenergy sources, which can produce energy with zero carbon residue are emerging as excellent clean energy source. For maximizing the efficiency and productivity of clean fuels via green & renewable methods, it's crucial to understand the configuration, sustainability and techno-economic feasibility of these promising energy alternates. The book series presents a comprehensive coverage combining the domains of exploring clean sources of energy and ensuring its production in an economical as well as ecologically feasible fashion. Series involves renowned experts and academicians as volume-editors and authors, from all the regions of the world. Series brings forth latest research, approaches and perspectives on clean energy production from both developed and developing parts of world under one umbrella. It is curated and developed by authoritative institutions and experts to serves global readership on this theme.

Neha Srivastava • P. K. Mishra
Editors

Technological Advancement in Algal Biofuels Production

Springer

Editors
Neha Srivastava
Department of Chemical Engineering and Technology
IIT (BHU) Varanasi
Varanasi, India

P. K. Mishra
Department of Chemical Engineering and Technology
IIT (BHU) Varanasi
Varanasi, India

ISSN 2662-6861 ISSN 2662-687X (electronic)
Clean Energy Production Technologies
ISBN 978-981-19-6805-1 ISBN 978-981-19-6806-8 (eBook)
https://doi.org/10.1007/978-981-19-6806-8

This Springer imprint is published by the registered company Springer Nature Singapore Pte Ltd.
The registered company address is: 152 Beach Road, #21-01/04 Gateway East, Singapore 189721, Singapore

Acknowledgements

The editors are thankful to all the academicians and scientists whose contributions have enriched this volume. We also express our deep sense of gratitude to our parents whose blessings have always prompted us to pursue academic activities deeply. It is quite possible that in a work of this nature, some mistakes might have crept into the text inadvertently and for these we owe undiluted responsibility. We are grateful to all authors for their contribution to this book. We are also thankful to Springer Nature for giving this opportunity to editors and the Department of Chemical Engineering & Technology, IIT (BHU) Varanasi, U.P., India, for all technical support. We thank them from the core of our heart.

Contents

About the Editors

Neha Srivastava has received her PhD in Biotechnology from the Department of Molecular and Cellular Engineering, SHIATS, India, in the area of bioenergy. She works as Research Scientist in the Department of Chemical Engineering and Technology, IIT (BHU) Varanasi, India. She has published more than 37 research articles in peer-reviewed journals of SCI impact factor. She has filed three patents with one technology transfer. She has 14 published books and two book series with renowned international publishers. She is working on bioprocess technology and biofuels production (microbial screening and enzymes; production and enhancement, biohydrogen production from waste biomass, bioethanol production).

P. K. Mishra is currently Professor and Head of the Department of Chemical Engineering & Technology, Indian Institute of Technology (BHU), Varanasi, India. He has authored/coauthored over 60 technical papers published in reputed national/international journals and supervised more than 20 doctoral students. He has received several awards and honors and has five patents with one Technology Transfer. He is Fellow of Institution of Engineers India. He has received several awards and honors at national/international levels.

Chapter 1
Biotechnological Approaches to Enhance Algae Biofuel Production

Umar Shahbaz, Sidra Zubair, Amna Younas, Xiao bin Yu, Nazra Fatima, Shahzal Babar, Samra Basharat, Asma Bibi, and Muhammad Iftikhar Hussain

Abstract Algae are aquatic species that may reproduce quickly and have over 3000 different breeds, making them more abrasive than terrestrial plants. They may be able to convert CO_2 from the air into oxygen and remove it from the breaking cells of algae plants, which is how they produce wonderful oil. A viable source of feedstock for biofuels, oleaginous microalgae have a number of advantages over terrestrial plants. Due to the lack of robust algal strains with increased lipid content and biomass and methodologies for economically viable oil extraction, the microalgal fuel business is still in its infancy. By carefully targeting important metabolic nodes, microalgal metabolic engineering demonstrates the huge potential to improve lipid

---

U. Shahbaz (✉)
Dipartimento di Genomica e Biologia delle Piante da Frutto, Fondazione Edmund Mach, San Michele all'Adige, TN, Italy

C3A-center agriculture food and environment, University the Trento, Trento, Italy
e-mail: Umar.shahbaz@fmach.it; https://www.fmach.it/

S. Zubair
Department of Bioinformatics and Biotechnology, Government College University, Faislabad, Pakistan

A. Younas
Center of Agribiotechnology and Biochemistry, University of Agriculture, Faislabad, Pakistan

X. b. Yu · S. Basharat · M. I. Hussain
Key Laboratory of Carbohydrate Chemistry and Biotechnology, Ministry of Education, Jiangnan University, Wuxi, China

N. Fatima
Institute of Atherosclerotic Disease, Xi'an Jiaotong University, Xi'an, Shaanxi Province, China

S. Babar
School of food science, Ministry of Education, Jiangnan University, Wuxi, China

A. Bibi
The Key Laboratory of Microbiology and Parasitology, Anhui School of Basic Medical Sciences and Laboratory Diagnostics, The First Hospital of Anhui Medical University, Hefei, China

N. Srivastava, P. K. Mishra (eds.), *Technological Advancement in Algal Biofuels Production*, Clean Energy Production Technologies,
https://doi.org/10.1007/978-981-19-6806-8_1

accumulation without compromising cell growth. The genetic enhancement of microalgae without impairing cellular biomass remains an underutilized option for large-scale biofuel production, despite recent advances in synthetic biology. To improve microalgae as a biofuel platform for the production of biohydrogen, alcohols generated from starch, substitutes for diesel fuel, and/or alkanes, we consequently present a thorough overview of numerous biotechnological techniques in this chapter.

Keywords Biofuel · Genetic engineering · Microalgae · Nano-additive · Bio-bricks · Biorefinery

Abbreviation

ABE	Acetone-butanol-ethanol
DHA	Docosahexaenoic acid
EPA	Eicosa pentaenoic acid
EST	Expressed sequence tags
FAME	Fatty acid methyl ester
GHG	Greenhouse gases
HBV	Hepatitis B virus
LED	Light-emitting diodes
LHC	Light antenna complexes
PC	Phosphatidylcholine viation
ROS	Reactive oxygen species
SCP	Single-cell proteins
TAG	Triacylglycerols

1.1 Introduction

Algae are the aquatic species that have the fastest capability to reproduce with over 3000 different breeds, therefore more adverse than land plants. They can have the capacity to extract CO_2 from the air and turned into oxygen and have the ability to yield great oil, i.e., the breaking cells of algal plants are extracted (Adeniyi et al. 2018). Renewable fuels are bioalgal fuels derivative of algal feedstock that can be related with its capability to abundantly photosynthesize. The capability to convert all feedstock's energy into various varieties of valuable biofuels is the main advantage (Demirbas 2010). Their other application adds the creation of energy cogeneration (electricity and heat) afterward oil extraction, deduction of CO_2 gases through industrial chimney (algal bio-fixation), bio-fertilizer, animal feeds, and further food products. The important and imaginable effects that algae exist and can survive in thrilling heat, scarcity, salinity, and radiation. Therefore, ecological conditions will affect the country's crop growing ways. Freshwater and marine algae, for example, Chlorophyceae (green algae), Cyanophyceae (blue-green algae), and Pyrrophyceae (fire algae), could be cultured as expected happening in the UK (Adeniyi et al. 2018). Phaeophyceae a brown alga with synthetic cultivation techniques of photobioreactors could be genetically modified. Aimed at the creation of fatty acid methyl ester (FAME), Chlorophyceae, Cyanophyceae, and Pyrrophyceae were recommended. Because of its excessive sugar content, for ethanol production, Phaeophyceae tend to be the greatest suitable algal feedstock (Islam et al. 2015). Owing to growing response for transportation fuel, worldwide, algae have appeared as an appropriate candidate due to their sustainable and renewable characters together with financial reliability to compete by international request aimed to conveyance fuels (García-Olivares et al. 2018). The algal biofuel conversion processes, like fermentation, transesterification, and hydro treatment, are economically expensive and more complex as compared to fossil fuels. The possible earth of optimum established are sustainability for the feedstock, or the enhanced possibility of products and innovative applications because of ranges (Culaba et al. 2020). Due to algal positive characters, it tends to be construed that this feedstock is the world's one of the furthermost valuable renewable and sustainable fuel resources that could show an important role in controlling environmental contamination (Bharathiraja et al. 2018). The major threat is the emission of CO_2 from varied fossil fuels through atmosphere. Worldwide economy in the past is out of petroleum funds, whereas it became rich and switch the unfamiliar trade marketplace in light by petroleum reserves (Mohsin et al. 2019).

To all these abovementioned challenges, the only viable solution for both the worldwide economy and greenhouse gas (GHG) emission of fuel formed from various materials (plant). However, renewable energy has different bases, for example, geothermal, breeze, and solar which probably won't be economically practicable as biofuels; still these GHG play a meaningful part of resolving the concerns of global warming. Biofuels are yields of various sustainable and biodegradable feedstock that can be changed of producing opportunities for advances in agriculture as a result of direct contribution to agriculture plants (Gielen et al. 2019).

1.2 Microalgal Species for Producing Biofuels

Recent biofuel manufacturing exploiting microalga biomass is not cost-effective. To make microalga yields and by-products economically feasible, the growing research is keen to observe new microalgal candidates (Koyande et al. 2019). Various advantages of microalgal species must gain methodical concept as to present a profitable source of extreme worth of products (chemicals) like antioxidant, polysaccharides, β-carotene, natural dyes, bioactive, docosahexaenoic acid, and efficient pigment, astaxanthin, algal extracts, antioxidants, and eicosapentaenoic acid. Those classes show a fundamental function of nutraceuticals, human food, cosmeceuticals, fodder, bioremediation, and pharmaceutical aquaculture (Koyande et al. 2019).

Table 1.1 Some major microalgal species, products, and applications (Mobin and Alam 2017)

Species/groups	Products	Areas of application
Arthrospira (Spirulina) platensis	Phycocyanin, biomass	Health food, cosmetics
Arthrospira (Spirulina)	Protein, vitamin B_{12}	Antioxidant capsule, immune system
Aphanizomenon flos-aquae	Protein, essential fatty acids, β-carotene	Health food, food supplement
Chlorella sp.	Biomass, carbohydrate extract	Animal nutrition, health drinks, food supplements
Chlorella vulgaris	Biomass, carbohydrate extraction	Health food, feed, food supplement

Economically viable is to produce algal biomass, production of extraordinary worth of byproduct pull out are utmost. Normally microalgal organisms are naturally single-cell photosynthetic autotrophic microscopic located at marine and garden-fresh environments. Production of numerous multipart complexes like carbohydrates, protein, and lipids, consuming basic elements situated in their surrounding environment (Trivedi et al. 2015) (Table 1.1). Microalgae are "photosynthetic plants" like microorganisms deprived from land pland possess by particular cell and organ types. For energy production, which take carbon from surrounding airborne. By consuming organic carbon, some microalgae yield energy. Microalgae have more than 300,000 species, and around 30,000 species stay renowned. The living locals are complex or may adjust rapidly in severe environmental circumstances (e.g., heat, UV irradiation, variable salinity, and nutrients) (Ramaraj et al. 2015). However, they can yield an increased variability of interesting subordinate metabolites (biological energetic) through unique structures and biological actions which is usually not found in other candidates. The microalgae also yield certain valuable bio goods comprising polysaccharides, carotenoids (specifically β-carotene), antioxidants, docosahexaenoic acid (DHA), astaxanthin, natural dye, eicosapentaenoic acid (EPA), functional or bioactive stains, and algal extracts (Sathasivam et al. 2019) (Fig. 1.1). On a commercial level, microalgal cultivation has begun five decades ago. The global marketplace worth of the microalgae are considered around US$ 2.5 billion created via health food sector. US$ 1.5 billion are yielding DHA or then US$ 700 million by way of aquaculture. Yearly microalgal production is almost 7.5 million tons.

1.3 Promising Microalgal Species and High-Value Applications

Microalgae are divided into four main collections: (1) cyanobacteria (blue and green algae), (2) chlorophytes (green algae), (3) rhodophytes (red algae), and then (4) chromophytes (wholly remaining algae). Every set consists of more than

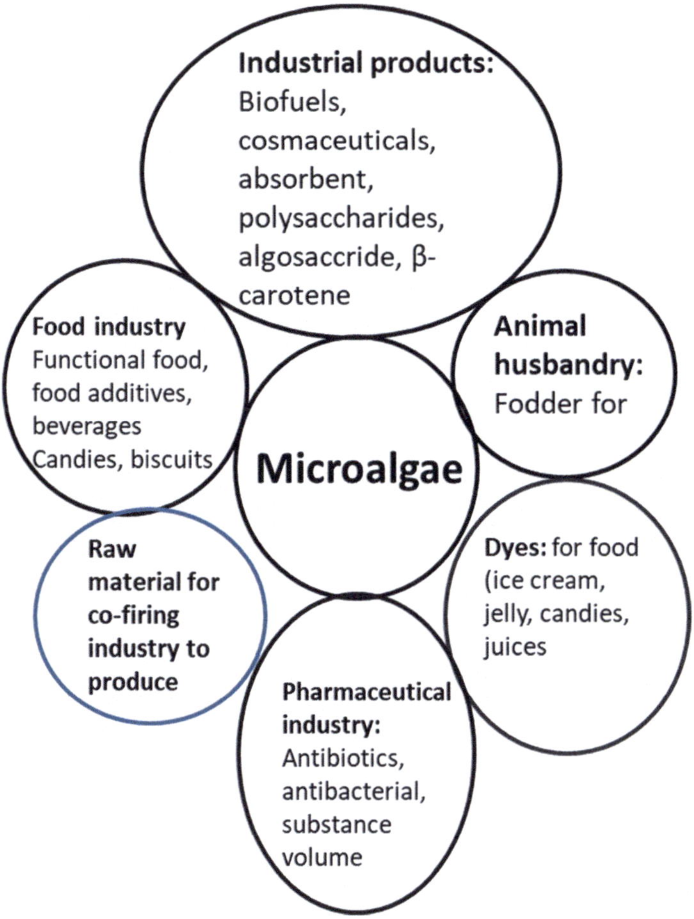

Fig. 1.1 Application of microalgae from numerous areas

hundreds of algae types, and altogether species have thousands or more of strains. For possible beneficial use, only a small variety has been studied. Bacillariophyceae (as well as diatoms), Cyanophyceae (blue-green algae), Chlorophyceae (green algae), and Chrysophyceae (including golden algae) stand generally used microalgae. Some are main microalgal types, their goods, or their application for biotechnological (Sathasivam et al. 2019).

Over the last two decades, four major microalgae are focused on biotechnological applications: (1) *Haematococcus pluvialis*, (2) *Chlorella vulgaris*, (3) *Dunaliella salina*, and (4) *Spirulina* (*Arthrospira*). The characteristics, composition, and production mechanism and, respectively, microalgal type is explained now at the following section (de Morais et al. 2015).

1.3.1 Spirulina

Spirulina (*Arthrospira*) is multicellular, blue and green algae, symbiotic, and filamentous that use nitrogen from the atmosphere. *Spirulina* has two unique shapes; disk-like or spiral rod. Phycocyanin (blue color) is the main photosynthetic pigment in *Spirulina*. The *Spirulina* microalgae likewise hold carotenoids and chlorophyll a. The presence of Phycoerythrin makes the color of microalgae pink or red. Autotrophic contain *Spirulina* then they replicate via binary fission (Mobin and Alam 2017). In 2014 the global manufacture of *Spirulina* sp. was 86,000 tons, as indicated by a report by FAO. It is some of cells considered algae used for extensive outside culture. The high pH (9–11) is best to grow with an extraordinary bicarbonate concentration. To culture *Spirulina*, paddle wheel-driven raceway ponds are used (Mobin and Alam 2017). In the ponds, water varies from 250 to 500 mm because of the microalgae density and seasons. Depth of water pond is also dependent upon pond size, optimal light absorption, and flow velocity by the microalgal culture. Temperature plays a vital role in *Spirulina* productivity (Mobin and Alam 2017). It grows well in between 35 and 37 °C. *Spirulina* filamentous mowing is informal. The starting focus to reached using an inclined gravity screen going on vibrating scree-filter (Zhang 2015). Its dewatering remains attempted through vacuum belt filter building a solid pate about 15%; further detail on *Spirulina* is that it is rich in protein, minerals, essential amino acids, vitamins, or required polyunsaturated fatty acid and pigments comprising Zeaxanthin, Myxoxanthphyl, and Phycocyanin. It carries (46–71%) proteins, (8–16%) carbohydrates, and (4–9%) lipids (dry weight). *Spirulina* vital amino acid contains leucine, valine, and also isoleucine. It contains comparatively extraordinary meditation of vitamin K, provitamin A, vitamin B_{12}, and also β-carotene. Fatty acid of *Spirulina* also includes Llinolenic and γ-Linolenic acid and $\Omega - 3$ with $\Omega - 6$ polyunsaturated fatty acids. DHA is the natural source for *Spirulina platensis* demonstrating up to 9.1% of the whole fatty acid (Hemantkumar and Rahimbhai 2019). It furthermore encloses an antioxidant-rich combination more than carotenoids. *Spirulina* mineral content hinge on water, grown and this one content of calcium, iron, and magnesium delivers great nutritional-value (Bensehaila et al. 2015). According to studies, *Spirulina* powder covers provitamin A (2.330 × 03 IU/kg), vitamin E (100 mg, 100/g), β-carotene (140 mg, 100/g), riboflavin B_2 (4.0 mg, 100/g), niacin B_3 (14.0 mg, 100/g), thiamine B_1 (3.5 mg, 100/g), vitamin K (2.2 mg, 100/g), vitamin B_6 (0.8 mg, 100/g), inositol (6 mg/g), vitamin B_{12} (0.32 mg, 100/g), biotin (0.005 mg, 100/g), folic acid (0.0 mg, 100/g), and pantothenic acid (0.1 mg, 100/g) (Ragaza et al. 2020).

1.3.2 Chlorella

Chlorella is spherical-shaped, single-cell (2–10 μm) in diameter, and phototrophic green microalgae without flagella. In its chloroplast, both chlorophyll a and chlorophyll b (green photosynthetic pigment) are present. It multiplies speedily and needs sunlight, water, CO_2, and a slight volume of minerals (Simosa 2016). *Chlorella* was grown in photobioreactor large round tanks and paddle wheel mixed open and circular-open ponds. Though making of microalgae used for aquaculture is mostly at smaller scales, or at most of situations, it is produced within (20–40 L) carboys and huge plastic bags (~1000 L in volume) (Kunjapur and Eldridge 2010). *Chlorella* is harvested by auto-flocculation or centrifugation. Following the collection of biomass sprayer exists drum dried and dust is wholesaled straightforwardly for used to proceed tablets. Chemical configuration of *Chlorella* indicates which comprises (12–28%) carbohydrate, (111–58%) protein, and (2–46%) lipids via dry weight. The usual configuration of *Chlorella* residue may be present at Borowitzka. It also encloses several vitamins like that provitamin A (55,500 IU/kg), vitamin E (1 mg 100/g), inositol (165.0 mg 100/g), thiamin B_1 (1.5 mg 100/g), biotin (191.6 mg 100/g), riboflavin B_2 (4.8 mg 100/g), vitamin B_6 (1.7 mg 100/g), vitamin B_{12} (125.9 mg 100/g), folic acid (26.9 mg 100/g), β-carotene (180 mg 100/g), and pantothenic acid (1.3 mg 100/g). *Chlorella*'s worldwide profitable market worth esteem is above a billion US dollars (Hemantkumar and Rahimbhai 2019).

1.3.3 Dunaliella

Dunaliella is edible, nutrient-rich, single-celled, flagellated, and extremophile green microalgae. *Dunaliella* is brought into being at countless fresh waters or marine water habitats. *Dunaliella salina* (*D. salina*) has gained ample care from scientists by the intense stages of antioxidant actions (Dolganyuk et al. 2020). It is the best source of β-carotenoids, which contains a high extent of β-carotene (above 14% of dry biomass) contrast with further well-known cradles. *Dunaliella* may be grown over 30% NaCl saturation though the ideal development salinity is 22% NaCl saturation. There is only a rare reported challenging type and predators for *Dunaliella*; thus, modest open pond culture is sufficient for it. *Dunaliella salina* (*D. salina*) is full-fledged at each shallow, extensive areas (5–200 ha), in light paddle wheel race way ponds and unstirred ponds which is more than 1000 m^2 in area (Dolganyuk et al. 2020). For *D. salina* using technique is the semi-continuous culture. It has been discovered that at lower salinity level is the best to grow *Dunaliella* closely (22% (w/v) NaCl) saturation however β-carotene content (nearly 33% NaCl). *Dunaliella* shows promising growth at lower salinity conditions, but there are risks of contamination by brine shrimp or even by *Artemia* or protozoa. There are several manufacturing tactics researched by Ben-Amotz for *Dunaliella*. The challenging culture was using nitrogen inadequacy, extreme salt concentration, or penetrating solar radiation which boost fruitful biomass of *Dunaliella* and then β-carotene fabrication (Pourkarimi et al. 2020). By using the centrifugation technique, harvesting of *Dunaliella* has been completed, further flocculating and employing centrifugation methods manipulating the cell membrane at hydrophobic nature. Certain complications coupled to the reaping of *D. salina* are:

1. Cells are of jaggedly to the comparable density like as culture medium.
2. Lower cell concentrations in the culture (typically less than 0.1 g dry weight per liter).
3. Cells are very faint because of no cell wall.
4. Instability of cell (uncertainty of cells are broken in the period of harvesting, so the rapidly dissolving of β-carotene) (Hosseini Tafreshi and Shariati 2009).

Immediately after being harvested, the biomass might be sprayed through drum dried, and the β-carotene may be detached straightly by means of hot oil or any additional solvents. The biochemical configuration of *Dunaliella* is protein (49–57%), lipids (6–8%), and carbohydrate (4–32%) of its total dry weight (Wu and Chang 2019).

1.3.4 Haematococcus pluvialis

Haematococcus pluvialis (*H. pluvialis*) present as unicellular bi-flagellate freshwater Chlorophyta microalga dispersed around the world. Under different stress conditions, the accumulation of a wide variety of strong antioxidant astaxanthin is renowned as *Haematococcus pluvialis* (more or less 2–3% on dry weight). A commercially producing organism is astaxanthin of *H. pluvialis* (Shah et al. 2016). Heterotrophic, photoautotrophic, or mixotrophic centered on progress state, internal or secure photobioreactors, or sweeping channel ponds are mostly recycled for *H. pluvialis* refinement. Standard photobioreactors castoff for refinement cover simmer column, tubular, plus airlift photobioreactors (Narala et al. 2016). A twice step agriculture approach tactics are generally adapted for marketable nearer efficiency. The initial step includes the mounting of algal green biomass at motile stage in a secured system (photobioreactors) and exposed skill (like that pond raceways), and the following step comprises manufacturing of astaxanthin covering aplanospores enlargement through the insufficiency nitrite or phosphate and then improved light intensity and temperature (Kim 2015). Then accretion of astaxanthin is squeezed via natural features, for example, salt concentration, nutritional stresses, light, and pH. Cellular structure of *H. pluvialis* differs remarkably among red-green phases of agriculture. Mowing is attempted by using a grouping of flaccid resolving besides following flotation and centrifugation. Mowing of biomass has got dried out by shower, chilling, sun-oriented drying, lyophilization, or cryodesiccation. To dry high-esteem *H. pluvialis* items, splash drying is the most fitting strategy (Shah et al. 2016). A widespread variety of procedures has been industrialized toward interrupting the *H. pluvialis* cell or recovering the intracellular metabolites. Then mechanical procedures (bead milling expeller pressing) are the most appropriate cell disruption technique. *H. pluvialis* cells are cuddled underneath extraordinary pressure toward breaking the dense pollen ultimately. Later in the disruption of cell walls, the biomass of *H. pluvialis* is essential to be treated within a couple of times to evade damage. However, super critical carbon dioxide (SC-CO_2) and solvent drawing out methods are well-known owing to their proficiency or compatibility. *H. pluvialis* is able to mass more or less 5% dried weight of astaxanthin and is well-thought-out as the good organic cause of the great worth of carotenoid colorant. Due to the huge cost of manufacture, synthetic astaxanthin leads the presently profitable market (yearly production of 130 tons creating a value of US$ 200 million) (Sanzo et al. 2018). Some additional microalgal species areas are as under studying for their appropriate for food or nutritional supplements bases. In the 1970s *Scenedesmus* was not found favorable for human or animal intake due to its huge cost (Ghani et al. 2020). In the 1990s, the dinoflagellate *Crypthecodinium cohnii* has been recognized as the utmost capable entrant for DHA creation. In heterotrophic culture, the cell bulks of *Crypthecodinium cohnii* have been attained up to around 40 g/L with a lipid percentage of 15–30%. The DHA value for this lipid is described to 25–30% (Santos-Sánchez et al. 2016).

1.4 Microalgae

Microalgae have gained considerable attraction worldwide currently, by their wide-spread usage potency in the renewable energy, biopharmaceutical industries, and nutraceutical. Microalgae are renewable, viable, cost-effective homes of biofuels, bioactive medicinal products, and food ingredients. Several microalgae kinds have been examined for their likely function as significant yields with extraordinary biological and pharmacological potentials (Subhadra 2010). As per a biofuel, microalgae are an entire alternative for melted remnant fuels with charge, renew-ability, and environmental interests. It has a considerable capability to change atmospheric CO_2 into valuable products such as carbohydrates, lipids, and additional bioactive metabolites (Hussain et al. 2021). For biopharmaceuticals and bioenergy, the microalga is a viable source, some challenges and problems are remaining, but they must be overcome to advance the equipment from experimental-phase to built-up level. For bioethanol production, pretreating biomass, dewatering algal culture for biomass production is the most essential and challenging feature for improving the growth rate and product synthesis (Tonnaer 2017). For biodiesel production, most microalgae species are appropriate, because of their high lipid contents of 50–70% and may reach 80% such as incase of microalgae *B. braunii*, which accumulate up to 80% of oil in its biomass. Polyunsaturated fatty acids are vital in tissue integrity and have beneficial health effects. Omega-3 and omega-6 fatty acids, in particular, are crucially essential for humans but the human body cannot produce them by itself. Many microalgae species (e.g., *Arthrospira platensis*, *Porphyridium cruentum*, and *Odontella*, *I. galbana*) have been explored for their capability to synthesize these fatty acids. *Pavlova lutheri* produces polyunsaturated fatty acids in large quantities. Microalgae are a rich source of various vitamins; *P. cruentum* produces a high quantity of vitamin C and E, as well as β-carotene (vitamin A); *Haslea ostrearia* is a rich source of vitamin E (tocopherol).

Microalgae have a broad spectrum of industrial applications. Mostly wastewater purification and biofuel production have been reported. Some industrial applications are high-value food, nutraceuticals, pharmaceutical products, health food for humans, fodder additives, polysaccharides, cosmeceuticals, antioxidants, dyes, food for aquaculture, and bioplastic production. Microalgal biomass is also used as a raw material for the co-firing industry to generate power because it has a high heating value other than biomass feedstock. *Spirulina* is used for cholesterol reduction and immune system enhancement. Sulfated polysaccharides of *Spirulina* are broadly used as an antiviral agent. Its tablets are being marketed since 1975 in Japan. *Chlorella* composes of vital substances: β-1,3-glucan. This compound is a free radical scavenger that enhances immune response, also responsible for lowering lipid in the blood. It is also effective in stomach ulcers, wound healing, antitumor activity, and removing toxins from the body. *Spirulina* has a preventive effect in hypercholesterolemia. *Chlorella* is effective on low blood sugar level, enhances hemoglobin concentration, and acts as hepato-protective agent. Astaxanthin from *Haematococcus* is utilized in aquaculture for coloring fish muscles (salmon fish). The antioxidant characteristics of astaxanthin prevent the production of inflammatory compounds and also prevent protein degradation, oxidative stress, and macular degeneration. *Spirulina*, *Dunaliella*, and *Chlorella* all are widely used in the food industry. Their biomass is also utilized for forming a variety of health products including tablets and capsules. Three microalgae are used for bread, noodles, candies, bean curd, and other common food with high health values. *Chlorella vulgaris* stimulates the synthesis of skin collagen; they also help in the regeneration of fibers and make the skin surface free from wrinkles.

Extract of *Spirulina* minimizes the age effects. Purified phycobiliproteins, a product of *Spirulina*, are used in cosmetics, food (colorant), antioxidant treatment, anti-inflammatory, photodynamics of various cancers, leukemia and tumor treatment, and florescent marker production.

1.5 Macroalgae

Algae are distributed in an extreme and diverse environment. Due to their high content in compounds, they are valuable with different biological activities, including both complex organic compounds, both primary and secondary metabolites. It is significant to observe that the majority of these substances include phytopigments (carotenoids and xanthophylls), polyunsaturated fatty acids, phenolic compounds, tannins, peptides, lipids, vitamins, enzymes, terpenoids, and others. Thus algae are viable and economical biomass sources of valuable compounds with potential applications in the pharmaceutical, nutraceutical, chemical, food, and cosmetic industries due to their biological active and regenerative characteristics (Fig. 1.2). Microalgae gained more and more value due to their usage in health aspects. They are promoting properties that can reduce the risks of many chronic diseases and help to extend the life span. They are also used for wastewater treatment or as natural

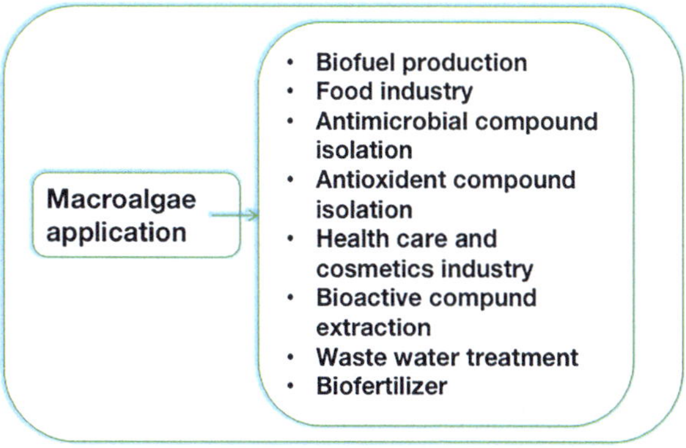

Fig. 1.2 The main application of macroalgae

fertilizer in agricultural areas, therefore improving the quality of products and reducing the need for chemical fertilizers. As a source, the potency of macroalgae of renewable energy is also of considerable interest. These aquatic organisms mitigate carbon dioxide emissions and nowadays are being used as feedstock to form "clean" or so-called third-generation biofuels.

1.6 Chemical Composition

Different marine macroalgae (seaweed) as a source of bioactive and essential compounds had the advantages to use an under-utilized sustainable natural reserve. It had been confirmed that secondary metabolites are made up of biomass (Biris-Dorhoi et al. 2020). The chemical composition has changed due to natural conditions (temperature, candlepower, sea water salinity, and growth habitat) and genetic modifications among species. The protein substance can go from 7% to 31% of total dry weight and a lipid substance varying from 2% to 13% of total dry weight. A big quantity of carbohydrates can range from 32% to 60% of dry weight (Aratboni et al. 2019). Macroalgae are considered a true natural source of vitamin A and E (tocopherol). The abundance of vitamin B_{12} advances the macroalgae created products concerning the dietary supplement for a vegetarian lifestyle taking a risk of deficiency of B_{12} (Biris-Dorhoi et al. 2020). The numerous microelements found in seaweeds are mostly supported by their mineral content; sodium, mechanism, and calcium, which present above 97%, you look after the general mineral content. Additional microelements like iron, zinc, manganese, and copper are discovered in minor quantities (ranging from 0.00 to 0.094) of seaweeds of dry weight (Arguelles 2020). Phenolic compound existence in macroalgae has taken great attraction due to their particular bio-activities and health supporting benefits incorporating antiallergic, antiproliferative, antioxidant, antimicrobial, antidiabetic, and neuro-protective properties (Santos et al. 2019). Secondary metabolite's presence in macroalgae is encouraging the regular defense system against several injuries, diseases, and environmental aggression. For an extended time, macroalgae are encouraged for their prospective function in protective cancer rate, tumor development, and even health upturn after chemotherapy or radiotherapy treatments (Biris-Dorhoi et al. 2020). Iodine can also be used as an anticancer response, due to its capability to start apoptosis in cancer cells. Similar characteristics are often ascribed to the omega-3 fatty acid like stearidonic acid and hexadecatetraenoic acid discovered in eatable marine algae-like *Ulva* and *Undaria* along with 40% of total fatty acids (Biris-Dorhoi et al. 2020; Fitton et al. 2008).

1.7 Laminarian

Alginate, fucoidan, and lots of other seaweed polysaccharides are proven to possess antitumor activities. A great quantity of polysaccharides (~65% polysaccharides in total dry weight) are often found in many seaweeds like *Ascophyllum*, *Ulva*, *Palmaria*, and *Porphyra* (Ghosh et al. 2021). Alginate filling the intestinal system additionally aids in boosting immunity and intestinal health, reducing the risk of cancer (Eliaz et al. 2007). Laminarian and fucoidan induce apoptosis to prevent cancer, but also unidentified seaweed polysaccharides are able to exhibit straight and unintended antitumor influence. *Sargassum latifolium* reserved cytochrome P450 IA and glutathione S-transferases and prevent cell viability and stimulating apoptosis. Another study about the antiviral potential of algal against foodborne viruses is getting interested about current years; recently the accessible data are still limited (Zorofchian Moghadamtousi et al. 2014). The most compounds from algae that are proved to possess antiviral potency are sulfated polysaccharides, including sesquiterpene hydroquinone, carrageenan, sulphoglycolipids, and fucoidan (Ahmadi et al. 2015). Polysaccharides derived from marine and their lower relative molecular mass oligosaccharides derivatives are displayed to have a kind of antiviral activities and also exercise antimicrobial and antioxidant influence (Zhu et al. 2021). The algal polysaccharides can overturn the DNA replication and leave the host cell colonization by the virus. For instance, the antiviral capability of polysaccharides from brown seaweed exposed substantial inhibiting action against the hepatitis B virus (HBV) and DNA polymerase, accordingly influencing its replication. The antiviral action of those polysaccharides is exerted through suppression of virus adhesion to the host cells (*U. pinnatifida*, bladder wrack, *Cystoseira indica*). Fucoidan inhibits syncytia formation with remarkable selectivity (IS50 > 2000) during the early phases of virus infectivity (0–60 min post–infection) (Gheda et al. 2016).

1.8 Algal Biofuel Production

There are an excellent revolution and challenges in biofuel production to exchange fossils fuels. The biological yield of biofuels is apprehending the world's market thanks to the restrictions of petroleum-based fuel. Researchers are more interested in the exploration of latest technologies for biofuel yield. Biomass for biofuel production is one among the alternatives thanks to its sustainability and renewability, low CO_2, fewer greenhouse gases, etc. The most issue regarding biomass usage is the efficiency to exchange complete fossil fuels (Khan et al. 2018). For biomass generation, land use causes ethical issues that are consequences in enhancing food prices. The most focus of scientists is to believe new technologies to beat energy needs, decrease the prices of fuel, and be ready to solve environmental issues too (Fribourg 2008). Photovoltaic technology systems are going to be directly employed by genetically engineered photosynthetic microorganisms of completely synthetic factories. Biofuels are considered the foremost promising within the short term as their market maturity is above other options. Global climate change concerns the buildup of greenhouse gases causing linked regarding the utilization of fossil fuels because it is the major energy source (Hannon et al. 2010). To beat the problems of global climate change, the one solution is to believe the potential of microorganisms to use renewable substrates for biofuel production, thanks to tremendous progress in several industries which are resulting in polluting the environment. The microbial technology gives the simplest solution with an environmentally friendly approach, by identifying the microbial strains, causative agents for the matter, and implementing the useful one for environmental bioremediation (Srivastava 2019). Microalgae have long been known as potential good sources for biofuel production due to their comparatively high oil level and speedy biomass production. Microalgae raise very timely as competed to continental crops; algal mass can grow on non-arable lands using non-potable saline or wastewater (Srivastava et al. 2020).

The microbial-related bioremediation process has benefited society by exploiting the metabolic abilities of microorganisms. Due to the depletion of worldwide petroleum and its value increases, biodiesel is becoming one among the simplest promising worldwide energy markets within the coming future. Growing pattern in the biofuel and high-rate biochemical yield decline the necessity for nonrenewable and artificial sources, to decrease the harmful environmental effects and advance biorefinery (Koenraad et al. 2015). It's likely that the biochemical market will increase from 2% to 20% interest by the year 2025 due to the main growing demands for bio-based products which have directed the R & D efforts to specialize in commercially oriented research styles (Cordova et al. 2020). At the present, biofuels have gotten significant potential among the financial and environmental disasters of fossil fuels. Fuel reproduction molecules as isoprenoids, fatty acid-derived molecules, hydrocarbons, and improved alcohols have a plus over their conventional complement due to engine compatibility, compatibility with present storage, higher density, and transport structure. Furthermore, biologically functional high-value substances containing isoprenoids and fatty acid-derived biomolecules are of environmental, biotechnological, pharmacological, agricultural, and also industrial significance (Adegboye et al. 2021). For the assembly of required titers for subsequent generation, there's a requirement for improvements in gene-splicing techniques that have assisted scientists to develop advanced robust strains (Bharadwaj et al. 2020). Biofuels also require starchy/sugary substrate or lipid-rich biomass for their successive conversion into advanced alcohols: butanol, isoprenoids, bioethanol, and fatty acid-derived substances (Mehmood et al. 2021). First-generation feedstock include food crops, for instance, sugarcane, barley, corn, beetroot, wheat, etc.; however it made the conversation of food deficiency (Hirani et al. 2018).

1.9 The Second-Generation Feedstock

Second generation focused intensely towards the raw and waste materials to account for the issues of first-generation feedstock, yet they're fundamentally expensive and laborious pretreatments for their decomposition into simpler components for their simpler succeeding adaptations to those products (Anto et al. 2020). Microalgae and cyanobacteria are photosynthetic microbes considered because third generation feedstock supply both higher lipid-based pieces of stuff and alcohols. Due to their biochemical composition, microalgae got a standing over lignocellulosic biomass, which contains lipids, protein contents, and carbohydrates (Abomohra and Elshobary 2019). Carbohydrate ingredients are viable to yield bioethanol and alcohols, while the lipid section is employed to supply biodiesels, isoprenoids, and extra lipid-based compounds. The restraints of cultivation, harvesting, and downstream processing are thanks to the commercial execution of third-generation feedstock (Laurens et al. 2017b). The deficiencies of the previous generations were alleged to be reported by GM microbes (designated as fourth generation) to support the growth proficiency and product yield. The earlier decade has understood a rise in extreme value biochemical productions and biofuels engaging microbial cell factories (Shuba and Kifle 2018). To reduce the greenhouse gases and to satisfy the worldwide energy burdens, technological advancement requires specialization in as follows:

1. To boost up the biochemical production, through improving the microbial cell factories.
2. For higher efficiency, optimize the existing production technologies.
3. Development of cell capability to yield biofuel-related designer molecules that advance fuel quality.

Combined methods could support in overwhelming the technological difficulties met during designing a functional, and reliable biofuel production pipeline. For the modernization of metabolic pathways, genome editing (like gene insertion and removal) has developed the reconstruction efforts because it advanced the metabolic engineering of both native and non-native hosts to yield renewable biofuels. By using of strain improvement technique, several approaches are used for the assembly of high titer, e.g., genetic modification, promoter engineering, pathways synthesis, process engineering, enzyme engineering, and competitive pathways blocking (Aro 2016).

Advancement in synthetic biology decreases the struggles, contrasting previously in developing microbial cell factories (Xia et al. 2019). Therefore, the key challenge is the partial number of existing obvious genes and their combinatorial control on metabolomics structure of the host organism. The modification consequence during a complicated phenotype-genotype results in opposing and accidental effects (Manzoni et al. 2018). The purpose for receiving developed titers from microbial cell factories is careful rebuilding and optimization of the metabolic pathway regulation in terms of organic phenomenon, enzyme activity, and metabolic influx (Lee et al. 2018). Difficulties are encountered during a different system of biology and gene-splicing methods to make fourth generation alongside the possible solutions to handle these challenges.

1.10 Heterologous Synthesis of Hydrocarbons

The international marketplace for *n*-butanol is growing at a prompt rate with a predictable market worth of 5 billion USD (Tao et al. 2014). Biological fermentation process (acetone-butanol-ethanol pathway) or petrochemical/chemical methods (for 7.0–8.0 billion USD/year) are usually used for the economic production of butanol. However, the dependence of chemicals on the price of oil makes this alternative unsuitable for the near future. Additionally, *n*-butanol synthesis is taken into account carbon neutrality and sustainability. It's produced naturally by *Clostridia* spp. through acetone-butanol-ethanol (ABE) pathway; however slow ratio of growth, complicated life cycle, complex nutrient need, and demanding genetic modifications constrain the assembly (biological) of butanol (Vogt and Richnow 2013). Furthermore, Clostridium spp. is incapable of directly using low-price substrates like hemicellulose, organic waste, and cellulose and must be dependent on the molasses and starchy materials whose accessibility has been limited by environmental dependence and competition with human food correspondingly (Kucharska et al. 2018). Special engineering strategies are adapted by using non-butanol-producing species, like *E. coli*, *B. subtilis*, *P. putida*, *S. cerevisiae*, and *Lactobacillus brevis* to extend marketable production targets of butanol. Toxicity induced by butanol is the primary major difficulty to become commercial yields. With the chaotropic action of 37.4 kJ/kg/M, butanol interrupts the macromolecular complexes of the host cell and then produces oxidative destruction which finally results in product-induced growth (Cray et al. 2015).

1.11 Production of Microalgae-Biofuels with Nano-Additives

For biofuel generation, microalgae are seen as potential feedback within the current age due to its energy-rich source, cheap culture approaches, inflated rate of growth, the prominent ability of CO_2 fixation, and O_2 accumulation to the atmosphere. Recently, research is constant regarding the improvement of microalgal biofuel technologies (Zhu et al. 2018) (Fig. 1.3). Application of nano-additives has been understand as a prominent improvement to mitigate this experience. At different stages from microalgal cultures, the nano-additive application to end-product use presented a solid possibility into the longer term (Anwar et al. 2020). Currently, the foremost complex technology is that the nanotechnology incorporation with the applications of bioenergy by the nano-energy zone possesses a durative on biofuel transformation mechanism and improvement of engine progress. This technology is described because of the design of a material or device on a nanoscale (10–9 m). To accomplish the biofuel product and develop the effectivity of biofuel consumption in diesel and petrol, nanotechnology has been introduced through nano-additives like nano-crystals, nano-droplets, nano-fibers, nano-magnets, etc. (Hossain et al. 2019).

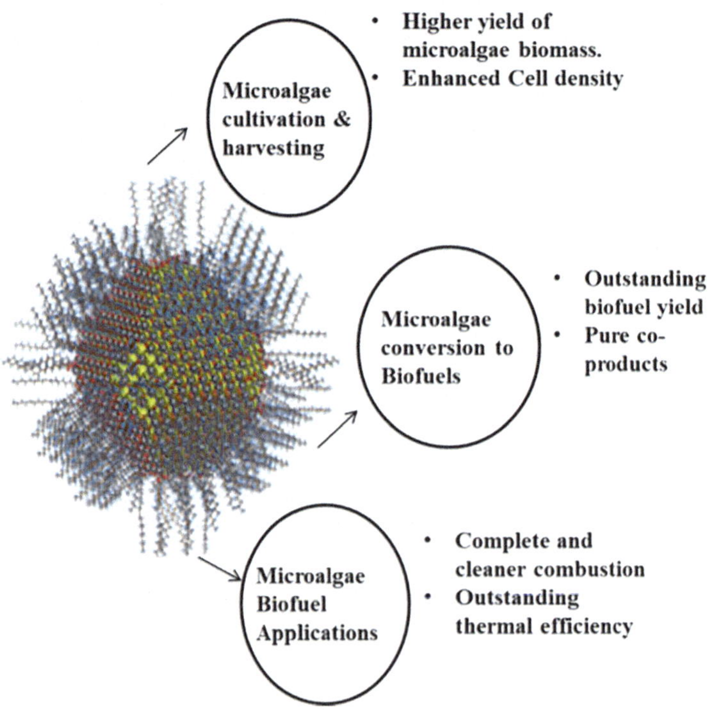

Fig. 1.3 Nano-additive application for the enhancement of microalgae cultivation to biofuel implementation

The thought of microalgae cultivation has appeared to the spotlight for biofuel production due to many positive outlooks like:

1. They don't clash with human and animal food chains.
2. Rich in proteins, carbohydrates, and oil contents.
3. Ability to sprout on aqueous media like wastewater, saline water, and water and assimilate nutrients from brackish water, highly contaminated water.
4. Low tide requirement.
5. Support the power to supply whole year naturally with sunlight existence.
6. Are often grown within the bare land (especially within the cold region), ponds, waste dump area, river, and municipal and industrial waste drainage.
7. Create a sustainable O_2 production system.
8. Use of CO_2 for photosynthesis respiration helps CO_2 elimination from atmosphere (Pal et al. 2019).

Furthermore, microalgae have a really short harvesting biological clock and yield promising biomass that generate greater yield of the specified biofuel. Interesting microalgae have an extraordinary quantity of lipids, protein, and carbohydrates, the only element of biofuel conversion. For the biofuel productions, nanotechnology applications are implemented, meanwhile when the present notorious approaches of usual microalgae culture, biofuel production includes some restrictions like uneven industrial-scale microalgae yield, increased cost harvesting and production, the energy utilization of biofuel construction from microalgae, and therefore the expansion of greenhouse emission concentration within the environment (Work 2014). For various phases of microalgae cultivation, nanotechnology applications are often entitled to microalgae-biofuel application fuel engines due to stability, durability, crystallinity, proficiency, recyclability, catalytic performance, high storing capacity, absorption, and eco-friendly characteristics (Hossain et al. 2019). Due to improvement, nanotechnology improved microalgal cultivation, the larger yield of several microalgae biofuels also as microalgae-biofuel implications in diesel engines and petrol. Numerous nano-materials, e.g., nano-tubes, nano-particles, nano-fibers, nano-sheets, and further nano-structures, are seen as active nanocatalysts in direct and indirect approaches in biofuel (e.g., bioethanol, biodiesel, biomethane, and others) for product improvement (Darwesh et al. 2021). However, magnetic nanoparticles were used as a transporter for enzyme control for biodiesel and bioethanol production effectivity. Due to potent paramagnetic properties, magnetic nanoparticles were also favored for methanogenesis to yield biomethane (Hossain et al. 2019). Bioenergy yields from lignocellulosic biomass (agriculture excesses) and waste of industries (slurry) furthermore as algae (macro and microalgae), with optimization at nanoscale that has just enforced on the instrument of nano-particles, characteristics of biomass, and various application on biomass progression (Kamla et al. 2021). The scientific assent is that by the process of photosynthesis respiration, algae convert O_2 and CO_2 and generate an enormous quantity of cellular energy contents into sugar, proteins, and lipid (Shuba and Kifle 2018). Development and industrialization threaten the present ecosystem due to the dumping of heavy metals, waste containing sulfur, nitrogen, phosphorous, etc. commonly with exhaling a high quantity of CO_2 to the free air (Hossain et al. 2019).

1.12 Nano-Additives for Microalgae Biomass Conversion to Biofuels

Among all microalgae biofuel, biodiesel got high importance because of the commercial and admired biofuel within the oil market. For the production of biodiesel, applications of basic and acidic nano-catalyst spheres can substitute chemical compounds as sodium methoxide by reacting with the oil and free fatty acids. The constructive effect of this nano-catalyst is that reaction occurs with pressure and coldness also as advance reduces the environmental bome by sodium methoxide (Hossain et al. 2019). Another study about biodiesel at industrial level confirmed that commercial CaO-NPs offered 91% biodiesel conversion efficacy throughout scaled-up catalytic transesterification. Microalgal development with sphere-shaped nanoparticles constituted of calcium, and sand silica compounds shown that cellular growth of microalgae boosted significantly exclusive of destroying harvesting also as biofuel production from oil (Akia et al. 2014). The research study about mesoporous silica nano-catalyst Ti-loaded SBA-15 signified ten times advanced free carboxylic acid (FCA) and water tolerance level than another catalyst for production of biodiesel from oil also as this nano-catalyst accomplished three times improved than other effective nano-catalysts titanium silicate-1 (TS-1) and titanium oxide silicate (TiO_2-S) (Chen et al. 2013). Furthermore, Ti-loaded Santa Barbara Amorphous-15 (SBA-15) nano-catalyst application reduced the chemical (alkaline catalyst, NaOH) cost of transesterification process for biodiesel production by recycling the nano-catalyst also as this method is more environment friendly (Hossain et al. 2019). Another study show that sulfate incorporated Ti-SBA also performed as a biocatalyst to vary oil to 100% esterified bio-lubricant. Bio-oil of microalgae with the aid of nano-particle aided to source of bio-lubricant from bio-oil. Another research reported that Niobia (N_2O_5) incorporated with SBA-5 applications on biodiesel production from biomass via esterification characterized a big development for microalgae-biodiesel yield (Chen et al. 2020). A kind of nano-particle, zeolite (an alumina silicate mineral), has been used as a billboard absorbent through transesterification process. Zeolites are able to absorb the unfavorable moisture content (4–6%) and generate pure glycerine as a co-product furthermore of biodiesel (Tran 2018).

1.13 Genetically Engineered Algae for Biofuels

For several years due to the revolution of the global level energy crisis, microalgae have been developed and are using as a replacement of feedstock for the yield of biofuels. Furthermore, for producing key chemicals, microalgae consider having huge strength for bio cell factories like recombinant enzymes, alcohol, hydrogen lipids, and protein (Jagadevan et al. 2018). Microalgae-dependent renewable energy is considered very economic due to the concept of high-value products such as the algal biorefinery approach. Genetic engineering is a new discipline that combine in a sequence of engineering and science to aid support construct and design novel-biological systems and to gain themes that are formulated rationally (Jagadevan et al. 2018). The tools and resources used for artificial gene network construction, nuclear manipulation, and reproduction of microalgae of genome-scale are limited. Microalgae of genetic engineering (GE) are appearing to be a biofuel synthesis commercial release without such ecological studies or public information to inspect the possible risks (Kumar et al. 2020). Green eukaryotic algae and cyanobacteria a blue-green algae are likely to spread-from ponds which are uncluttered and on the subordinate scale with smaller possibilities from unopened photo bioreactors. Cyanobacteria are especially problematic to detect due to the danger of horizontal gene exchange with discrete microbes (Day and Stanley 2012). Before novel, genetically engineered algae transfer to the environment; environmental and major biosafety dangers should usually be addressed by experts' teams like ecologists. For the synthesis of biofuels, biologists assumed that microalgae will be reformed by using a vision or perception from artificial biology and the innovative process of producing biofuels in the form of genetically engineered organisms (GEO) (Ancillotti et al. 2016) (Fig. 1.4). High-throughput sequencing of hybridization, DNA sequencing, metagenomics, and accelerated evolution. Genetic engineering is currently expanding, becoming more accessible, and becoming more effective. Nemours strains of algae have been selected worldwide and collected for raw genetic material (Park and Kim 2016). The pathways and related genes are vital for the biosynthesis of biodiesel in *Dunaliella tertiolecta*, a non-model aquatic flagellate (Rismani-Yazdi et al. 2011). In the microalgae, the composition of DNA which enhances the rate of growth, and Boston the nitrogen efficiency uses, have been studied for Nemours patents. In photobioreactors and in open ponds, the growing algae are very expensive. By research, it is summarized that there is still much need to do innovation including genetically engineered microalgae cultivation (Barry et al. 2016). The novel characters can be accomplished by methods of non-transgenic therefore value assessment, to increase the performance of wild strains, recombinant DNA is used. To introduce microalgae at the commercial level, Sapphire Energy Corporation works together with Monsanto company to investigate innovative genes which converse extra growth and additional positive traits (Snow and Smith 2012). There are spin-off applications in crop plant. The company Joule Unlimited has patented a mix of foreign microbe genes and genetically engineered cyanobacteria that makes more ethanol. Novel GEOs are

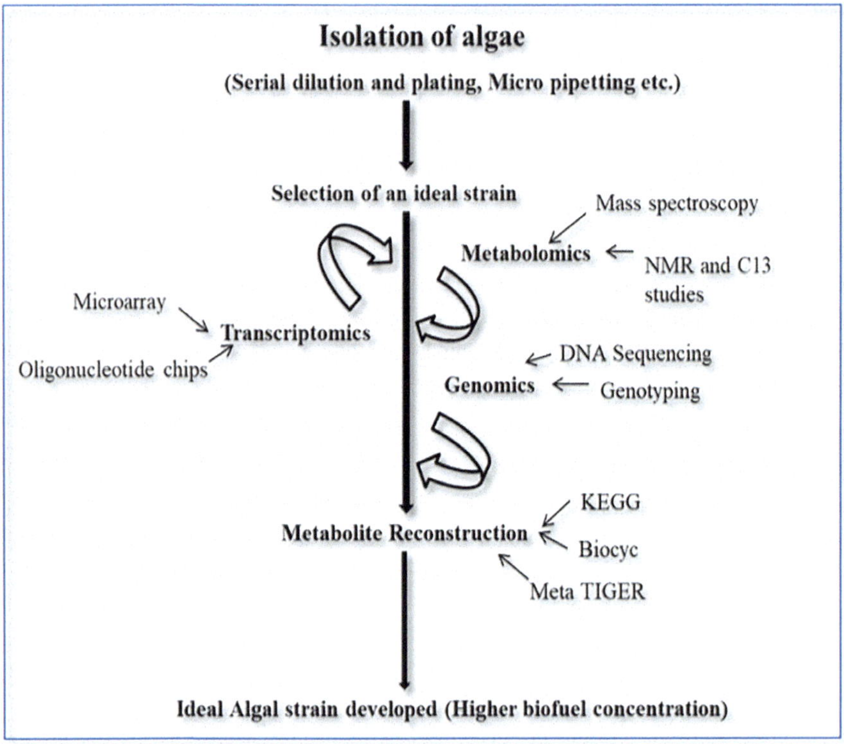

Fig. 1.4 Pictorial representation of the overall process toward biofuel production in microalgae using synthetic biology approach selection of an ideal strain redirecting metabolism to maximize the synthesis of the targeted biofuel

represented within the atmosphere. Moreover, major biosafety dangers should be addressed to invest cause of harm and to the potential stage of exposure (Snow and Smith 2012). To GE algae endurance outside of open lakes or encased bioreactors is essential standard source for hazard investment. The problem has arisen that genetically engineered algae discover lab-created strains must unable to subsist within the rough specially if tamed or kept forming huge volumes of industrial type of by-products. If an unproposed release era to prissiest, the concerned GE algae choose that their developing descendants may have that would allow them to survive and continue in natural surroundings (Henley et al. 2013). Such engineered traits, for example, resistance to severe temperatures or improved growth, may help GE algae survive and develop in a suitable ecosystem.

Significantly more had to think about the condition of lab than their regular habitats (Joutey et al. 2013).

1.14 Algal Biorefinery Products

There is no comparison between the economic suitability of microalgae biofuels with existing technologies but no focus on the innovation of high-value co-products from microalgae to develop the foundation of economy on biorefinery (Patil et al. 2008). Biorefinery is an organization where high-rate products and energy fuel are produced, for example, lipids, pigments, protein, antioxidants, vitamins, and carbohydrates are formed through different processes by biomass (Chew et al. 2017). Lipids, carbohydrates, and proteins are abundant in microalgae, and the absolute amounts of these biological compounds varied among microalgal strains. It has been described in research that *C. reinhardtii*, cyanobacteria, *A. variabilis*, and *Anabaena cylindrica* green algae strains of algal are famous to produce hydrogen in presence of sunlight (Aratboni et al. 2019). The organisms listed above can extract protons and electrons from water and hydrogen. This is done with the help of the chloroplast (Hyd A2 and A1) hydrogenase, which speeds up the process. Research shows that the metabolism of C. reinhardtii's respiratory system was changed by finishing strength competition with hydrogenase for an electron. This made the plant's natural rate of biohydrogen production faster (Jagadevan et al. 2018). Moreover, the hexose uptake protein expression in a production of hydrogen *C. reinhardtii* mutant (Stm6) shows 150% enhancement in the ability to formats H_2 (Doebbe et al. 2007). Because of the upgrading research which is investigating the fourth generation of biofuels, numerous studies on the production of bioethanol by implementing algal strains like *Desmodesmus* sp., *Tetraselmis suecica*, and *Porphyridium cruentum* have been stated. Through recombination system of double homologous, the strain of algal PCC 6803 *Synechocystis* sp. formed, and this is able to do CO_2 photoautotrophically changing to bioethanol, by means of improved theoretical growth of ethanol per gram of 0.696 as in contrast to glucose ethanol per gram 0.51 through *S. cerevisiae* (Doebbe et al. 2007). There is a need to do concentration on the research of the production of bioethanol which is derived from algae to make this method economically feasible. Various researches directed the biobutanol production from specific microalgae like its strain which is full of carbohydrates. CL-MI, *Neochloris aquatica* acid is a pretreated biomass of biodiesel residues of microalgae and JSC-6 *Chlorella vulgaris* biomass (Behera and Varma 2016; Doebbe et al. 2007). Iso-butyralaldehyde with iso-butanol production a strain of algal *Synechococcus elongatus* was studied for the production of butanol. A higher ratio of iso-butyraldehyde produce with the GE *S. elongates* than those researches from lipid, hydrogen, or ethanol production with algae or cyanobacteria through the upregulation of ribulose bisphosphate oxygenase/carboxylase (Rodriguez and Atsumi 2014). The major difficulties to introduce biobutanol's large product at the commercial level are the other by-products that impurify the production of the yield of pure butanol. Biodiesel has similar chemical attributes as fossil fuels and consider a good alternative. Various algal species like *Dunaliella*, *Chaetoceros*, *Nannochloropsis*, *Scenedesmus*, *Pseudochlorococcum*, and *Botryococcus* are famous to concentrate a high ratio of native lipids (Kolesinska et al. 2019). Algal strains' metabolic pathways can format 16–20.5% fatty acids *n*-Carbon as precursors for biodiesel growth. Formation of fatty acid for the synthesis of biodiesel has happened through *Nannochloropsis* sp. Carotenoids and biohydrogen are other by-products. Seventy percent of pigments and 45 g/100 g lipids dry biomass were extracted when supercritical CO_2 fluid extraction is used. For the production of biodiesel, another research was performed to investigate on a high lipid-forming microalgae (*Euglena sanguinea*) (Kolesinska et al. 2019). It was confirmed that this biodiesel will not cause any disturbance in the engine because of unsaturated availability in fatty acids of C: 18:1 along with SFA (C24:0-C22:0, -C16:0, -C18: 0) in it. Hence, it could be used in automobiles without any change in engine structure. Flux analysis, in particular, can give an effective technique of prediction in biology systems by measuring carbon flux during accumulation, growth, and carbon fixation altogether (Nagy and Tiuca 2017). Calvin cycle's coefficient data is controlled by enzyme flux; it helps in the growth and fixation of carbon. Metabolic subsequent map and metabolic ^{13}C flux data derived from algae can originate the targets and steps which are involved in lipid metabolism (Behera et al. 2021). These data show that pathways and enzymes which they modified are exert significant and its rate-limiting control on the high metabolism. In the present, researchers are investigating on the pathways of metabolic engineering for enlargement of fatty acid chain length (Park and Nicaud 2020). This modification will help to enhance the growth of algal biodiesel production in the future. Nemours researches have been done on the genes of microalgal encoding enzymes that are involved in the preparation of high-value carotenoids like *C. reinhardtii* type of microalgae but cannot prepare keto-carotenoid, and gene of β-caroteneketolase through In *C. Haematococcus pluvialis* was discovered. *C. reinhardtii* is working on developing a novel ketocarotenoid. From *P. tricornutum* of *C. vulgaris* of nitrate reductase gene promoter and transformed terminator and the recombinant strain was shown to be capable of producing high-value proteins (Aratboni et al. 2019). For biopolymer production, microalgae can be used as expression systems, for example, poly-3-hydroxybutyrate which is a key precursor for biodegradable plastic production. Over the original strain, the strain that was designed with a facility for revocable opening to high flux of carbon was able to produce much higher levels of 3-hydroxybutyrate (Jagadevan et al. 2018). As a protective matrix for hydrated biofilms, most microalgae generate extracellular polymeric compounds in their immediate habitats. Anti-coagulant, anti-viral, drag reducers, antioxidants, and biolubricant are examples of high-value applications for these compounds. Consumed biomass of microalgal and lipid-extracted algae (LEA) is a true replacement of the manufacturing of different products as it comprises 30–60% carbon residual which is present as fermentable sugar (Fanesi et al. 2019).

Lipid-extracted alga has been usually utilized through an anaerobic mechanism as methanation's substrate. Another research which was conducted on product-extracted algal samples like protein and lipid extracted enhancement was noted in biogas production. Pre-treated algal has higher production while using cumulative methane. The yield of methane increases while using anaerobic digestion of lipid-excused biomass than the exhausted biomass of non-lipid (Rabii et al. 2019). It is also used in the fermentation of butanol. It is also used in the fermentation of butanol. Lipid extracted algal may also be efficiently transformed into liquid fuel, most of the alkanes through hydrothermal upgrading and liquefaction processes like by hydrocracking and hydrotreating. 69.5% of energy efficiency is noted overall at a higher heating value (Ramirez et al. 2015). In another research on *Scenedesmus acutus*, it was observed that it can assimilate nitrogen from lipid extracted algal residues and was capable of conversing nitrate within media of culturing media, therefore adding recovering of nutrients. The phosphorous/nitrogen limit in microalgae hints at a discrepancy in the productivity of liquid, and the acetyl-CoA carboxylase gene expression was used to verify this (Mahmud et al. 2021).

1.15 Microalgae Biofuel Engineering

Because of the infected technique of harvesting and low yield because of less effective design of photobioreactor and inadequate efficiency of photosynthetic, it is compulsory to maintain balance between liquid production and biomass content (Xu et al. 2020). The genetic engineering tool's aim is to increase the structure of enzymes involved in storage and reduction of lipid catabolism mechanisms as well as the synthesis of lipid, which has huge potential for engineering vital metabolisms (Mahmud et al. 2021). The interactions between the many enzymes that algae uses to produce lipids. In recent years, ACCase acetyl-CoA carboxylase overexpression, gene knockout of acyl-CoA synthase, oxidation of fatty acid (acyl Co oxidase), camitineacyl transferase, fatty acyl-CoA dehydrogenase, and acyl-CoA synthase, and acyl-CoA synthase have become more popular. Genetic manipulation is difficult since editing techniques are species specific and cannot be used interchangeably due to strategies including defensive-like restriction and methylation enzymes, codon usage, nucleic acid uptake, and cell porosity.

There are just a few completely annotated microalgal genomes accessible, if used with advancements in technology of sequencing. In the future, there is predicted to be a substantial increase in this type of data (Lv et al. 2013).

1.16 In Microalgae: "The Synthetic Biology"

The core of artificial biology is capable to build synthetic monitoring circuits that can resist cellular performance depending upon use defined (input/signal/stimuli) which produce the most wanted output like as biofuel chemicals and proteins in the creation of necessary changes in metabolism (Mukherji and Van Oudenaarden 2009). The most focus is being devoted to the use of large-scale microalgae through sustainable or strong energy from microalgal biotechnology, where feedback is given at (a) During metabolic engineering, improve their photosynthetic efficiency to make more oil and improve mass cultures by increasing the rate at which carbon is taken up. (b) Using carbon flux as a fuel source, energy-rich molecules are produced. (c) Strong and committed algal cells are spreading out which can hold out low cost with large-scale farming causing in minor carbon footprint of the produced chemicals and lower operational cost (Jagadevan et al. 2018). The function of mock biology via the building artificial biological schemes by a blending of engineering policy through biotechnological utensils, grounded on genetic, metabolic, or regulatory statistics collected by trials. It is a field of biology that concentrated on re-engineering or rebuilding the metabolic alleyways by means of starter of genetic units then detected such as "biological circuits." Then crucial purpose is redesign microalgae that obtain the original task. The rebuilding of applicable biochemical lanes (metabolite products are vital) addicted to incomplete and complete copied replacements necessity of a steady assemblage beside toward assimilation of heterologous DNA segments into masses and chassis-strains (Konig et al. 2013). The diverse says gene that can be contrived and manipulated for renewal are (a) digestion, ligation of DNA wreckages, (b) in vivo homologous recombination, and (c) in vitro homologous recombination of fragments (Finnigan and Thorner 2015).

1.17 "Bio-bricks" or Circuits as the Synthetic Elements

Definite challenges require to be essential to commercialize transgenic microalgae. For genetic engineering, there are poor molecular apparatus, and there is a minimum level of heterologous gene faces from the nucleus. By the emergence of this narrative genomic instrument, synthetic biology is increasing rapidly, and the concept of "biobricks" in microalgal systems may be established (Moses et al. 2017). Uniform DNA segments of DNA having a common interface are known as bio-bricks that can be collected into biological methods. Such types of portions are exchangeable units like the organizers, ribosomes binding sites (RBS), terminators, trans-elements, or several directing molecules which are utilized to stimulate the genetics of microalgae and eventually their metabolism (Røkke et al. 2014). The genetic tool omics method play a vital role in configuring corridor and re-construction at microalgae by supporting to recognize the metabolism inside the entire structure, i.e., stimulated via response onward and backward coils which disturb the output. Get the high quantity of omics data liquid-inducing conditions (stress stimulus) are channelized therefore prominent the effective instruction of metabolisms through the application of mock biology not only mere overexpression of enzyme coding genes but also pointing the regulatory networks (Warner and Suggett 2016).

1.18 Single-Cell Protein

Single-cell proteins (SCP) are termed as bio proteins, biomass, and microbial proteins, which is a substance of the dried cell. SCP is formed via microorganisms, for example, yeast, bacteria, fungi, and algae. Therefore fungi and bacteria are the main manufacturers of the protein. Increasing efficiency of proteins by that foundation was primarily owing to their speedy progress level, and comparatively their chemical environment has complex protein level. More classes of algae are recycled as they are cultivated in the water environment (Warner and Suggett 2016). Furthermore, with the high content of protein, single-cell protein also has high contents of vitamins, nucleic acids, fats, minerals, and carbohydrates. Moreover, single-cell protein has high levels of amino acids, which include lysine, threonine, and methionine. These proteins are the best resource of algae than other exclusive options, e.g., soybean meals and fish. The conclusion has appeared that single-cell protein (SCP) could be merely a substitute for other traditional and exclusive (plants and animals) protein bases inside humans (Warner and Suggett 2016). Protein foods found on microbial algal biomass have the energy to fulfill this necessity, and these microbes are thought out as protein-rich supplements and ingredients in animals and human diets. This is also a good substitute because it doesn't require a vast area for growth and proliferation (Wells et al. 2017). Contrasting plant productions are also free of atmosphere and seasonal deviation, and it must be created all through the year. Furthermore, it couldn't produce greenhouse gases in surroundings as per plants' protein sources (Wells et al. 2017) (Table 1.2). The most crucial thing to decrease the production of single-cell protein is to choose the economical and applicable substrates or biodegradable agro-industrial by-products for the microorganisms as a nutrient basis to yield and grow tons of proteins (Spalviņš 2020). Due to this, in the past diverse substrates were used. Few common substrates are yam peel, apple pomace papaya waste, potato peel, pineapple waste, etc. It's important to select the surplus products for propagation of single-cell protein-forming microorganisms (Spalviņš 2020). For industrial purposes and study, the accessibility of microorganisms is not a big problem, and numerous strains of bacteria, fungi, algae, and in the lab yeast can be cultured in different ways. The center is now algal sources for

Table 1.2 List of substrates for various algal species (substrate for algae to produce SCP)

Algae	Substrate
Spirulina species	Carbon dioxide
Chlorella salina	Alkaline waste effluent
Caulerpa racemosa	Carbon dioxide
Spirulina maxima	Sunlight and carbon dioxide
Chlorella species	Carbon dioxide
Sargassum	Carbon dioxide and sunlight
Dunaliella	Carbon dioxide and sunlight
Laminaria	Carbon dioxide and sunlight
Diatoms and *Chlorella*	Carbon dioxide and sunlight

single-cell protein (SCP). Most kinds of microalgae are nurtured for human and animal utilization or generally have strong protein substances raised up to 70%. The protein from algae is primarily source of fats like chlorophyll, mineral salts vitamins, fatty acids, and omega-3. They hold less concentration of nucleic acid 3–8%. African and Mexican natives near texcoco harvested an algal species *Spirulina* (Yaakob et al. 2014). When it got dried, then it was used in human diets. As a feed source, *Scenedesmus* and *Chlorella* biomass have been used. Improved protein content, simple cultivation, fast growth, and appropriate utilization of solar energy have vital subsidy in which they are broadly recognized as feed ingredients all through the world. Green algae are noble antioxidant (Mobin et al. 2019). The interest looks more motivated due to high crude protein content, by the potential production of docosahexaenoic acid (DHA) and omega-3 fatty acid (i.e., eicosapentaenoic acid (EPA)), and carotenoids termed as astaxanthin an antioxidant, and pink pigment for salmon and shrimp and immune stimulant, e.g., by the strain of *Aurantiochytrium* (Matos 2017).

1.19 Strategies to Improve the Yield of Microalgal Biomass and Lipids for Biofuel Feedstock

The yielding of biofuels, related to further sources of microalgae, has massive feedstock, because of little effect on food security, low eco-friendly, and algal genuine yield. Hence commercializing the algal biofuels needed more profits (Rawat et al. 2013). Enhancing the economics of consuming microalgae as biofuels is life-threatening to the achievement of tactics of biomass and lipid production. Spending of genetic engineering and mutagenesis are the necessary methodologies for the collection of appropriate algal strains (Mondal et al. 2017). Approaches are added to improve lipid production and biomass of microalgae like co-cultivation and genetic of microalgae with yeast and bacteria and also phytochromes and metabolic engineering. The best technique to stable the lipid production and algal biomass is to improve culture system and hybrids (combination of photobioreactors and open ponds). The usage of discarded water of agri-industrial to sowing microalgae is cooler method to decrease in production cost. These types of methods help in bio elimination of waster CO_2 and bioremediation of undesirable water. To meet the source lodging, current and future energy anxieties have triggered researchers to discover other energy sources for proliferation in fossil fuels. It is expected that the present supply of fossil fuels would be sufficient until 2030. In 107 years, 35 years, and 37 years, respectively, fossil fuels such as oil, coal, and gas will be depleted. The People's Republic of China, India, Malaysia, Indonesia, the Philippines, or Thailand is major biofuel producers. Conventional feedstock such as cooking oil, oil crops, and animal fats are utilized but not enough to meet demands to produce biofuels. Algal biofuels are an attractive aspect just toward healthier ecological act, mainly concerning lower greenhouse gas (GHG) emissions but not in food completion. The revenue of oil from microalgae is 300 times greater than soya bean and 23 times greater than oil of palm per hectare of cultivation. Fresh water for human consumption wouldn't be attained by fully fledged microalgae in waste and saline water. The waste remunerations are the main approaches to take deep attention in the manufacturing of algal biofuels by integration of expertise that includes in by-product industries leveraging environmental protection. Driven of renewable fuels from microalgae comprising biohydrogen generated via photo biological process, biodiesel resulting after microalgal oil, or methane derived from anaerobic ingestion of the algal-biomass. Energy assets in microalgae, i.e., triacylglycerols (TAG), are mined from algal cells and transformed to biodiesel via transesterification. After that great construction cost required to commercialize the algal biofuels is the key barrier for standby fossil fuels. Before consuming microalgae as a thriftily practicable feedstock for biofuels, some procedural matters must be stunned such as low biomass production, less lipid content, and relaxed development rate which are required. Fabricating lipids, to develop straight plans first to boost impressive nutrients stress as well as operating environmental influence like light, salinity, and temperature. Fresh policies are co-culturing of microalgae by new microorganisms with toting of phytohormones (plant regulators) too; chemical additives are capable tactics. For the perfection of strains in algae meant for biofuel production, metabolic and genetic engineering are potent gears.

1.19.1 Improvement of Microalgal Strains for Biofuel Production

There are 44,000–72,500 species of microalgae that have been found (Guiry 2012). Phycoprospecting and repeating of microalgae are desirable characteristics based on growth rate, and lipid production is thriving in robust strains of biofuel feedstock. Thirty-three tropical isolates of thraustochytrids were isolated from mangrove leaves by the Malaysian team for lipid production and fatty acid profiles. The peak of insulates of genus *Aurantiochytrium* and their yields of lipid oscillated as 0.27–70.86 mg/L/day with projecting fatty acid of 16:0 (Aratboni et al. 2019). Alienated strains of species *Desmodesmus* from Taiwan are curtain for lipid output and thermo-tolerance. The growing of required traits of algae in salinities (Chu 2017). To tolerate a wide variety of aptitude of 12 traits of *Tetraselmis* which is detached by natural saline in western Australia. Little strains of *Tetraselmis* were subjected to euryhaline, which resulted in increased biomass and a high level of lipid efficiency, therefore strength for biofuel efficiency (D'Adamo et al. 2014). For yielding biodiesel efficiency, whatever water is used, so it is economically and environmentally feasible to use sea water or inland saline ground water. Besides, there is less chance of adulteration in algal culture in cultured saline medium with the next organisms (Hirooka et al. 2020). The possible platform is the mutagenesis for the strain of microalgae to improve lipid and biomass efficiency. For lipid yielding, there is the usage of UV-mutagenized strain of *Scenedesmus* species, after oxidation exposes the H_2O_2. Catalases and a major enzyme in fatty acid biosynthesis' activity that is acetyl-CoA carboxylase boost expressively in the modified mutant (Sivaramakrishnan and Incharoensakdi 2017). Enlarging the biomass and lipid productivity by enlarging the mutant process of *Chlorella pyrenoidosa* is positively useful in the atmospheric temperature plasma mutagenesis process, and it is on working for yielding biomass and lipid production (Arora et al. 2020). The 33 °C temperature and pH 9.0 has kept then mutant (II-H6) can show great inborn stability and grow optimally. Through adaptive laboratory evolution development are usage for favorable microalgal strains by a screening of exact cultural conditions (Chu 2017). *Crypthecodinium cohnii* is a glucose-tolerant strain paralleled with wild strain that exhibits lipid growth by accomplished laboratory progress and also increases outputs in *Chlamydomonas reinhardtii* which is mutants of low starch (Sun et al. 2018). There is the intense spreading of extremophilic microalgae for lipid yield. In impartial lipids, the collection of polar algae fatty acyl chains, specifically C16 polyunsaturated series of fatty acids, C18:1ω9 with C118:3ω3. The six microalgae were mostly selected from the topsoil of semi-arid areas in northwest China for lipid and biomass output. *Chlorella* species 11 and *Monoraphidium dybowskii Y2* are two microalgae mined from the soil crust of biological desert, which cause high efficiency of lipid and biomass when grown in outdoors at 140-1 PBR (Aratboni et al. 2019) (Chu 2017).

1.19.2 Manipulation of Nutrients with Environmental Factors

The foremost purpose of microalgae for feedstock is to harmonize biomass and lipid efficiency.

To increase the lipid and biomass productivity is utilized to manipulated nutrient levels (e.g., phosphorous and nitrogen), and ecological features like light, salinity, and temperature are conventional edges (Yaakob et al. 2021).

1.19.3 Nutrient Stress

The foremost purpose of microalgae for feedstock is to harmonize biomass and lipid efficiency. To increase the lipid and biomass productivity is utilized to manipulated nutrient levels (e.g., phosphorous and nitrogen), and ecological features like light, salinity, and temperature are conventional edges (Remmers et al. 2018). Utilization of nitrogen acclimation inhibitor compounds to provide an alternate method to recreate the similar circumstances as *C. reinhardtii* have the state of nitrogen deficiency. Reproduction of neutral lipid composition *C. reinhardtii* caused by exposure to inhibitors (methionine sulfoximine) subsequent behavior connected to nitrogen deficiency that was only active for three days (Bono Jr et al. 2013). After the starch synthesis, most of the nitrogen limits were suppressed by the bounding of TAG (triacylglycerol) that appeared in *T. obliquus*. As shown in *Nannochloropsis oceanica*, nitrogen starvation is connected with degradation of lipid membrane that gathered phospholipids and TAG (Goncalves et al. 2016). A proteomic-based method is used to illuminate metabolic changes in nitrogen scarcity which is found in tropical *Chlorella* species. Another tactic is found to increase lipid and biomass productivity by combining high glucose with low nitrogen level in microalgae. Cultured under heterotrophic conditions increase the lipid content of *Chlorella sorokiniana* by low $NaNO_3$ and high glucose levels. To simulate lipid buildup in microalgae is the tactic to limit the culture of nutrients like sulfur (S) and phosphorous (P). For example, under phosphorous constraint (0.1 mg^{-1}), *Scenedesmus* sp. LXI could accumulate lipids up to 53% of its biomass; the lipid productivity/unit volume of culture was not increased (Yun et al. 2020). Now *C. Reinhardtii*, sulfur deficiency continued to be the primary cause of TAG accumulation by directing metabolic carbon flow away from protein synthesis and toward TAG synthesis (Shahid et al. 2020). Malnourishment of all elements can be accomplished instantly, to pretend lipid binding is the method for microalgae in nutrient-deficiency medium. By increasing, lipid content in microalgae is achieved by stress to the nutrient (Sun et al. 2018). The key control is the low biomass production as it distresses the overall lipid efficiency. Some approaches are used for lipid outputs, first stage, to yield the maximum amount of biomass by the flourished to microalgae under optimal conditions (Sajjadi et al. 2018). Some approaches are used for lipid outputs, first stage, to yield the maximum amount of biomass by the flourished to microalgae under optimal conditions. In the second stage, cultures will be exposed to stress conditions. Anxiety circumstances will be endangered when simulating growth and lipid fusion. These two stages were active to control lipid outputs in *Chlorella vulgaris*. In the initial phase, when nitrogen stress was provided, it changed the circumstances of grown microalgae which grow under nutrient-rich medium (Cui et al. 2017). Under such manipulation, at least twofold increased the lipid synthesis. Lipid throughput of *Ankistrodesmus falcatus* better by 36.5–45.5% under two-phase cultivation conditions are provided (Álvarez-Díaz et al. 2014). Beforehand being transported to nitrate and phosphate limited medium, first algae were cultivated in glucose-supplemented condition, which enhanced tenfold lipid yield of *T. obliquus* found by the studies. For high algal biomass production, it must confirm that the CO_2 level is not low enough in ambient air. 1–5% of CO_2 is frequently supported for maximal microalgal growth, and 5–15% of CO_2 is mostly aerated for laboratory cultures (Panahi et al. 2019). Most inventions are that high in CO_2 cause high in lipid efficiency. Besides, high CO_2 levels (30–50%) could stimulate the growth of PUFA and lipids in *T. obliquus* SJTU-3 and *C. pyrenoidosa* SJTU-2. Research finds that the quantity of highest CO_2 bio-fixation rate and biomass concentration was achieved once 100% CO_2 was found for microalgae developments (Show et al. 2017).

1.19.4 Light Intensity and Wavelength

For biofuel fabrication, it is necessary to enhance the production of lipids and biomass in microalgae. The highly attractive approach is painted, light-emitting diodes (LED) and the use of dyes among the modern approaches. Showing the microalgal culture in blue (450–475 nm) and red (630–675 nm) light increases the absorption of chlorophyll colorants and improves the functionality of photosystems I and II (Ra et al. 2016). By increasing knowledge, it comes to know that lipid addition in *T. obliquus* increases, when LED light exposure at various frequencies and also *Cyclotella cryptica* development at 450 nm (yellow wavelength) causes lipid content spread up to fourfold (Choi et al. 2015). Microalgal cells are protected from photoinhibition and photooxidation through filters made from nanoscale covering, by IR and UV rays that cause oxidative pressure (Michael 2015). The light strainer expressively boosted algal biomass up to 13–34% and biomass efficiency up to 70–100%. The filters increase the microalgal growth, but the high rate of manufacturing of filters arises an issue (Michael 2015). The additional strategy is the organic dyes which enhance the outputs in microalgae bounding, and by adding the organic dyes, it increases growth two- to threefold in lipid production of *Chlorella vulgaris* (Ramanna et al. 2017). To modifying the solar spectrum and fascinating the surplus, solar radiation is useful by using fluorescent paint solutions. Red paint accomplished the highest biomass yield (1.7 g/L) although blue paint solution is consuming to yield high lipid content (30% dry weight) (Seo et al. 2015).

1.19.5 Temperature Fluctuations

Temperature disturbs not only lipid yield but also fatty acid profiles in microalgae. Fatty acid unsaturation distresses by variation in temperature. Balancing membrane fluidity in cell development required affinity to increase fatty acid un-\saturation at a lower temperature (Morales et al. 2021). Instantly at low-temperature unsaturated fatty acid content becomes doubled in *Chlorella ellipsoidea*. Furthermore, it results in greater unsaturation in its phosphoglycerides due to low-temperature strain of microalgal confined complex amounts of α-linoleic acid (Aratboni et al. 2019). When it is imperiled to boost culture temperature, it enhances lipid yield and growth amount understudy of *Nannochloropsis* saline. Temperature variation can change the lipid profile in algae development; features of algal biodiesel are predictable to be modified under diverse seasons and climates (Bounnit et al. 2020).

1.19.6 Addition of Phytohormones

Lipid efficiency in *Monoraphidium* species rises by adding melatonin (phytohormone) coupled with photoinduction. Reactive oxygen species (ROS) levels and lipid biosynthesis-related enzyme activities are linked and affect the rise of lipid bounding (Cui et al. 2021). According to research phytohormones belongings on metabolism in algae, mainly about lipid biosynthesis. The attractive method is to surge biomass, and lipid yield is influenced by a number of phytohormones which are very little and less in cost (Chu 2017). Adding fluvic acid encouraged lipid bounding in *Monoraphidium* species through gene expression, intracellular ROS yield, activities of acetyl-CoA carboxylase, malic enzyme, and phosphoenolpyruvate carboxylase (Che et al. 2017).

1.19.7 Metabolic and Genetic Engineering

Using a molecular strategy to increase biomass and lipid accumulation, makes biofuel. Expressed sequence tag (EST) databases established for few microalgae, nuclear, mitochondrial, and chloroplast genomes of numerous species have been sequenced (Chu 2017). The genome sequence of microalgae by biofuel feedstock is possible, like *Chlorella vulgaris*, *Nannochloropsis* sp., and *Phaeodactylum tricornutum* are not present. The growth of synthetic biology tools, named "biobricks" such as transcriptional terminators, ribosomes binding sites, and organizers, has progressed the development in genetic engineering of microalgae. *Chlamydomonas reinhardtii* is the model organism of choice for molecular engineering aimed at biofuel throughputs as its genome has been sequenced and then availability of tool set for microalgae (Young et al. 2020). Particle bombardment, glass bead agitation, micro injection, electroporation, and Agrobacterium tumefaciens-mediated transformation are a few techniques that can be used to change algae. Successful conversion was demonstrated on *C. reinhardtii* (Doron et al. 2016). In the occurrence of DNA, glass beads, and polyethyleneglycol, the cell wall-free microalga was agitated. Interaction between the nuclear and chloroplast genomes of *P. tricornutum* and *C. sorokiniana*, extremely active cell after DNA-coated metal particle bombardment (Che et al. 2017). To adjust saturation and length of fatty acids done are arose for lipid metabolism by adding lipid biosynthesis and catabolism which is targeted by altering the microalgal for biofuel outputs (Radakovits et al. 2010). Sequenced microalgal genomes contain homologous genes that regulate lipid metabolism. Transgenic techniques were used on higher plants and algae. Possible approach for increasing lipid production in microalgae by overexpressing important enzymes involved in the processing of fatty acids. With further research, instead of lipid pathways, the carbohydrate has focused by overexpressing the glucose-6-phosphate dehydrogenase (G6PD) from the pentose phosphate pathway; the output of NADPH was increased that run to make greater the production of lipids up to 55.7% dry weight in *P. tricornutum* (Blatti et al. 2013). Chloroplast contain G6PD and also presence of increasing in amount and size of lipid bodies, representing the buildup of neutral lipids. Additional skills are the destruction of lipid catabolism for lipid bounding of microalgae (Chu 2017). In an order to targeted knock-down of a multifunctional phospholipase/lipase/acyltransferase enzymes superior lipid production deprived of disturbing the growth of *T. pseudonana* (Aratboni et al. 2019). Hence, the reserved lipid catabolism might disturb biomass efficiency and proliferation. In particular varieties of microalgae, for cell division, the catabolic passageways are vital in providing the energy substrates. Through inducible agents, lipid synthesis genes are controlled in their expression. During the static phase, microalgal cultures must acquire high density, while promoters must be inspired to improve the expression of lipid synthesis genes. Microalgae recognized some pretend promoters, comprising one with copper-responsive elements in *C. reinhardtii* and other promoter responsive to nitrate in diatoms (De Bhowmick et al. 2015). To increase lipid content yield in microalgae is to delaying metabolic pathways which binds the energy-rich storage compound like starch (Ge et al. 2017). For instance, the starch-less mutant of *C. pyrenoidosa* was originated to integrate higher levels of lipids through nitrogen deficiency while showing high progress rate and biomass yield than wild type. The starch-deficient strains in *C. reinhardtii* appeared when disturbing the genes which code for ADP-glucose pyrophosphorylase or isoamylase (Radakovits et al. 2010). Microalgal fatty acids require a chain length between 14 and 20, mainly 16:0, 16:1, and 18:1. Chain lengths of numerous fatty acids are acyl-ACP thioesterases which regulate enzymes that hold some forms of variation of organisms, precise for certain fatty acid lengths (Glencross 2009).

1.19.8 Photosynthetic Efficiency

Designed to enhance algal biomass by capturing the most light, its necessary process increases biofuel production. To advance photosynthetic proficiency is to greater the fascinated level of photosynthetically active pigments. *Acaryochloris marina* has been revealed to absorb light in the nearby infrared region for chlorophyll, demonstrating the strength of transmission of microalgae by the ability for biofuel production (Carvalho et al. 2011). Due to non-photochemical reduction by diminishing the size of antenna in photosystem to lessen the chance of damages energy in microalgae. Microalgae developed inside the lab to have light antenna complexes (LHC) that catch the lights in dim lighting conditions (Borowitzka 2016). Somehow, microalgae once grown in height irradiance maximum of photons immersed will be lost as fluorescence and heat to protect against photodamage. Total of 80–90% of energy loss from the system by the inefficient fluorescence and heat loss (Mussgnug et al. 2007).

1.20 Conclusion

The first stage is to increase enough biomass by photosynthesis for good growth of biomass of algae needing CO_2 sufficiency, nutrients, optimal temperature, as well as the high-performance strains that affect the enzyme kinetics. Research of these factors yet depends on cultures of laboratory-scale (Zhou et al. 2014). Various engineering aspects of the culture system must be tuned in addition to photosynthetic and biological optimal output. *C. reinhardtii* was employed at bench level to systematically investigate the conditions which improve the production of lipid and its growth using experimental approaches with modified integrated computational. By this approach, different parameters are set for the availability of carbon and nitrogen substrate. Following this model, it was discovered that using 0.0742 g/L nitrogen, 0.005–2.1906 g/L acetate, beginning biomass inoculum, a 32.85% increase in algal oil productivity could be achieved, using artificial intelligence (AI), namely, evolutionary and statistical learning, as well as neural learning-based methodologies in an innovative method to improving and managing productivity and costs in algal biofuel production (Boyle et al. 2017). Bayesian clustering and Nave models, closest neighbor, and hidden Markov models are some of the statistical approaches that can be used in AI. One study found that yeast cells may be induced to release TAG into the medium via random mutagenesis (González-Díaz et al. 2008). The use of ABC transporters, which are involved in the mediating of wax release in plants, is another technique to promote lipid secretion from microalgae. Before metabolic engineering may be used to increase the synthesis of lipids in microalgae, the principles of metabolism of lipid in microalgae must be addressed. Analysis of lipidomic using advanced technologies like UPLC/Q-TRAP will help researchers learn more about the assimilation of TAG and lipid metabolism in microalgae (Radakovits et al. 2010). This method was used to do an extensive profile of polar glycerolipids in *Nannochloropsis oceanica*, involving 112 species, to investigate the changes of such lipids in response to N shortage. Through oleic acyl desaturation, phosphatidylcholine (PC) was used as a linoleic and linolenic acyl donor. Through oleic acyl desaturation, phosphatidylcholine (PC) served as a linoleic and linolenic acyl donor and PC converted acyl 16:0 and 18:1 to TAG under N stress conditions (Servaes et al. 2015). The valorization of microalgal products is dependent on a biorefinery idea, which should be the way ahead to enhance algal biofuel economics production. The biorefinery system's main goal is to produce a wide range of products with little or no waste creation (Laurens et al. 2017a). An intriguing technique would be to boost the yield of TAG, a high-value lipophilic compound like carotenoids and LC-PUFA, in prospective microalgae in real time. Such yields share not only storage sinks and biosynthetic precursors but also the stimulation of the molecules' maker, which is frequently dependent on common environmental influences (Schüler et al. 2017). Methods with a high throughput, such as triggered fluorescence sorting of cells, will be useful in separating producers triple capable of simultaneously collecting large amounts of TAG, LCPUFA, and carotenoids (Han et al. 2020). Furthermore, transcriptomics comparisons between wild type and triple producers could also be used to recognize products of gene involved in the biomolecules stimulation. To boost production of lipid and algal biomass, as well as to improve the economics of biofuels whose productivity is dependent on microalgae, multi-pronged techniques including many methodologies should be used. As a result, optimal scale pilot plant testing is needed to confirm the outcomes of laboratory-scale experiments.

References

Abomohra AE-F, Elshobary M (2019) Biodiesel, bioethanol, and biobutanol production from microalgae. In: Alam M, Wang Z (eds) Microalgae biotechnology for development of biofuel and wastewater treatment. Springer, Singapore, pp 293–321

Adegboye MF, Ojuederie OB, Talia PM, Babalola OO (2021) Bioprospecting of microbial strains for biofuel production: metabolic engineering, applications, and challenges. Biotechnol Biofuels 14(1):1–21

Adeniyi OM, Azimov U, Burluka A (2018) Algae biofuel: current status and future applications. Renew Sust Energ Rev 90:316–335

Ahmadi A, Zorofchian Moghadamtousi S, Abubakar S, Zandi K (2015) Antiviral potential of algae polysaccharides isolated from marine sources: a review. Biomed Res Int 2015:825203

Akia M, Yazdani F, Motaee E, Han D, Arandiyan H (2014) A review on conversion of biomass to biofuel by nanocatalysts. Biofuel Res J 1(1):16–25

Álvarez-Díaz PD, Ruiz J, Arbib Z, Barragán J, Garrido-Pérez C, Perales JA (2014) Lipid production of microalga Ankistrodesmus falcatus increased by nutrient and light starvation in a two-stage cultivation process. Appl Biochem Biotechnol 174(4):1471–1483

Ancillotti M, Rerimassie V, Seitz SB, Steurer W (2016) An update of public perceptions of synthetic biology: still undecided? NanoEthics 10(3):309–325

Anto S, Mukherjee SS, Muthappa R, Mathimani T, Deviram G, Kumar SS et al (2020) Algae as green energy reserve: technological outlook on biofuel production. Chemosphere 242:125079

Anwar M, Fayyaz A, Sohail N, Khokhar M, Baqar M, Yasar A et al (2020) CO2 utilization: turning greenhouse gas into fuels and valuable products. J Environ Manag 260:110059

Aratboni HA, Rafiei N, Garcia-Granados R, Alemzadeh A, Morones-Ramírez JR (2019) Biomass and lipid induction strategies in microalgae for biofuel production and other applications. Microb Cell Factories 18(1):1–17

Arguelles ED (2020) Evaluation of nutritional composition and in vitro antioxidant and antibacterial activities of Codium intricatum Okamura from Ilocos Norte (Philippines). Jordan J Biol Sci 13(3):375

Aro E-M (2016) From first generation biofuels to advanced solar biofuels. Ambio 45(1):24–31

Arora N, Yen H-W, Philippidis GP (2020) Harnessing the power of mutagenesis and adaptive laboratory evolution for high lipid production by oleaginous microalgae and yeasts. Sustainability 12(12):5125

Barry A, Wolfe A, English C, Ruddick C, Lambert D (2016) 2016 National algal biofuels technology review. DOE, Washington DC

Behera BK, Varma A (2016) From algae to liquid fuels. In: Microbial resources for sustainable energy. Springer, New York, NY, pp 123–180

Behera B, Unpaprom Y, Ramaraj R, Maniam GP, Govindan N, Paramasivan B (2021) Integrated biomolecular and bioprocess engineering strategies for enhancing the lipid yield from microalgae. Renew Sust Energ Rev 148:111270

Bensehaila S, Doumandji A, Boutekrabt L, Manafikhi H, Peluso I, Bensehaila K et al (2015) The nutritional quality of Spirulina platensis of Tamenrasset, Algeria. Afr J Biotechnol 14(19): 1649–1654

Bharadwaj SV, Ram S, Pancha I, Mishra S (2020) Recent trends in strain improvement for production of biofuels from microalgae. In: Microalgae cultivation for biofuels production. Elsevier, Amsterdam, pp 211–225

Bharathiraja B, Sudharsana T, Jayamuthunagai J, Praveenkumar R, Chozhavendhan S, Iyyappan J (2018) Biogas production–a review on composition, fuel properties, feed stock and principles of anaerobic digestion. Renew Sust Energ Rev 90(April):570–582

Biris-Dorhoi E-S, Michiu D, Pop CR, Rotar AM, Tofana M, Pop OL et al (2020) Macroalgae—a sustainable source of chemical compounds with biological activities. Nutrients 12(10):3085

Blatti JL, Michaud J, Burkart MD (2013) Engineering fatty acid biosynthesis in microalgae for sustainable biodiesel. Curr Opin Chem Biol 17(3):496–505

Bono MS Jr, Ahner BA, Kirby BJ (2013) Detection of algal lipid accumulation due to nitrogen limitation via dielectric spectroscopy of Chlamydomonas reinhardtii suspensions in a coaxial transmission line sample cell. Bioresour Technol 143:623–631

Borowitzka MA (2016) Algal physiology and large-scale outdoor cultures of microalgae. In: The physiology of microalgae. Springer, New York, NY, pp 601–652

Bounnit T, Saadaoui I, Rasheed R, Schipper K, Al Muraikhi M, Al Jabri H (2020) Sustainable production of Nannochloris atomus biomass towards biodiesel production. Sustainability 12(5): 2008

Boyle NR, Sengupta N, Morgan JA (2017) Metabolic flux analysis of heterotrophic growth in Chlamydomonas reinhardtii. PLoS One 12(5):e0177292

Carvalho AP, Silva SO, Baptista JM, Malcata FX (2011) Light requirements in microalgal photobioreactors: an overview of biophotonic aspects. Appl Microbiol Biotechnol 89(5): 1275–1288

Che R, Huang L, Xu J-W, Zhao P, Li T, Ma H, Yu X (2017) Effect of fulvic acid induction on the physiology, metabolism, and lipid biosynthesis-related gene transcription of Monoraphidium sp. FXY-10. Bioresour Technol 227:324–334

Chen S-Y, Mochizuki T, Abe Y, Toba M, Yoshimura Y (2013) Production of high-quality biodiesel fuels from various vegetable oils over Ti-incorporated SBA-15 mesoporous silica. Catal Commun 41:136–139

Chen S, Wu T, Zhao C (2020) Conversion of lipid into high-viscosity branched bio-lubricant base oil. Green Chem 22(21):7348–7354

Chew KW, Yap JY, Show PL, Suan NH, Juan JC, Ling TC et al (2017) Microalgae biorefinery: high value products perspectives. Bioresour Technol 229:53–62

Choi Y-K, Kumaran RS, Jeon HJ, Song H-J, Yang Y-H, Lee SH et al (2015) LED light stress induced biomass and fatty acid production in microalgal biosystem, Acutodesmus obliquus. Spectrochim Acta A Mol Biomol Spectrosc 145:245–253

Chu W-L (2017) Strategies to enhance production of microalgal biomass and lipids for biofuel feedstock. Eur J Phycol 52(4):419–437

Cordova LT, Butler J, Alper HS (2020) Direct production of fatty alcohols from glucose using engineered strains of Yarrowia lipolytica. Metabol Eng Commun 10:e00105

Cray JA, Stevenson A, Ball P, Bankar SB, Eleutherio EC, Ezeji TC et al (2015) Chaotropicity: a key factor in product tolerance of biofuel-producing microorganisms. Curr Opin Biotechnol 33:228–259

Cui H, Meng F, Li F, Wang Y, Duan W, Lin Y (2017) Two-stage mixotrophic cultivation for enhancing the biomass and lipid productivity of Chlorella vulgaris. AMB Express 7(1):1–11

Cui N, Xiao J, Feng Y, Zhao Y, Yu X, Xu J-W et al (2021) Antioxidants enhance lipid productivity in Heveochlorella sp. Yu. Algal Res 55:102235

Culaba AB, Ubando AT, Ching PML, Chen W-H, Chang J-S (2020) Biofuel from microalgae: sustainable pathways. Sustainability 12(19):8009

D'Adamo S, Jinkerson RE, Boyd ES, Brown SL, Baxter BK, Peters JW, Posewitz MC (2014) Evolutionary and biotechnological implications of robust hydrogenase activity in halophilic strains of Tetraselmis. PLoS One 9(1):e85812

Darwesh O, Eida M, Matter I (2021) Environmental nanobiotechnology: microbial-mediated nanoparticles for sustainable environment. In: Lateef A, Gueguim-Kana EB, Dasgupta N, Ranjan S (eds) Microbial nanobiotechnology: principles and applications. Springer, Singapore, p 145

Day JG, Stanley MS (2012) Biological constraints on the production of microalgal-based biofuels. In: The science of algal fuels. Springer, New York, NY, pp 101–129

De Bhowmick G, Koduru L, Sen R (2015) Metabolic pathway engineering towards enhancing microalgal lipid biosynthesis for biofuel application—a review. Renew Sust Energ Rev 50: 1239–1253

Demirbas A (2010) Fuels from biomass. Biorefineries: For biomass upgrading facilities. Springer, London, pp 33–73

Doebbe A, Rupprecht J, Beckmann J, Mussgnug JH, Hallmann A, Hankamer B, Kruse O (2007) Functional integration of the HUP1 hexose symporter gene into the genome of C. reinhardtii: impacts on biological H2 production. J Biotechnol 131(1):27–33

Dolganyuk V, Belova D, Babich O, Prosekov A, Ivanova S, Katserov D et al (2020) Microalgae: a promising source of valuable bioproducts. Biomolecules 10(8):1153

Doron L, Segal NA, Shapira M (2016) Transgene expression in microalgae—from tools to applications. Front Plant Sci 7:505

Eliaz I, Weil E, Wilk B (2007) Integrative medicine and the role of modified citrus pectin/alginates in heavy metal chelation and detoxification-five case reports. Forsch Komplementmed 14(6): 358–364

Fanesi A, Paule A, Bernard O, Briandet R, Lopes F (2019) The architecture of monospecific microalgae biofilms. Microorganisms 7(9):352

Finnigan GC, Thorner J (2015) Complex in vivo ligation using homologous recombination and high-efficiency plasmid rescue from Saccharomyces cerevisiae. Bio-protocol 5(13):e1521

Fitton JH, Irhimeh M, Teas J (2008) 14 Marine algae and polysaccharides with therapeutic applications. In: Barrow CJ, Shahidi F (eds) Marine nutraceuticals and functional foods. CRC Press, Boca Raton, FL, p 345

Fribourg HA (2008) Biomass energy-food or fuel-a global perspective. NACTA J 52:40–57

García-Olivares A, Solé J, Osychenko O (2018) Transportation in a 100% renewable energy system. Energy Convers Manag 158:266–285

Ge S, Champagne P, Plaxton WC, Leite GB, Marazzi F (2017) Microalgal cultivation with waste streams and metabolic constraints to triacylglycerides accumulation for biofuel production. Biofuels Bioprod Biorefin 11(2):325–343

Ghani N, Shahzadi N, Sadaf S, Ullah I, Ali E, Iqbal J et al (2020) Isolation of several indigenous microalgae from Kallar Kahar Lake, Chakwal Pakistan. Iran J Biotechnol 18(3):e2214

Gheda SF, El-Adawi HI, El-Deeb NM (2016) Antiviral profile of brown and red seaweed polysaccharides against hepatitis C virus. Iranian J Pharmaceut Res 15(3):483

Ghosh T, Singh R, Nesamma AA, Jutur PP (2021) Marine polysaccharides: properties and applications. In: Inamuddin, Ahamed MI, Boddula R, Altalhi T (eds) Polysaccharides: properties and applications. Wiley, New York, NY, pp 37–60

Gielen D, Boshell F, Saygin D, Bazilian MD, Wagner N, Gorini R (2019) The role of renewable energy in the global energy transformation. Energy Strat Rev 24:38–50

Glencross BD (2009) Exploring the nutritional demand for essential fatty acids by aquaculture species. Rev Aquac 1(2):71–124

Goncalves EC, Wilkie AC, Kirst M, Rathinasabapathi B (2016) Metabolic regulation of triacylglycerol accumulation in the green algae: identification of potential targets for engineering to improve oil yield. Plant Biotechnol J 14(8):1649–1660

González-Díaz H, González-Díaz Y, Santana L, Ubeira FM, Uriarte E (2008) Proteomics, networks and connectivity indices. Proteomics 8(4):750–778

Guiry MD (2012) How many species of algae are there? J Phycol 48(5):1057–1063

Han X, An Y, Zhou Y, Liu C, Yin W, Xia X (2020) Comparative transcriptome analyses define genes and gene modules differing between two Populus genotypes with contrasting stem growth rates. Biotechnol Biofuels 13(1):1–21

Hannon M, Gimpel J, Tran M, Rasala B, Mayfield S (2010) Biofuels from algae: challenges and potential. Biofuels 1(5):763–784

Hemantkumar JN, Rahimbhai MI (2019) Microalgae and its use in nutraceuticals and food supplements. In: Vítová M (ed) Microalgae-from physiology to application. IntechOpen, London, p 10

Henley WJ, Litaker RW, Novoveská L, Duke CS, Quemada HD, Sayre RT (2013) Initial risk assessment of genetically modified (GM) microalgae for commodity-scale biofuel cultivation. Algal Res 2(1):66–77

Hirani AH, Javed N, Asif M, Basu SK, Kumar A (2018) A review on first-and second-generation biofuel productions. In: Kumar A et al (eds) Biofuels: greenhouse gas mitigation and global warming. Springer, New York, NY, pp 141–154

Hirooka S, Tomita R, Fujiwara T, Ohnuma M, Kuroiwa H, Kuroiwa T, Miyagishima S-Y (2020) Efficient open cultivation of cyanidialean red algae in acidified seawater. Sci Rep 10(1):1–12

Hossain N, Mahlia T, Saidur R (2019) Latest development in microalgae-biofuel production with nano-additives. Biotechnol Biofuels 12(1):1–16

Hosseini Tafreshi A, Shariati M (2009) Dunaliella biotechnology: methods and applications. J Appl Microbiol 107(1):14–35

Hussain F, Shah SZ, Ahmad H, Abubshait SA, Abubshait HA, Laref A et al (2021) Microalgae an ecofriendly and sustainable wastewater treatment option: biomass application in biofuel and bio-fertilizer production. A review. Renew Sust Energ Rev 137:110603

Islam MA, Rahman MM, Heimann K, Nabi MN, Ristovski ZD, Dowell A et al (2015) Combustion analysis of microalgae methyl ester in a common rail direct injection diesel engine. Fuel 143: 351–360

Jagadevan S, Banerjee A, Banerjee C, Guria C, Tiwari R, Baweja M, Shukla P (2018) Recent developments in synthetic biology and metabolic engineering in microalgae towards biofuel production. Biotechnol Biofuels 11(1):1–21

Joutey NT, Bahafid W, Sayel H, El Ghachtouli N (2013) Biodegradation: involved microorganisms and genetically engineered microorganisms. Biodegrad Life Sci 1:289–320

Kamla M, Sushil A, Karmal M (2021) Nanotechnology: a sustainable solution for bioenergy and biofuel production. J Nanosci Nanotechnol 21(6):3481–3494

Khan MI, Shin JH, Kim JD (2018) The promising future of microalgae: current status, challenges, and optimization of a sustainable and renewable industry for biofuels, feed, and other products. Microb Cell Factories 17(1):1–21

Kim S-K (2015) Handbook of marine microalgae: biotechnology advances. Academic Press, London

Koenraad VH, Vandamme P, Vervaecke S, Anita VL (2015) Maldi-TOF MS of microbial mixtures: impressions of its usability for culture-independent analyses of microbial diversity in food ecosystems. Chall Complex 2015:162

Kolesinska B, Fraczyk J, Binczarski M, Modelska M, Berlowska J, Dziugan P et al (2019) Butanol synthesis routes for biofuel production: trends and perspectives. Materials 12(3):350

Konig H, Frank D, Heil R, Coenen C (2013) Synthetic genomics and synthetic biology applications between hopes and concerns. Curr Genom 14(1):11–24

Koyande AK, Show P-L, Guo R, Tang B, Ogino C, Chang J-S (2019) Bio-processing of algal bio-refinery: a review on current advances and future perspectives. Bioengineered 10(1):574–592

Kucharska K, Rybarczyk P, Hołowacz I, Łukajtis R, Glinka M, Kamiński M (2018) Pretreatment of lignocellulosic materials as substrates for fermentation processes. Molecules 23(11):2937

Kumar G, Shekh A, Jakhu S, Sharma Y, Kapoor R, Sharma TR (2020) Bioengineering of microalgae: recent advances, perspectives, and regulatory challenges for industrial application. Front Bioeng Biotechnol 8:914

Kunjapur AM, Eldridge RB (2010) Photobioreactor design for commercial biofuel production from microalgae. Ind Eng Chem Res 49(8):3516–3526

Laurens LM, Chen-Glasser M, McMillan JD (2017a) A perspective on renewable bioenergy from photosynthetic algae as feedstock for biofuels and bioproducts. Algal Res 24:261–264

Laurens LM, Markham J, Templeton DW, Christensen ED, Van Wychen S, Vadelius EW et al (2017b) Development of algae biorefinery concepts for biofuels and bioproducts; a perspective on process-compatible products and their impact on cost-reduction. Energy Environ Sci 10(8): 1716–1738

Lee H-M, Vo PN, Na D (2018) Advancement of metabolic engineering assisted by synthetic biology. Catalysts 8(12):619

Lv H, Qu G, Qi X, Lu L, Tian C, Ma Y (2013) Transcriptome analysis of Chlamydomonas reinhardtii during the process of lipid accumulation. Genomics 101(4):229–237

Mahmud K, Panday D, Mergoum A, Missaoui A (2021) Nitrogen losses and potential mitigation strategies for a sustainable agroecosystem. Sustainability 13(4):2400

Manzoni C, Kia DA, Vandrovcova J, Hardy J, Wood NW, Lewis PA, Ferrari R (2018) Genome, transcriptome and proteome: the rise of omics data and their integration in biomedical sciences. Brief Bioinform 19(2):286–302

Matos ÂP (2017) The impact of microalgae in food science and technology. J Am Oil Chem Soc 94(11):1333–1350

Mehmood MA, Shahid A, Malik S, Wang N, Javed MR, Haider MN et al (2021) Advances in developing metabolically engineered microbial platforms to produce fourth-generation biofuels and high-value biochemicals. Bioresour Technol 337:125510

Michael CI (2015) Utilizing optical light filters and biofilm based cultivating to enhance microalgal growth, Iowa State University

Mobin S, Alam F (2017) Some promising microalgal species for commercial applications: a review. Energy Procedia 110:510–517

Mobin SM, Chowdhury H, Alam F (2019) Commercially important bioproducts from microalgae and their current applications–a review. Energy Procedia 160:752–760

Mohsin M, Rasheed A, Sun H, Zhang J, Iram R, Iqbal N, Abbas Q (2019) Developing low carbon economies: an aggregated composite index based on carbon emissions. Sustain Energy Technol Assess 35:365–374

Mondal M, Goswami S, Ghosh A, Oinam G, Tiwari O, Das P et al (2017) Production of biodiesel from microalgae through biological carbon capture: a review. 3. Biotech 7(2):1–21

de Morais MG, Vaz BS, de Morais EG, Costa JAV (2015) Biologically active metabolites synthesized by microalgae. BioMed Res Int 2015:835761

Morales M, Aflalo C, Bernard O (2021) Microalgal lipids: a review of lipids potential and quantification for 95 phytoplankton species. Biomass Bioenergy 150:106108

Moses T, Mehrshahi P, Smith AG, Goossens A (2017) Synthetic biology approaches for the production of plant metabolites in unicellular organisms. J Exp Bot 68(15):4057–4074

Mukherji S, Van Oudenaarden A (2009) Synthetic biology: understanding biological design from synthetic circuits. Nat Rev Genet 10(12):859–871

Mussgnug JH, Thomas-Hall S, Rupprecht J, Foo A, Klassen V, McDowall A et al (2007) Engineering photosynthetic light capture: impacts on improved solar energy to biomass conversion. Plant Biotechnol J 5(6):802–814

Nagy K, Tiuca I-D (2017) Importance of fatty acids in physiopathology of human body. In: Catala A (ed) Fatty acids. IntechOpen, London

Narala RR, Garg S, Sharma KK, Thomas-Hall SR, Deme M, Li Y, Schenk PM (2016) Comparison of microalgae cultivation in photobioreactor, open raceway pond, and a two-stage hybrid system. Front Energy Res 4:29

Pal P, Chew KW, Yen H-W, Lim JW, Lam MK, Show PL (2019) Cultivation of oily microalgae for the production of third-generation biofuels. Sustainability 11(19):5424

Panahi Y, Khosroushahi AY, Sahebkar A, Heidari HR (2019) Impact of cultivation condition and media content on Chlorella vulgaris composition. Adv Pharmaceut Bull 9(2):182

Park ST, Kim J (2016) Trends in next-generation sequencing and a new era for whole genome sequencing. Int Neurourol J 20(Suppl 2):S76

Park Y-K, Nicaud J-M (2020) Metabolic engineering for unusual lipid production in Yarrowia lipolytica. Microorganisms 8(12):1937

Patil V, Tran K-Q, Giselrød HR (2008) Towards sustainable production of biofuels from microalgae. Int J Mol Sci 9(7):1188–1195

Pourkarimi S, Hallajisani A, Alizadehdakhel A, Golzary A (2020) Factors affecting production of beta-carotene from Dunaliella salina microalgae. Biocatal Agric Biotechnol 29:101771

Ra C-H, Kang C-H, Jung J-H, Jeong G-T, Kim S-K (2016) Effects of light-emitting diodes (LEDs) on the accumulation of lipid content using a two-phase culture process with three microalgae. Bioresour Technol 212:254–261

Rabii A, Aldin S, Dahman Y, Elbeshbishy E (2019) A review on anaerobic co-digestion with a focus on the microbial populations and the effect of multi-stage digester configuration. Energies 12(6):1106

Radakovits R, Jinkerson RE, Darzins A, Posewitz MC (2010) Genetic engineering of algae for enhanced biofuel production. Eukaryot Cell 9(4):486–501

Ragaza JA, Hossain MS, Meiler KA, Velasquez SF, Kumar V (2020) A review on Spirulina: alternative media for cultivation and nutritive value as an aquafeed. Rev Aquac 12(4): 2371–2395

Ramanna L, Rawat I, Bux F (2017) Light enhancement strategies improve microalgal biomass productivity. Renew Sust Energ Rev 80:765–773

Ramaraj R, Dussadee N, Whangchai N, Unpaprom Y (2015) Microalgae biomass as an alternative substrate in biogas production. International Journal of Sustainable and Green Energy. Spec Issue Renew Energy Appl Agric Field Natl Resour Technol 4(1-1):13–19

Ramirez JA, Brown RJ, Rainey TJ (2015) A review of hydrothermal liquefaction bio-crude properties and prospects for upgrading to transportation fuels. Energies 8(7):6765–6794

Rawat I, Kumar RR, Mutanda T, Bux F (2013) Biodiesel from microalgae: a critical evaluation from laboratory to large scale production. Appl Energy 103:444–467

Remmers IM, Wijffels RH, Barbosa MJ, Lamers PP (2018) Can we approach theoretical lipid yields in microalgae? Trends Biotechnol 36(3):265–276

Rismani-Yazdi H, Haznedaroglu BZ, Bibby K, Peccia J (2011) Transcriptome sequencing and annotation of the microalgae Dunaliella tertiolecta: pathway description and gene discovery for production of next-generation biofuels. BMC Genomics 12(1):1–17

Rodriguez GM, Atsumi S (2014) Toward aldehyde and alkane production by removing aldehyde reductase activity in Escherichia coli. Metab Eng 25:227–237

Røkke G, Korvald E, Pahr J, Øyås O, Lale R (2014) BioBrick assembly standards and techniques and associated software tools. In: Chandran S et al (eds) DNA cloning and assembly methods. Springer, New York, NY, pp 1–24

Sajjadi B, Chen W-Y, Raman AAA, Ibrahim S (2018) Microalgae lipid and biomass for biofuel production: a comprehensive review on lipid enhancement strategies and their effects on fatty acid composition. Renew Sust Energ Rev 97:200–232

Santos SA, Félix R, Pais A, Rocha SM, Silvestre AJ (2019) The quest for phenolic compounds from macroalgae: a review of extraction and identification methodologies. Biomolecules 9(12):847

Santos-Sánchez N, Valadez-Blanco R, Hernández-Carlos B, Torres-Ariño A, Guadarrama-Mendoza P, Salas-Coronado R (2016) Lipids rich in ω-3 polyunsaturated fatty acids from microalgae. Appl Microbiol Biotechnol 100(20):8667–8684

Sanzo GD, Mehariya S, Martino M, Larocca V, Casella P, Chianese S et al (2018) Supercritical carbon dioxide extraction of astaxanthin, lutein, and fatty acids from Haematococcus pluvialis microalgae. Mar Drugs 16(9):334

Sathasivam R, Radhakrishnan R, Hashem A, Abd-Allah EF (2019) Microalgae metabolites: a rich source for food and medicine. Saudi J Biol Sci 26(4):709–722

Schüler LM, Schulze PS, Pereira H, Barreira L, León R, Varela J (2017) Trends and strategies to enhance triacylglycerols and high-value compounds in microalgae. Algal Res 25:263–273

Seo YH, Lee Y, Jeon DY, Han J-I (2015) Enhancing the light utilization efficiency of microalgae using organic dyes. Bioresour Technol 181:355–359
Servaes K, Maesen M, Prandi B, Sforza S, Elst K (2015) Polar lipid profile of Nannochloropsis oculata determined using a variety of lipid extraction procedures. J Agric Food Chem 63(15): 3931–3941
Shah M, Mahfuzur R, Liang Y, Cheng JJ, Daroch M (2016) Astaxanthin-producing green microalga Haematococcus pluvialis: from single cell to high value commercial products. Front Plant Sci 7:531
Shahid A, Rehman AU, Usman M, Ashraf MUF, Javed MR, Khan AZ et al (2020) Engineering the metabolic pathways of lipid biosynthesis to develop robust microalgal strains for biodiesel production. Biotechnol Appl Biochem 67(1):41–51
Show PL, Tang MS, Nagarajan D, Ling TC, Ooi C-W, Chang J-S (2017) A holistic approach to managing microalgae for biofuel applications. Int J Mol Sci 18(1):215
Shuba ES, Kifle D (2018) Microalgae to biofuels: 'Promising' alternative and renewable energy, review. Renew Sust Energ Rev 81:743–755
Simosa AE (2016) Factors affecting algal biomass growth and cell wall destruction
Sivaramakrishnan R, Incharoensakdi A (2017) Enhancement of lipid production in Scenedesmus sp. by UV mutagenesis and hydrogen peroxide treatment. Bioresour Technol 235:366–370
Snow AA, Smith VH (2012) Genetically engineered algae for biofuels: a key role for ecologists. Bioscience 62(8):765–768
Spalviņš K (2020) Single cell protein and single cell oil production from agro industrial by products
Srivastava RK (2019) Bio-energy production by contribution of effective and suitable microbial system. Materials Sci Energy Technol 2(2):308–318
Srivastava RK, Shetti NP, Reddy KR, Aminabhavi TM (2020) Biofuels, biodiesel and biohydrogen production using bioprocesses. A review. Environ Chem Lett 18(4):1049–1072
Subhadra BG (2010) Sustainability of algal biofuel production using integrated renewable energy park (IREP) and algal biorefinery approach. Energy Policy 38(10):5892–5901
Sun X-M, Ren L-J, Zhao Q-Y, Ji X-J, Huang H (2018) Microalgae for the production of lipid and carotenoids: a review with focus on stress regulation and adaptation. Biotechnol Biofuels 11(1): 1–16
Tao L, He X, Tan EC, Zhang M, Aden A (2014) Comparative techno-economic analysis and reviews of n-butanol production from corn grain and corn stover. Biofuels Bioprod Biorefin 8(3):342–361
Tonnaer E (2017) Microalgae cultivation for the production of algae jet fuel in the Netherlands-a feasibility study
Tran NN (2018) Optimization of the production of biodiesel from recycled grease trap waste
Trivedi J, Aila M, Bangwal D, Kaul S, Garg M (2015) Algae based biorefinery—how to make sense? Renew Sust Energ Rev 47:295–307
Vogt C, Richnow HH (2013) Bioremediation via in situ microbial degradation of organic pollutants. In: Schippers A et al (eds) Geobiotechnology II. Springer, Berlin, pp 123–146
Warner ME, Suggett DJ (2016) The photobiology of Symbiodinium spp.: linking physiological diversity to the implications of stress and resilience. In: Goffredo S, Dubinsky Z (eds) The Cnidaria, past, present and future. Springer, Berlin, pp 489–509
Wells ML, Potin P, Craigie JS, Raven JA, Merchant SS, Helliwell KE et al (2017) Algae as nutritional and functional food sources: revisiting our understanding. J Appl Phycol 29(2): 949–982
Work VH (2014) Metabolic and physiological engineering of photosynthetic microorganisms for the synthesis of bioenergy feedstocks: development, characterization, and optimization. Colorado School of Mines
Wu W, Chang J-S (2019) Integrated algal biorefineries from process systems engineering aspects: a review. Bioresour Technol 291:121939
Xia PF, Ling H, Foo JL, Chang MW (2019) Synthetic biology toolkits for metabolic engineering of cyanobacteria. Biotechnol J 14(6):1800496

Xu Z, Wang H, Cheng P, Chang T, Chen P, Zhou C, Ruan R (2020) Development of integrated culture systems and harvesting methods for improved algal biomass productivity and wastewater resource recovery–a review. Sci Total Environ 746:141039

Yaakob Z, Ali E, Zainal A, Mohamad M, Takriff MS (2014) An overview: biomolecules from microalgae for animal feed and aquaculture. J Biol Res Thess 21(1):1–10

Yaakob MA, Mohamed RMSR, Al-Gheethi A, Ravishankar GA, Ambati RR (2021) Influence of nitrogen and phosphorus on microalgal growth, biomass, lipid, and fatty acid production: an overview. Cells 10(2):393

Young R, Haines M, Storch M, Freemont PS (2020) Combinatorial metabolic pathway assembly approaches and toolkits for modular assembly. Metab Eng 63:81

Yun H-S, Kim Y-S, Yoon H-S (2020) Characterization of Chlorella sorokiniana and Chlorella vulgaris fatty acid components under a wide range of light intensity and growth temperature for their use as biological resources. Heliyon 6(7):e04447

Zhang X (2015) Microalgae removal of CO2 from flue gas. IEA Clean Coal Centre, London

Zhou W, Chen P, Min M, Ma X, Wang J, Griffith R et al (2014) Environment-enhancing algal biofuel production using wastewaters. Renew Sust Energ Rev 36:256–269

Zhu S, Wang W, Feng J, Shang C, Wang Z, Liu S (2018) 2. Foundations in microbiology. In: Yuan Z (ed) Microbial energy conversion. De Gruyter, Berlin, pp 19–108

Zhu B, Ni F, Xiong Q, Yao Z (2021) Marine oligosaccharides originated from seaweeds: source, preparation, structure, physiological activity and applications. Crit Rev Food Sci Nutr 61(1): 60–74

Zorofchian Moghadamtousi S, Karimian H, Khanabdali R, Razavi M, Firoozinia M, Zandi K, Abdul Kadir H (2014) Anticancer and antitumor potential of fucoidan and fucoxanthin, two main metabolites isolated from brown algae. Sci World J 2014:768323

Chapter 2
The Use of Omics Technologies, Random Mutagenesis, and Genetic Transformation Techniques to Improve Algae for Biodiesel Industry

Ali Osman Adiguzel

Abstract Biodiesel as a transportation fuel has gained interest because of fossil fuel depletion and global warming. Algae, with the ability to grow fast in a variety of conditions, are a sustainable choice for biodiesel production. The most critical challenge is insufficient lipid yield in algal sources and associated high biorefinery costs. It is well-known that some algal strains accumulate relatively high lipid under environmental or nutritional stress that hinder algal growth. Genomic, transcriptomic, and proteomic analyses of stressed algal strains help identify and characterize essential molecules that induce lipid accumulation. Random mutagenesis, in addition to the development of algal strains with high lipid content for the biodiesel industry, provides clues for further genetic engineering attempts if combined with omics technologies. Another powerful genetic engineering tool to improve algae growth, lipid content, and stress tolerance is genetic transformation. In the present chapter, first, the commonly used omics methods and the use of omics data to improve algae species with potential in the biodiesel industry are summarized. In addition, random mutagenesis and its applications in strain improvement for the biodiesel industry are examined. Finally, genetic transformation methods and vector construction strategies are reviewed as well as up-to-date reports on the progress in algal transformation for biodiesel production.

Keywords Biodiesel · Genomics · Transcriptomics · Proteomics · Random mutagenesis · Genetic transformation

A. O. Adiguzel (✉)
Department of Molecular Biology and Genetics, Science Faculty, University of Ondokuz Mayıs, Samsun, Turkey

N. Srivastava, P. K. Mishra (eds.), *Technological Advancement in Algal Biofuels Production*, Clean Energy Production Technologies,
https://doi.org/10.1007/978-981-19-6806-8_2

2.1 Introduction

Humanity is faced with the energy crisis and environmental problems. The increase in the world population is predicted to reach nine billion in 2050 (Fayyaz et al. 2020). Moreover, the current energy demand has been estimated to increase by about 50% per 20 years (Stephenson et al. 2011). The current scenario necessitates the use of renewable energies from natural sources such as solar, wind, water, and biomass. Algae are influential producers of many compounds, including amino acids, proteins, pigments, carbohydrates, hormones, and lipids. Their ability rapidly grows by solar energy and converts CO_2 into biomass (Khan and Fu 2020). They also contribute to the reduction of environmental pollution via the degradation of some toxic chemicals (Sivakumar et al. 2012). Their water footprint is low as it can be grown in industrial wastewater and seawaters. Therefore, algae have gained increasing interest as a renewable feedstock for methane, electricity, biohydrogen, syngas, bioethanol, and biodiesel production (Zhang et al. 2021a; Kannan and Donnellan 2021; Kaloudas et al. 2021).

Biodiesel, one of the important renewable energies, is traditionally produced using sunflower oil (Antolın et al. 2002), palm oil (Al-Widyan and Al-Shyoukh 2002), rapeseed oil (Peterson et al. 1996), and residual oil (Bouaid et al. 2007). Alternatively, naturally lipid-rich algal strains of *Chlamydomonas*, *Chlorella*, *Isochrysis*, *Nannochloropsis*, *Neochloris*, *Scenedesmus*, and *Schizochytrium* can be used as cheap and cleaner feedstock in the biodiesel industry (Razeghifard 2013). Many of them have limited use in industrial-scale biodiesel production because considerable lipid productivity is not always achieved during the rapid growth of algal strains (Ng et al. 2017). It is well-known that different cultivation methods induce the lipid productivity of algal strains for a long time (Salama et al. 2019). Molecular analysis of changes in algal cells under different conditions provides clues about key molecules involved in lipid production. Furthermore, sequential application of omics and conventional genetic engineering approaches facilitate the improvement of yield, quality, and accessibility. They also help to reduce the toxic effects of biofuels, production cost, and process time. Schematic of the algal improvement concept for the biodiesel industry is shown in Fig. 2.1. The present chapter summarizes methods in genomics, transcriptomics, proteomics, random mutagenesis, and genetic transformation. Furthermore, applications of omics, random mutagenesis, and genetic transformation are presented.

2.2 Algal Omics

"Omic" studies focus on quantifying and characterizing DNA, RNA, and protein pools extracted from cells (Guarnieri and Pienkos 2015). Detailed investigation of these biological molecules, their functions, and expression patterns with "omic" technologies, containing genomics, transcriptomics, and proteomics, is essential to

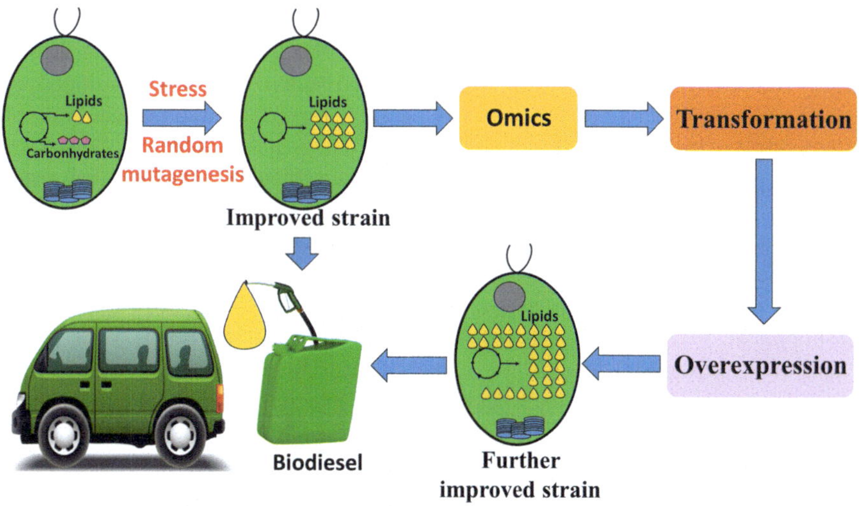

Fig. 2.1 Improvement of algal strains for the production of biodiesel

improve the production of algal biofuels by genetic engineering approaches (Anand et al. 2019). In addition, large data sets from "omic" technologies provide understanding metabolism and physiology of algal cells and predicting biofuel yields of species under different conditions. An application of omics technologies on algal species is known as "algomics."

2.2.1 Genomics

Genomics identifies nucleotide sequences, genes, gene arrangement, repeated elements, regulatory elements, and cDNAs. Genome information helps discover novel metabolic pathways, clarify conserved behaviors across algal species, and determine the types, numbers, and functions of proteins produced by organisms. In the first step of the genomic studies, DNA is sequenced (Reijnders et al. 2014). There are many different sequencing techniques like pyrosequencing, sequencing by ligation, and real-time sequencing. However, next-generation sequencing (NGS) on different technology platforms such as Roche GS FLX Titanium XL+, Life Technologies Ion torrent, Illumina/solexa HiSeq2000/2500, Illumina MiSeq, and Helicos Heliscope (Buermans and Den Dunnen 2014) is currently preferred because of its fast, sensitive, and high capacity. In the second step, bioinformatic databases and tools including Newbler (Roche), Velvet (https://www.ebi.ac.uk/~zerbino/velvet/), CLC Genomic Workbench (Qiagen), Augustus (http://bioinf.uni-greifswald.de/augustus/), Eugene (http://eugene.toulouse.inra.fr/), BLAST (https://blast.ncbi.nlm.nih.gov/Blast.cgi), Algal Functional Annotation Tool (http://pathways.mcdb.ucla.

edu), UniProtKB/Swiss-Prot (https://www.expasy.org/resources/uniprotkb-swiss-prot), GeneMark (http://exon.gatech.edu/GeneMark/), Pfam (https://pfam.xfam.org/), and KEGG (https://www.genome.jp/kegg/pathway.html) are exploited to genome assembly, predict the function of genes, and elucidate the probable role of hypothetical proteins.

Algal genomic studies first began to sequences of *Chlamydomonas reinhardtii* genome (Shrager et al. 2003). Since then, 200 formerly genome sequencing project (https://www.ncbi.nlm.nih.gov/genome) have been completed for different purposes, especially the classification, identification, and characterization of algal strains (Cock et al. 2010; Curtis et al. 2012; Moore et al. 2012; Mock et al. 2017; Guillory et al. 2018; Shoguchi et al. 2018; Takahashi et al. 2018; Benites et al. 2019; Zhang et al. 2020a, b). However, 31 of them provide the required basic and current information with both assembled and annotated genes for genetic engineering of algae with potential in biofuel production listed in Table 2.1.

2.2.1.1 Application of Genomics in the Improvement of Algal Lipid Biosynthesis

So far, various genes and gene products involved in biofuel production have been identified with genetic information from the nuclear genome along with the organelles to facilitate genetic engineering. Radakovits et al. (2012) investigated the potential of *N. gaditana* in biofuel production by comparative genomic data. Data revealed that the key enzymes associated with the biosynthesis of fatty acids and TAG were acetyl-CoA synthetase/carboxylase, malonyl-CoA:ACP-trans-acylase, 3-oxoacyl-ACP synthase/reductase, enoyl-ACP reductase I, oleyl-ACP hydrolase, acyl-ACP desaturase, omega-6 fatty acid desaturase delta-12, stearoyl-CoA desaturase, glycerol-3-phosphate dehydrogenase (GPDH)/acyltransferase, 1-acylglycerol-3-phosphate O-acyltransferase, phosphatidic acid phosphatase, diacylglycerol kinase, diacylglycerol acyltransferase, monogalactosyldiacylglycerol synthase, and UDP-sulfoquinovose synthase. In another study, whole-genome (65.35 Mb) *Scenedesmus quadricauda* LWG002611, a possible candidate for biofuel production, was sequenced, assembled, and annotated (Dasgupta et al. 2018). A total of 13,514 and 16,739 genes were identified by de novo and reference-guided assembly, respectively. As a result of the study, it was determined that the single palmitoyl-acyl carrier protein thioesterase gene plays a significant role in the production of higher amounts of palmitate (16:0) and lesser amounts of other fatty acids. In a study conducted by Sahoo et al. (2020), putative hypothetical proteins play an important role in biofuel synthesis and can be defined by comparing genomes from algae classified in different taxa and further bioinformatic analysis based on homology modeling approach. A comparative genomics study showed that genes responsible for ATP-citrate lyase, phosphoenolpyruvate carboxylase, and malic dehydrogenase are vital for the accumulation of lipid in 26 microalgal species classified in *Chlorophyta*, *Cryptophyta*, *Dinoflagellate*, *Haptophyta*, *Heterokonta*, and *Rhodophyta* (Sahoo et al. 2020).

Table 2.1 List of algal strains whose genomes have been sequenced

Species	Genome size (Mb)	Chromosome number	GC%	Predicted gene number	References
C. reinhardtii	121	17	64	15,143	Merchant et al. (2007)
Hemiselmis andersenii CCMP644	0.632425	3	25.53	525	Lane et al. (2007)
Ostreococcus lucimarinus CCE9901	13.2	21	60.44	7651	Palenik et al. (2007)
Phaeodactylum tricornutum CCMP2561	27.40	33		10,402	Bowler et al. (2008)
Micromonas pusilla	21.96	17	65.7	10,248	Worden et al. (2009)
Cyanidioschyzon merolae 10D	16.73	20	54.81	4803	Rossoni et al. (2019)
Volvox carteri f. nagariensis	137.68	14	55.3	14,437	Prochnik et al. (2010)
Chlorella variabilis NC64A	46.16	12	65.5	9780	Blanc et al. (2010)
Nannochloropsis gaditana CCMP526	29.00		39	3557	Radakovits et al. (2012)
Guillardia theta CCMP2712	87.15	3	52.9	24,923	Curtis et al. (2012)
Coccomyxa subellipsoidea C-169	48.83	20	52.9	9915	Blanc et al. (2012)
Monoraphidium neglectum	69.48		64.2	16,763	Bogen et al. (2013)
Emiliania huxleyi CCMP1516	167.68		64.5	38,549	Read et al. (2013)
Porphyridium purpureum	20.8221	8–10	55.7	8355	Bhattacharya et al. (2013)
Nannochloropsis granulata CCMP525	27.65		50.1		Wang et al. (2014)
Ostreococcus tauri OT2009	12.9	20		7699	Blanc-Mathieu et al. (2014)
Auxenochlorella protothecoides 0710	22.9	≥6	63	7039	Gao et al. (2014)
Saccharina japonica	537	15	49.1	18,733	Ye et al. (2015)
Dunaliella salina	343.7		48.9	16,683	Polle et al. (2017)
Micractinium conductrix SAG 241.80	60.8	≥13	67.3	9349	Arriola et al. (2018)
Chlorella sorokiniana UTEX 1602	59.4	≥13	64.1	9587	Arriola et al. (2018)
Raphidocelis subcapitata NIES-35	51.16		70.0	13,429	Suzuki et al. (2018)
Skeletonema costatum (Ariake8)	46.9		45.1	16,449	Ogura et al. (2018)

(continued)

Table 2.1 (continued)

Species	Genome size (Mb)	Chromosome number	GC%	Predicted gene number	References
Synechococcus elongatus PCC 11801	2.7		54.9	2793	Jaiswal et al. (2018)
Chlorella vulgaris 211/11P	40	14	61	10,724	Cecchin et al. (2019)
Picochlorum renovo	14.4		46.2	8902	Dahlin et al. (2019)
Desmodesmus armatus	116.3		56.6		Knoshaug et al. (2020)
Tetradesmus acuminatus	119.24		55		Astafyeva et al. (2020)
Messastrum gracile SE-MC4	60.83		68.27		Teh et al. (2019)
Tribonema minus UTEX B ZZ1240	158.35		57.0	18,441	Mahan et al. (2021)
Trichormus variabilis	7		41.39	5751	Chen et al. (2021)

2.2.2 *Transcriptomics*

Transcriptomics is RNA-based systematic approach aimed to shed light on the function of genes through expression patterns of cells under different conditions. Transcriptomics analyses compare the responses of the algal strains to varying environmental conditions (de Carvalho et al. 2019). It bridges the gap between genomics and proteomics (Dong and Chen 2013). The genomic data to identify the key genes for enhanced lipid productivity is not always useful for all algal strains. It generally ensures the understanding of the carbon route. Therefore, further transcriptome analyses are often required (Larsson et al. 2021).

Expressed RNA molecules at a specific moment referring to transcriptome can be investigated through expressed sequence tags (ESTs), microarrays, and RNA sequencing (RNA-seq) techniques. ESTs are single-pass sequencing of cDNA clones randomly selected from library (Hatey et al. 1998). mRNAs with poly-A tail are extracted from the target cell and then reverse transcribed to cDNAs. Subsequently, cDNAs were cloned to an appropriate vector to construct a library. Randomly picked clones from the library are sequenced (200–800 bp) using universal primers that bind to specific sites on the vectors at both ends of the insert. Microarray quantifies the mRNAs through hybridization to cDNA fragments (or oligonucleotides in situ synthesized) spotted on the surface of solid material like glass microscopy slides. mRNAs from control and sample are labeled with Cy3 (green fluorescent dye) and Cy5 (red fluorescent dye) and applied on an array. Each labeled nucleotide attached to each point is quantified by various software such as Chipster (Kallio et al. 2011), Maanova (Wu et al. 2003), Base (Saal et al. 2002),

GenePix Pro (Bengtsson and Bengtsson 2006), and TM4 (Saeed et al. 2006) after the array was scanned by laser emission. RNA-seq is the most recent and developed transcriptome analysis technique. It provides identification of both coding and noncoding RNAs as well as mapping of the transcriptome (Lowe et al. 2017). One of the important advantages of RNA-seq is that it does not need any reference during analysis. First, total RNA is fragmented into short pieces ranging in length from 200 to 500 base. Fragments are used to synthesis of the first strand of cDNA. Then, a complementary strand was synthesized. Double strand DNA are subjected to end repairing, ligated to adaptors for prepare sequencing library. Fragments are sequenced with NGS (Kukurba and Montgomery 2015). RNA-seq reads can be aligned using various tools, including GSNAP (Wu and Nacu 2010), MapSplice (Wang et al. 2010), RUM (Grant et al. 2011), STAR (Dobin et al. 2013), and TopHat (Trapnell et al. 2009). Aligned reads are assembled by either reference annotation or de nova reconstruction approaches and quantified using different tools such as Cufflinks (Trapnell et al. 2010), iReckon (Mezlini et al. 2013), and FluxCapacitor (Griebel et al. 2012).

2.2.2.1 Application of Transcriptomics in the Improvement of Algal Lipid Biosynthesis

Several transcriptome studies aimed to identify significant genes, molecules, and novel approaches to enhance biofuel production by algae have been reported. In one of them, transcriptome analysis of *Chlorella pyrenoidosa* revealed that the *accA* (acetyl-CoA carboxylase; ACCase), *me g6562* (malic enzyme; ME), *me g4297*, *rbcS* (RuBisCO large subunit), and *pepc g8086* (phosphoenolpyruvate carboxylase; PEPC) genes triggered lipid biosynthesis under nutrient deficiency (nitrogen, phosphorus, and iron), while rbcL (RuBisCO large subunit), *me g3137*, and pepc g6833 genes decreased the yield (Fan et al. 2014). Transcriptomes of *Neodesmus* sp. UTEX 2219-4 exposed to various stress have been sequenced and compared (Chang et al. 2016). They suggested that the ACCase is the main enzyme to increase the biosynthesis of fatty acids. Likewise, transcriptional upregulations have also been reported for several algae under conditions that promote lipid accumulation (Wan et al. 2014; Liu et al. 2017). It is possible that upregulation of ACCase is an essential condition for some algae to enhance their lipid content. However, in certain other algal strains, upregulations of ACCase at the transcript level were not invariably followed by increased lipid accumulation. Valenzuela et al. revealed that the ACC precursors like acetyl-CoA are more significant than the ACCase for TAG accumulation in *P. tricornutum* (Valenzuela et al. 2012). Similar findings have also been reported by Liang et al. for *Nannochloropsis* sp. (Liang et al. 2013). Some other transcriptome studies suggested that the various genes responsible for phosphatidic acid phosphatase (PAP), which catalyzes the dephosphorylation of phosphatidic acid resulting in diacylglycerol (DAG), or diacylglycerol acyltransferase (DGAT), which catalyzes the DAG to triglycerides, had an impact on cellular lipid biosynthesis in some *Chlorella* (Huang et al. 2016; Qu and Miao 2021), *Nannochloropsis* (Wang et al. 2014; Fattore et al. 2021), *Scenedesmus* (Sharma and Chauhan 2016; Yang et al. 2018), *Chromochloris* (Mao et al. 2020; Zhang et al. 2021a, b, c), *Ettlia* (Sturme et al. 2018), and *Tribonema* (Wang et al. 2017) species. Some transcriptome studies have revealed that upregulation of various transcription factors rather than enzymes involved in metabolic pathways increases lipid biosynthesis (Courchesne et al. 2009; Xing et al. 2021a, b). Shang et al. (2016) reported an increase in the expression level of *wri1* and coded a transcription factor known as WEINKLED1, during lipid accumulation of *Dunaliella parva* under nitrogen limitation and *Dunaliella tertiolecta* exposed to triethylamine stress (Chen et al. 2020). Correlation between wri1 and lipid accumulation has been shown in *Scenedesmus dimorphus* and *Scenedesmus quadricauda* with a comparative transcriptome study conducted by Sharma and Chauhan (2016). Transcriptome analysis is also used to identify the downregulated genes during lipid biosynthesis. A comparison of the transcriptomes of *Scenedesmus acutus* TISTR8540 revealed that numerous lipase genes were downregulated under nitrogen limited condition, which promotes lipid accumulation (Sirikhachornkit et al. 2018). Similarly, it has been reported that downregulation of lipolytic genes in *Monoraphidium neglectum* had a positive impact on lipid accumulation (Jaeger et al. 2017). In *C. subellipsoidea*, downregulation of PEP carboxylase enhances biomass by altering the carbon flux direction (Peng et al. 2016). As seen in the applications mentioned above, transcriptomic researches identify the possible targets for future genetic engineering approaches to improve the production of algal biofuels.

2.2.3 Proteomics

mRNAs are not always translated into protein for reasons such as rapid degradation or ineffective translation of mRNA. In addition, alternative splicing, post-translational modifications, and the necessity of some proteins to form complexes with other molecules to gain activity cause an increase in protein diversity in cells (Schubert et al. 2017; Li et al. 2017; Dupree et al. 2020). Therefore, expressed mRNAs do not reflect the active protein profile of cells. Thus, the combination of transcriptomics and proteomics is necessary to understand the molecular basis of stress-induced lipid accumulation in algae. Algal proteomics determines the quantitative changes in protein profiles under different stress conditions leading to fatty acid or carbohydrate accumulation (Ndimba et al. 2013; Beaulieu 2019). Thus, target molecules for further improvement by genetic engineering are identified.

Proteomics refers to identifying the protein repertoire of cells at different conditions and/or time (Tyers and Mann 2003). It also quantifies the proteins and presents insight into their function in biochemical pathways. Proteomic studies can be carried out using different approaches, including gel-based, mass spectrometry (MS)-based (Noor et al. 2021), stable isotope labeling by amino acids in cell culture (SILAC)-based (Chen et al. 2015), isotope-coded affinity tags (ICAT)-based (Shiio and Aebersold 2006), isobaric tags for relative and absolute quantification (iTRAQ)-

based (Zhang et al. 2018), tandem mass tag (TMT)-based (Myers et al. 2018), dimethyl labeling-based and gel-free approaches, and isobaric-labeling-based (Rai et al. 2016). Among them, gel-, MS-, and iTRAQ-based approaches have become a prime method in proteome analysis of algal cells. Especially gel-based approach is a visual approach for mapping variations in protein expression that is extremely straightforward to use.

Success in the extraction of proteins from algal cells is a crucial step for application of both gel- and MS-based approaches. Although the algal proteins are conventionally extracted by physical (ultrasonic homogenization, osmotic shock, grinding, and bead milling), chemical (NaOH, HCl, phenol, TCA-acetone, TRIzol) and enzymatic (Lysozyme, Autolysin, cellulase, κ-carrageenase, β-agarase, xylanase) extraction methods, there is no ideal protein extraction method for studying the algal proteome due to the variability of cell wall structure and the obstruction of protein extraction by other macromolecules and secondary metabolites (Bleakley and Hayes 2017; Karthikaichamy et al. 2017; Sierra et al. 2017). Therefore, species-specific optimization is needed before the algal proteomic studies. On the other hand, methods using organic solvent are preferred for protein extraction from algae employed in biodiesel production. Thus, de-lipidation of protein solution is achieved. Another major challenge, which comprises concentration, desalting, or partial fractionization of proteins, is often overcome by ultrafiltration. Ultrafiltration (UF) is a technique of removing relatively small particles and dissolved molecules from protein lysates using membranes (Chernokalskaya et al. 2004). It also uses for buffer exchange. Contaminants with relatively low molecular weight are passed through the membranes of nominal molecular weight cut-off (NMWC) using either stirred ultrafiltration cells (SUC) based on nitrogen gas pressure or centrifugal ultrafiltration devices (CUD) based on centrifugation force. CUD is more preferable in proteomic studies due to its fast, cost-effective, and easy handle properties. The protein concentration of cell lysate or concentrated protein sample is estimated by Bradford (Bradford 1976) or Lowry methods (Lowry et al. 1951).

The general workflow of conventional gel-based proteomic studies is shown in Fig. 2.2. First, proteins are separated by sodium dodecyl sulfate-polyacrylamide gel electrophoresis (SDS-PAGE) or two-dimensional-polyacrylamide gel electrophoresis (2D-PAGE) (Görg et al. 2004). 2D-PAGE is the most preferred technique for the separation of proteins due to its resolution. Proteins are separated in the horizontal direction according to their isoelectric points (pI) on commercially available immobilized pH gradient (IPG) strips by isoelectric focusing (IEF) in the first dimension (Bjellqvist et al. 1982; Issaq and Veenstra 2008). IEF consist of rehydration, focusing, and equilibration steps (Görg et al. 2006). Rehydration buffer containing urea (5–8 M), thiourea (2 M), detergent (CHAPS, CHAPSO or NP-40 at 0.5–4%), reducing agent (dithiothreitol or dithioerythritol at 20–100 mM), and traces of bromophenol blue is loaded with algal protein sample and then poured on a channel of the tray. Rehydration is performed at 50 V at 20 °C for 10–12 h after the IPG strip places on the channel. Focusing is conducted at linear, rapid, or slow voltage ramping mode. Resistance of the IPG strip and content of the rehydration mixture are taken into consideration when selecting the voltage ramping

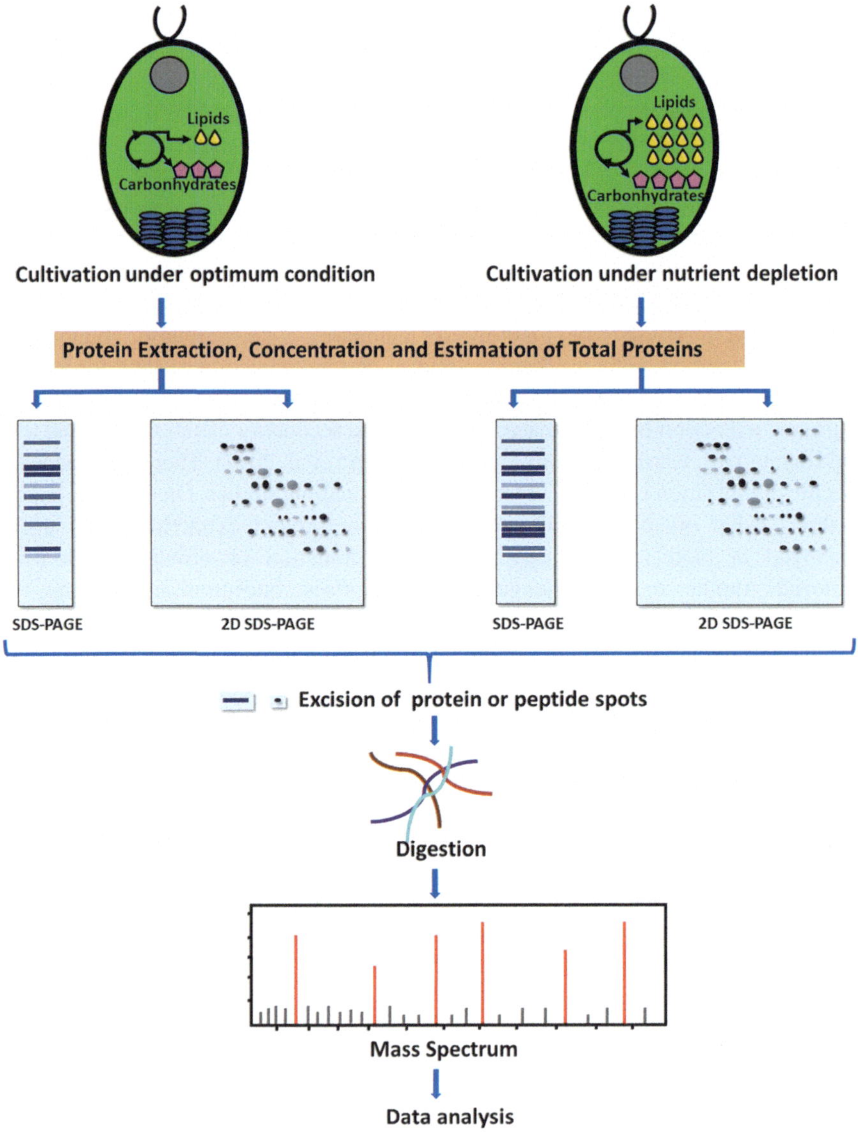

Fig. 2.2 Schematic overview of conventional gel-based proteomic studies

mode. Proteins migrate to their pI at the end of focusing. Subsequently, the IPG strip is equilibrated so that proteins may properly interact with SDS in the second dimension. Equilibration is accomplished by sequential incubating the IPG strip in 50 mM Tris-HCl (pH 8.8) containing 6 M urea, 2% SDS, 20% glycerol, 2% DTT, and the same buffer with 5% iodoacetamide instead of DTT for 15 min (Song et al. 2013). Proteins on the IPG strip are separated horizontally according to their molecular mass by SDS-PAGE. Separation is performed in 10% acrylamide gel using Tris-Glycine Buffer system (pH 8.3) with SDS (Laemmli 1970). The separated proteins are visualized by different methods. Autoradiography is a highly sensitive visualization method. However, it is expensive and harmful. The most practical, low-cost, and widely used method is the Coomassie staining, but its sensitivity is relatively low, with a detection limit in the range of 200–500 ng peptide per spot (Görg et al. 2004). Silver (Mortz et al. 2001), SYPRO Ruby (Cong et al. 2008), and deep purple staining (Mackintosh et al. 2003) is the method that allows the detection of higher than 0.1 ng, 1 ng, and 64 pg protein or peptide spots, respectively.

Conventional gel-based methodologies require the comparison of at least two distinct gels to uncover protein differences between cells grown under different conditions (Kim and Cho 2019). Inhomogeneity of gels limits spot matching during image analysis. Alternatively, differential gel electrophoresis (DIGE) enables the separation of two protein samples, labeled with two cyanine dyes (Cy3 and Cy5) in one polyacrylamide gel (Ünlü et al. 1997; Meleady 2018). Two protein samples separated simultaneously under the same conditions are visualized by two sequential imaging at 532 nm/580 nm and 633 nm/670 nm excitation/emission wavelength.

Comparing and analyzing two-dimensional images follows the excision of gel pieces, destaining, and in-gel digestion. The destaining procedure may differ according to the staining method. The gel pieces that are opaque as a result of destaining are partially dried and then saturated with a protease such as trypsin, Glu-C, Lys-C, and Asp-N. Trypsin, which cleaves proteins into small fragments with arginine or lysine at C-terminus, is the most preferred protease in proteomic analysis. It is highly specific, stable in various buffer systems, inexpensive, highly active in gels, and easy to inhibit. In addition, tryptic digestion of proteins yields double-charged peptides of suitable molecular weight for MS analysis.

Finally, tryptic peptides from each spot are subject to MS analysis which identified the peptides according to their mass-to-charge ratio (*m*/*z*). MS instruments contain three main components: a source of ions, a mass analyzer, and a detection system (Gafken and Lampe 2006). Matrix-assisted laser desorption and ionization-time of flight (MALDI-TOF) MS is a commonly used MS instrument in conjunction with 2D-PAGE in algal proteomics (Guerrera and Kleiner 2005). Briefly, the mixture of tryptic peptides and matrix consists of small organic molecules with chromophores which are exposed to pulsed laser beam. Photon energy absorbed by matrix chemicals is transferred to peptides for ionization. The *m*/*z* value of ions are determined by the help of their flight time through a tube of specified length.

Similar to the gel-based approach, MS-based approaches start by extraction of proteins from algal cells or their organelles. Analysis of proteome can be done "bottom-up" or "top-down" strategies (Zhang et al. 2013). MS analysis of peptides from the enzymatic digestion of proteins separated by chromatographic methods is known as the bottom-up strategy. On the other hand, top-down strategy refers to the MS analysis of intact proteins directly (Sickmann et al. 2003). The most common instrument is liquid chromatography-tandem mass spectrometry (LC-MS/MS) in algal proteomic studies that performed MS-based approach. The iTRAQ is quite similar to the MS-based approach with the bottom-up strategy. The main difference between them is labeling of proteolytic peptides prior to chromatographic separation. The iTRAQ-based approach exploits an isobaric reagent containing carbonyl mass balancing group, *N*-methyl piperazine reporter group, and *N*-hydroxy succinimide ester group, which reacts with each lysine side chain and N-terminus group of a peptide (Chiu et al. 2015).

2.2.3.1 Application of Proteomics in the Improvement of Algal Lipid Biosynthesis

Numerous proteomic analyses have been carried out to determining of up- and/or downregulated proteins in various oleaginous algae exposed the different stress conditions to induce lipid accumulation. In one of the early proteome studies, it is determined that enolase, glyceraldehyde-3-phosphate dehydrogenase (GAPDH), phosphoglycerate kinase (PGK), fumarate hydratase (FH), citrate synthase (CS), O-acetyl-serine lyase (OAS-L), aspartate aminotransferase (AST), and ATP sulfurylase (ATPS) are closely related with lipid biosynthesis in nitrogen-deprived *Isochrysis galbana* (Song et al. 2013).

Li et al. (2013a) have reported that the acyl-CoA dehydrogenase, which play a significant role in β-oxidation of fatty acids, was downregulated in *Chlorella protothecoides* cultivated under nitrogen deficiency. The same research group has revealed that the expression of glycerol-3-phosphate acyltransferase (GPAT), 3-hydroxyacyl-[acyl-carrier-protein] dehydratase, and acyl-CoA-binding protein was increased by the Cu(II) stress in *C. protothecoides*, while malate dehydrogenase and succinyl-CoA synthetase subunit alpha were downregulated (Li et al. 2013b). Quantitative proteomic profiling of *C. sorokiniana* has been displayed that the lipase was upregulated for lipid biosynthesis (Ma et al. 2013a, b). A study investigating the response of *C. protothecoides* to nitrogen deficiency and phosphor repletion showed that the ME was a key enzyme for lipid biosynthesis (Li et al. 2014). Similar findings have also been reported for *C. vulgaris* (Guarnieri et al. 2013; Li et al. 2015). Proteomic analysis of *Chlorella* UTEX29 under nitrogen starvation has been shown that the GPAT and ACCase GAPDH play a significant role in triacylglycerol accumulation (Goncalves et al. 2016). Nitrogen limitation also impacted lipid accumulation in *Chlorella* sp. FC2 by upregulation of hydroxyacyl-ACP dehydrogenase and enoyl-ACP reductase enzymes (Rai et al. 2017).

Several proteomics studies were performed to investigate the nitrogen deprivation associated with lipid biosynthesis in *Chlamydomonas* species. Choi et al. (2013) indicated that a low CO_2-inducible protein (P1) could be enhanced lipid accumulation in *C. reinhardtii*. In another study, nitrogen deprivation increased the expression of long-chain acyl-CoA synthetase and activates fatty acids for biosynthesis of complex lipids, in *C. reinhardtii* (Wase et al. 2014). Proteome analysis of *C. reinhardtii*, randomly mutagenized by EMS, has been displayed that ATP synthase beta subunit is the significant protein for improving lipid biosynthesis (Lee et al. 2014).

P. tricornutum, an oil-rich diatom, is one of the most studied species at the proteome level. Proteomic analysis of *P. tricornutum* has been revealed that DGAT, phospholipase C, ribulose-phosphate 3-epimerase, plastidic ATP synthase subunit beta, acyl-carrier proteins, and malonyl-CoA:ACP transacyclase are upregulated under nitrogen deprivation supporting the lipid biosynthesis (Yang et al. 2014; Longworth et al. 2016; Remmers et al. 2018). Additionally, it has been determined the increase in expression of GAPDH, PEPC, and pyruvate kinase when *P. tricornutum* was cultivated under dark stress to enhance the lipid biosynthesis (Bai et al. 2016).

Numerous studies have focused on identifying proteins that increase lipid biosynthesis and accumulation in various *Nannochloropsis* species under nitrogen deficiency. A proteomic analysis conducted by Tran et al. (2016) has revealed the enhancement in FAME caused by the increasing of lipid droplet surface protein. Interestingly, it has been shown that the ACCase was downregulated in *Nannochloropsis oculate*. Recently, it was reported that pyruvate dehydrogenase (PDH), GAPDH, GPAT, PEPC, and some lipid degrading enzymes like a lysophospholipase II were upregulated in *Nannochloropsis oceanica* under nitrogen starvation (Chen et al. 2019a; You et al. 2020).

A report published by Shang et al. (2017) pointed out that the downregulation of β-oxidation-associated enzymes such as enoyl-CoA hydratase and 3-hydroxyacyl-CoA dehydrogenase induced lipid biosynthesis in *D. parva* similar to in *C. protothecoides* (Shang et al. 2017). On the other hand, it was also detected that the upregulation of GPAT and lipases might contribute to lipid biosynthesis.

2.3 Random Mutagenesis

Random mutagenesis (RM) has been used to improve lipid biosynthesis in algae for a long time. Mutagenesis combined with omics technologies allows identifying the key regulatory elements, proteins, and enzymes that cause an increase in yield. The superiority of RM is that it does not require overall knowledge of the biochemical processes in cells, expensive devices, and technical expertise. RM also allows the analysis of numerous genomic variants derived from a wild-type algal strain.

Wild-type algal strains can be randomly mutated by chemical and/or physical mutagens. The most common chemical mutagen used to improve lipid biosynthesis in algae is ethyl methane sulfonate (EMS) which causes GC to AT and AT to GC transition mutations through alkylation of DNA (Loechler et al. 1984; Beranek 1990). In addition, methyl nitro nitroso guanidine (MNNG), 5′fluorodeoxyuridine monophosphate (5′FDU), and *N*-methyl-*N*′-nitro-*N*-nitrosoguanidine (NTG) have also been used effectively for mutagenesis. Indeed, the success rate of a chemical mutagen relies on algal strains and optimization of process parameters such as exposure time, concentration of mutagen, and temperature (Chatterjee and Walker 2017). The most effective mutagenesis occurs when a mutagen provides a large frequency of mutations with minimal lethality, in the range of 10–50% survival

(Lawrence 1991). Physical mutagens include UV light, laser beams, heavy-ion beams, X-rays, γ-radiations, and atmospheric and room temperature plasma (ARTP). Exposure of algal strains to physical mutagens may result in point mutations, frameshifts, gene deletions, rearrangement, chromosome breaks, and exchanges (Ikehata and Ono 2011; Desouky et al. 2015). UV light, which causes damage in DNA directly or indirectly through the induction of reactive oxygen species (ROS), is the favorite physical mutagen (Wardman 2009; Vignard et al. 2013).

Workflow of a mutagenesis experiment initiates with the cell recovery through centrifugation of fresh algae culture into the logarithmic phase. Cells are washed with distilled water to remove culture debris and then suspense in an isotonic solution or buffer. Cell suspension is exposed to mutagen under optimum condition according to appropriate lethality (Bose 2014). Exposed cells are spread on agar plates. Finally, mutant colonies are selected according to some characteristics such as lipid content, FAME content, and biomass yields.

2.3.1 Application of Random Mutagenesis in the Improvement of Algal Lipid Biosynthesis

So far, some strains of *Chlorella*, *Nannochloropsis*, *Scenedesmus*, and *Desmodesmus* have been randomly mutated by physical mutagens to enhance cell lipid content and productivity. It has been reported that UV-mutated *C. sorokiniana* (Vigeolas et al. 2012), *Scenedesmus obliquus* (Vigeolas et al. 2012), and *D. armatus* (Smalley et al. 2020) demonstrated 1.5-fold, 2.2-fold, and 2.0-fold improvement in lipid content, respectively. UV mutation of *N. oculate* UTEX LB 2164 has facilitated 1.3-fold increase in total fatty acid (Moha-León et al. 2019). It has been demonstrated that TAG accumulation in UV-mutated *Chlorella* sp. KAS603 was 1.8- to 2.5-fold higher than that of the wild strain (Manandhar-Shrestha and Hildebrand 2013). UV light has also improved the *C. sorokiniana* 7-11-05 biomass by increasing photosynthetic performance (Cazzaniga et al. 2014). Some researchers have used some other physical mutagens for the development of algal strains. The study, conducted by Choi et al. (2014), has been revealed that FAME content of *Scenedesmus dimorphus* increased 1.4-fold by ARTP mutagenesis. Lipid productivity of *C. pyrenoidosa* FACHB-8 has been successfully improved (16.85%) using the same method. Xing et al. (2021a, b) have been demonstrated that the helium-neon laser beam-induced mutation in *Chlorella pyrenoidosa* strain FACHB 31 caused a 2.2-fold increase in lipid content. In addition, *N. oceanica* IMET1 mutant with enhanced lipid productivity (28%) has been obtained by random mutagenesis caused by heavy-ion beams (Ma et al. 2013a, b).

EMS-induced mutagenesis of *Nannochloropsis* sp. (Doan and Obbard 2012), *C. reinhardtii* (Lee et al. 2014), *C. pyrenoidosa* NCIM 2738 (Sachdeva et al. 2016), *C. minutissima* MCC-27 (Mehtani et al. 2017), and *P. tricornutum* (Yi et al. 2018) has resulted in 2.5-fold, 2-fold, 1.6-fold, 1.42-fold, and 1.627-fold increase in lipid content, respectively. More than 50% improvement in lipid productivity has been observed when the same approach was applied on *Synechocystis* PCC 6803 (Patel et al. 2016) and *Chlorella* sp. 6-4 (Noochanong et al. 2018). Perin et al. (2015) reported that the EMS mutagenesis conduced to increase light use efficiency and thus enhanced the lipid yield in *N. gaditana* 849/5. Meanwhile, lipid yield in *Chlorella saccharophila* has been improved by increasing biomass productivity induced by EMS mutagenesis (Smita et al. 2020).

Apart from the approaches mentioned above, two-round mutagenesis has been applied in a few algal strains to enhance lipid content. It has been reported that the mutation with UV light and EMS has been enhanced lipid content (threefold) of *Nannochloropsis salina* CCAP849/3 (Beacham et al. 2015). Mutant *C. vulgaris*, generated by two-round EMS, has accumulated 1.67-fold higher total lipid than that of its wild strain (Sarayloo et al. 2018).

2.4 Transformation

Genetic transformation refers to the insertion of the particular DNA into the genome for expression. A typical transformation process for improving algal strains consists of the following steps: (1) cultivation of algal strain, (2) recovery of cells, (3) transformation, (4) selection of success transformant, and (5) analyses of transgene (s) expression level (Fig. 2.3).

Foreign DNA can be introduced into the nucleus, chloroplast, or mitochondria genomes in algae. Although expression of a foreign gene in the nucleus offers some advantages in terms of post-translational modifications, inducibility, folding, and targeting to a specific region of the cell, it often results in low and/or unstable expression of the gene due to biased codon usage, silencing, and positional effect. The presence of multiple copies of the mitochondrial genome in algal cells limited the expression of the transgene(s) and the generation of stable transformants (Walker

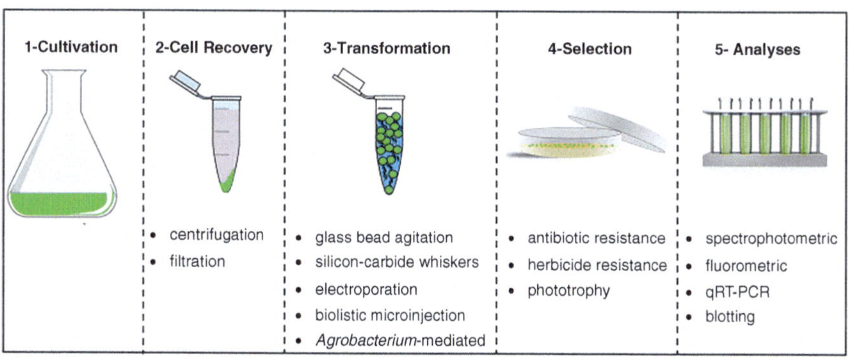

Fig. 2.3 Workflow of a transformation experiment for improving algal strain

et al. 2005). On the other hand, transgene(s) expression in chloroplasts is generally higher than those expressed in the nucleus and mitochondria. The main advantage of chloroplast transformation is the integration of DNA into targeted specific loci within the chloroplast genome through homologous recombination. Thus, silencing and unstable transgene(s) expression arising from the positional effect is avoided (Siddiqui et al. 2020). Chloroplast transformation provides the transcription of several genes under a single promotor as a polycistron (Day and Goldschmidt-Clermont 2011). Moreover, chloroplast facilitates the correct folding of recombinant proteins by post-translational modification, such as the formation of disulfide bond between protein subunits. Therefore, many researchers have focused on chloroplast transformation to improve lipid biosynthesis in algal strains.

Various transformation methods are used to introduce foreign DNA into the target genome in algal cells. These include glass bead agitation, silicon-carbide whiskers, electroporation, biolistic microparticle bombardment, microinjection, and *Agrobacterium*-mediated transformation (Doron et al. 2016). The efficiency of the transformation method varies based on algal species due to differences in their cell wall or cell membrane structures. Electroporation, biolistic microparticle bombardment, and microinjection are available for both nuclear and chloroplast transformation, while the others are only used for nuclear transformation.

Glass bead agitation is the oldest and simplest transformation method for algae (Kindle 1990). It is low cost as it does not require special equipment and devices. Briefly, a test tube containing algal cell suspension, exogenous DNA, polyethylene glycol (PEG), and glass beads in 200–600 μm diameter is agitated vigorously by a vortex, and then transformants are detected. The method generally requires the enzymatic pre-treatment of the cells to weaken or degrade the algal cell wall.

Another cheap and practical method is the silicon-carbide whiskers (Kaeppler et al. 1990; Dunahay 1993). Application of the method is quite similar to glass bead agitation. It proceeds by agitation of exogenous DNA and algae cells in the presence of silicon carbide fibers ranging from 5 to 500 μm in length and 0.5–1 μm in diameter. The fact that fibers cause respiratory damage in algal cells limits the use of the method (Ortiz-Matamoros et al. 2018).

The most prevalent transformation method for algae is electroporation. Electroporation is a method performed with an electric pulse, which creates micropores on the surface of the cell without causing the cell's death and thus enables the passage of DNA into the cell (Weaver and Chizmadzhev 1996; Chen and Lee 2019). Electroporation outperforms others in terms of short process times, high cell viability rates, and low heating effects. On the other hand, a special device called an electroporator is required for its application. Electroporation can be applied on macro-, micro-, or nano-scale using cuvettes, micro-channels, micro-capillaries, nano-channels, and micro-wells (Garcia et al. 2016; Zhao et al. 2016). Successful electroporation is achieved by strain-specific optimization of various parameters, including the number of pulses, pulse duration, voltage, resistance, capacitance, exogenous DNA concentration, and cell amount (Qin et al. 2012). Applied voltage, resistance, and capacitance for algal strains range from 0.05 kV/cm to 12.5 kV/cm, from 100 Ω to 1000 Ω, and from 25 μF to 50 μF, respectively (Mosey et al. 2021). In general, one pulse is adequate for algal strains. The use of an osmosis solution containing mannitol, sorbitol, and glycerol or addition of osmolytes into medium improves the transformation efficiency by reducing salt stress (Guo et al. 2013; Kotnik et al. 2015). Transformation methods with electroporation have been established for some strains of *Spirulina* (Toyomizu et al. 2001), *Chlorella* (Wang et al. 2007; Run et al. 2016), *Monoraphidium* (Jaeger et al. 2017), *Acutodesmus* (Muñoz et al. 2018), *Nannochloropsis* (Gan et al. 2018; Chen and Hu 2019; Jackson et al. 2019), *Chlamydomonas* (Muñoz et al. 2018; Kim 2018; Kim et al. 2019a, b), *Coccomyxa* (Kania et al. 2020), *Ettlia* (Lee et al. 2020a), *Picochlorum* (Krishnan et al. 2021), *Chaetoceros* (Yin and Hu 2021), and *Desmodesmus* (Douchi et al. 2021).

Biolistic microparticle bombardment is a physical method that directly delivers exogenous DNA with gold or tungsten particles to the nucleus or chloroplast genome of algae. Although equipment and device costs are high, its successful application in different types of algae makes the method preferable. For application of the method, exogenous DNA is initially precipitated onto particles using calcium chloride and spermidine. Subsequently, DNA-coated microparticles (microprojectiles) are bombarded toward algal cells using a gene gun system, mainly including a gas acceleration tube, rupture disk, microcarrier, and bombardment chamber. The burst of the rupture disk, blocking the gas accelerator tube, causes the release of helium gas pressure toward the microcarrier onto which microprojectiles (Williams et al. 1991). The successful application of the method for algal species depends on optimizing of some parameters such as gas pressure, particle size, microprojectile travel distance, and particle type. The gas pressure and particle size for the transformation of algal strains have been established to be within the range of 450–2000 psi and 5–9 cm, respectively (Teng et al. 2002; Zhang et al. 2020a, b; Douchi et al. 2021). Moreover, many researchers have been reported that the effective microprojectile travel distance is 6 cm for algal strains (Dahlin et al. 2019; Jallet et al. 2020; Douchi et al. 2021).

Microinjection, another physical transformation method, delivers the gene(s) to the target genome of the immobilized cell through a glass micropipette under a microscope (Crossway et al. 1986). Despite high efficiency, it is rarely employed for algal transformation because it is time-consuming and expensive (Jinturkar et al. 2011).

Agrobacterium-mediated transformation is based on transferring T-DNA, a part of Ti plasmid in *Agrobacterium tumefaciens*, into the plant genome with the aid of virulence (Vir) proteins (Simon et al. 2015). The method is also applied in algae with a process similar to that in plants. The process is typically performed by simultaneously incubating algal cells and *A. tumefaciens* harboring a plasmid with exogenous DNA. Unlike plants, incubation is performed in the presence of phenolic compounds such as acetosyringone, coumarin, cinnamic acid, and vanillin to induction of virulence (*vir*) genes (Cha et al. 2011; Srinivasan and Gothandam 2016). Moreover, the use of the method for marine algal strains requires careful selection of medium that supports both algal strain and *A. tumefaciens* growth simultaneously (Rathod et al. 2013). *Agrobacterium*-mediated transformation has been applied to *C. reinhardtii* (Kumar et al. 2004; Pratheesh et al. 2014), *Dunaliella bardawil* (Anila

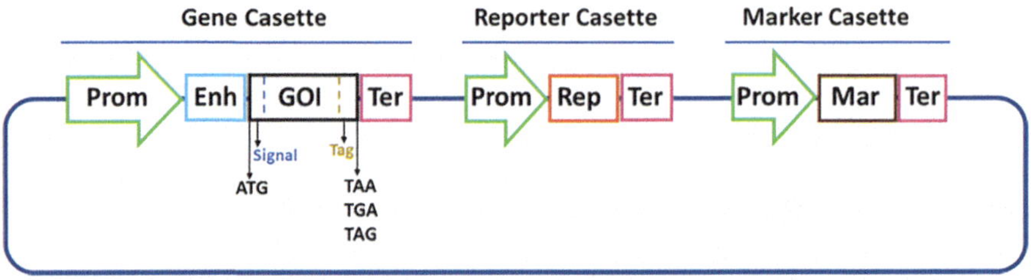

Fig. 2.4 Illustration of vector construct including promotor (**Prom**), enhancer element (**Enh**), gen of interest (**GOI**), terminator (**Ter**), reporter gene (**Rep**), and marker (**Mar**)

et al. 2011), *Schizochytrium* sp. TIO1101 (Cheng et al. 2012), *Haematococcus pluvialis* SAG 19-a (Kathiresan et al. 2009), *C. vulgaris* UMT-M1 (San Cha et al. 2012), *Parachlorella kessleri* (Rathod et al. 2013), *Ankistrodesmus* sp. (Sanitha et al. 2014), *Scenedesmus bajacalifornicus* (Sanitha et al. 2014), *D. salina* (Simon et al. 2016), *Dictyosphaerium pulchellum* (Bashir et al. 2018), *I. galbana* (Prasad et al. 2014), *D. armatus* (Douchi et al. 2021), *C. sorokiniana* (Sharma et al. 2021), *D. tertiolecta* (Norzagaray-Valenzuela et al. 2018), and *P. purpureum* (Prasad et al. 2019).

The stability and density of exogenous DNA expression are strongly influenced by vector construct. A vector harbors gene cassette that typically consists of exogenous DNA, promoter, enhancers, and terminator. A typical vector construct is shown in Fig. 2.4.

The selection of the promoter is one of critical issues due to its significant role on the transcriptional activities of a gene. Promoter availability has been extensively studied to obtain a success genetic transformation system. The efficiency of the promoters varies according to the target genome and algal species. In earlier, researchers focused on the promotors of viruses that naturally infect algae or plants. Various studies have been revealed that the promoter of cauliflower mosaic virus 35S (CaMV35S), Simian Virus 40 (SV40), and cytomegalovirus (CMV) is employed for expression of transgene(s) in some algae, including *Chlorella* (Hawkins and Nakamura 1999; Wang et al. 2007; Sharma et al. 2021), *Chlamydomonas* (Ruecker et al. 2008), *Kappaphycus* (Kurtzman 1991), *Porphyra*, *Dunaliella* (Tan et al. 2005), *Platymonas* (Cui et al. 2010), *Haematococcus* (Teng et al. 2002; Kathiresan et al. 2009), and *Symbiodinium* (Pasaribu and Jiang 2021). Heterologous promotors of *A. tumefaciens* nopaline synthase gene (*nos*), maize ubiquitin-Ω gene (*ubi*), and rice actine1 gene (*act1*) have also been employed in some algae to facilitate exogenous gene expression (Ortiz-Matamoros et al. 2015; Run et al. 2016; Zhou et al. 2021). However, it has been indicated that inadequate recognition and regulation are the main disadvantages of heterologous promoters (Lu et al. 2021). Alternatively, many exogenous and orthologous promoters have been successfully enhanced transcription and expression level of the exogenous gene. Among them, promoters of rubisco small subunit 2 (RBCS2), fucoxanthin chlorophyll binding protein (FCP), heat shock protein 70 (HSP70), photosystem I complex protein (PSAD), tubulin (TUB), RuBisCO (RBCL), and nitrate reductase

(NR) genes have been widely used in transformation of some algal strains including *C. reinhardtii* (Lumbreras et al. 1998; Dementyeva et al. 2021), *Chlorella ellipsoidea* (Kim et al. 2002), *P. tricornutum* (Xie et al. 2014; Xue et al. 2015), *Thalassiosira pseudonana* (Poulsen et al. 2006), *C. pyrenoidosa* (Fan et al. 2015), and *N. gaditana* (Radakovits et al. 2012). Furthermore, some researchers favor hybrid promoters, based on HSP70A acting as a transcription activator. HSP70A/RBCS2 is the most popular among hybrid promoters (Rasala et al. 2014; Shahar et al. 2020; Pivato et al. 2021; Klaitong et al. 2021). Some studies have been revealed that various elements such as untranslated regions (UTRs) and introns (from *RbcS2*, *Als*, *pds*, etc.) can be added into the gene cassette to improve the transcriptional effect of promoters (Liu et al. 2014; Young and Purton 2016).

The vector construct also consists of marker and reporter genes. Marker genes assist the discriminate success transformants from others. Algal transformants can be selected with two strategies: conferring (1) resistance to antibiotic or herbicide and (2) phototrophy of non-photosynthetic mutants. Genes of aminoglycoside 3′-adenyl transferase (*aadA*), glycopeptide binding protein (*ble*), chloramphenicol acetyltransferase (*cat*), erythromycin esterase type II (*ereB*), aminoglycoside 3′--phosphotransferase 7 (*aph7*), and aminoglycoside 3′-phosphotransferase A6 (*aphA6*) confer resistance to spectinomycin/streptomycin, zeocin, chloramphenicol, erythromycin, hygromycin, and kanamycin/amikacin, respectively (Esland et al. 2018). Neomycin phosphotransferase II (*npt2*) ensures neomycin, kanamycin, and geneticin selection. Bar/pat genes isolated from *Streptomyces hygroscopicus* and *Streptomyces viridochromogenes* inactivate herbicides with phosphinothricin (PPT). Genes of glyphosate acetyltransferase (*gat*) and acetohydroxyacid synthase (*ahas*) are the other possible markers that confer to herbicide resistance (Bashir et al. 2016). Photosynthesis-based selection is carried out using genes of β subunit of ATP synthase (*atpB*) (Yan et al. 2016), photosystem II protein D1 (*psbA*) (Michelet et al. 2011; Scranton et al. 2015), H subunit of photosystem II (*psbH*) (Wannathong et al. 2016), the structural component of nitrogenase Fe protein (*nifH*) (Cheng et al. 2005), and *N*-acetyl ornithine aminotransferase (*arg*) (Remacle et al. 2009). In the design of vector construct for algal transformation, the widely used reporters, which ensure the determination of expression level, are β-glucuronidase (GUS) (Zou et al. 2018) and green fluorescent protein (GFP) (Chungjatupornchai et al. 2016; Klaitong et al. 2021).

2.4.1 Application of Transformation in the Improvement of Algal Lipid Biosynthesis

Algal strains have been successfully improved by genetic transformation to obtain a lipid-rich feedstock in the biodiesel industry. The primary genetic transformation strategy is overexpression of a specific gene(s), encoded enzyme(s), and transcription factor(s). Some of the attempts made toward overexpression of genes in algal strains are shown in Table 2.2. *C. reinhardtii* is one of the most studied algal species

Table 2.2 List of algal strains improved by genetic transformation

Algal strain	Transformation method	Overexpressed molecule(s)	Promotor	Marker/ reporter	Target	Increase	Reference
P. tricornutum	Electroporation (1.5 kV, 25 μF, and 400 Ω)	DGAT2	FCP	–/–	Neutral lipid	35%	Niu et al. (2013)
C. reinhardtii	Electroporation	DGAT2	GPD	–/–	TAG	Ninefold	Hung et al. (2013)
P. tricornutum	Electroporation (1.5 kV, 25 μF, and 400 Ω)	GPDH	FCP	*cat*/–	Neutral lipid	60%	Yao et al. (2014)
C. reinhardtii	Agrobacterium-mediated	DOF-type transcription factor	CaMV35S	*npt2*/–	Total lipid	Twofold	Ibáñez-Salazar et al. (2014)
P. tricornutum	Electroporation (1.5 kV, 25 μF, and 400 Ω)	ME	FCP	*cat*/–	Total lipid	2.5-fold	Xue et al. (2015)
N. salina	Biolistic microparticle bombardment (0.6 μm, gold)	Transcription factor NsbHLH2	TUB	*ble*/–	FAME	33%	Kang et al. (2015)
P. tricornutum	Electroporation (1.5 kV, 25 μF, and 400 Ω)	PNPLA3 ortholog	FCP	*cat*/–	Total lipid productivity	70%	Wang et al. (2015)
N. oceanica	Electroporation (11 kV, 50 μF and 600 Ω)	Endogenous Δ12-desaturase	LDSP	*aph7*/–	TAG	Un-specified	Kaye et al. (2015)
C. reinhardtii	Electroporation	Phosphorus stress response 1 (PSR1)	HSP70A-RBCS2	*ble*/–	TAG	Twofold	Ngan et al. (2015)
N. oceanica	Electroporation (2.2 kV, 50 μF and 600 Ω)	DGAT2	HSP20	–/–	Neutral lipid	69%	Li et al. (2016)
S. obliquus	Electroporation (5 kV, 50 μF and 600 Ω)	DGAT2	CaMV35S	*aph7*/ GUS	Lipid content	127.8%	Chen et al. (2016)
Mortierella alpina	Electroporation	FADS6	TrpCter	–/–	Eicosapentaenoic acid	26.2-fold	Shi et al. (2016)
N. oceanica	Electroporation (2.2 kV, 50 μF and 600 Ω)	DGAT1	UBI	*ble*/–	TAG	2.4-fold	Wei et al. (2017)
N. salina		AtWRI1	TUB	*ble*/–	Total lipid	44.7%	Kang et al. (2017)
N. oceanica	Electroporation (11 kV, 50 μF and 600 Ω)	DGAT2 (NoDGTT5)	LDSP	–/LUC	TAG	Un-specified	Zienkiewicz et al. (2017)

C. reinhardtii		fatty acyl-ACP thioesterase	CaMV35S	*aph7*/−	Lipid content	56%	Tan and Lee (2017)
C. reinhardtii	Glass bead	LPAAT	Hsp70A-RBCS2	*ble*/−	Lipid content	44.5%	Wang et al. (2018)
C. reinhardtii	Glass bead	GPDH	Hsp70A-RBCS2	*ble*/−	Lipid content	67.5%	Wang et al. (2018)
Schizochytrium sp.	Electroporation (7.5 kV, 50 μF, 50 Ω)	Malonyl-coa: ACP transacylase	TEF-1	*ble*/−	Total lipid	39.6%	Li et al. (2018a)
P. tricornutum	Electroporation (0.5 kV, 25 μF, 400 Ω)	ACCase	RBCL	−/GFP	Neutral lipid	1.77-fold	Li et al. (2018b)
P. tricornutum	Electroporation (1.5 kV, 25 μF, and 400 Ω)	HsPNPLA3	FCP	cat/−	TAG	1.55-fold	Wang et al. (2018)
C. vulgaris	Glass bead	DOF-type transcription factor	Hsp70A-RBCS2	*ble*/−	Neutral lipid	1.5-fold	Tokunaga et al. (2019)
C. reinhardtii	Glass bead	Acetyl-CoA carboxylase	HSP70A-RBCS2	−/mCherry	Fatty acids	1.6-fold	Chen et al. (2019b)
C. reinhardtii	Electroporation	ME2	HSP70A-RBCS2	*aph7*/−	Lipid content	19.9%	Kim and Cho (2019)
Neochloris oleoabundans	Electroporation (6 kV)	LPAT, GPAT, and DGAT	CaMV35S	*aph7*/−	Total fatty acids	1.2-fold	Muñoz et al. (2019)
C. reinhardtii	Electroporation (0.75 kV, 25 μF, 200 Ω)	Fructose-1,6-bisphosphate aldolase 1	TUB	*psa*/−	Total lipid	Un-specified	Lee et al. (2020b)
P. tricornutum	Biolistic microparticle bombardment (0.6 μm, gold)	DGAT2 and ectopic expression of a Δ5-elongase	FCP	*ble*/−	TAG	1.8-fold	Haslam et al. (2020)
P. tricornutum	Electroporation (0.5 kV, 25 μF, 400 Ω)	DGAT1	CMV	*ble*/GFP	TAG and total lipid	Approximately twofold	Zhang et al. (2021a, b, c)

(continued)

Table 2.2 (continued)

Algal strain	Transformation method	Overexpressed molecule(s)	Promotor	Marker/ reporter	Target	Increase	Reference
C. ellipsoidea	Electroporation (6 kV, 001–0.2 s duration, 2^{10} frequencies, 2 mm distance)	Transcription factor leafy cotyledon1 (LEC1)	UBI	*npt2*/–	Total lipid	21.69–30.45%	Liu et al. (2021)
N. salina	Electroporation (1.2 kV)	NADP-dependent ME	TUB	*ble*/–	FAME	53%	Jeon et al. (2021)
N. oleoabundans	Electroporation (1 kV, 25 μF, 200 Ω)	LPAAT1 and DGAT2	HSP70-RBCS2	–/–	Total lipid, TAG and Total lipid productivity	1.6-, 2.1-, and 1.9-fold	Chungjatupornchai and Fa-Aroonsawat (2021)

in genetic applications applied for enhancing lipid accumulation. Various genes encoded DGAT2, overexpression of DOF-type transcription factor, phosphorus stress response 1 (PSR1), acyl-ACP thioesterase, lysophosphatidic acyltransferase (LPAAT), GPDH, acetyl-CoA carboxylase, ME2, and fructose-1,6-bisphosphate aldolase 1 successfully transformed and overexpressed in *C. reinhardtii*. It has been reported that the overexpression of DGAT2, GPDH, ME, phospholipase domain-containing protein 3 (PNPLA3) ortholog, ACCase, HsPNPLA3, and DGAT1 increased lipid accumulation up to 2.5-fold in *P. tricornutum*. In addition, efficient transformation methods have been developed for species from *Nannochloropsis* to improve their potential in biodiesel industry.

2.5 Conclusion

The enormous accumulation of knowledge from the "omics" reveals the entirety of genes, transcripts, and proteins and their interactions with each other. Omics, combined with strain improvement studies performed by cultural methods and random mutagenesis, allow the full understanding of the lipid production and accumulation mechanisms in algal cells. Thus, it helps predict the effect of gene (s) overexpression on the yield without laboratory experiments. Genetic transformation enables algae to gain many important features that are not naturally present and cannot be obtained by conventional methods. In the course of more than a decade, many direct and indirect transformation methods have been developed for different algal strains. However, the discovery of new oil-rich algae species and the identification of molecules involved in lipid metabolism in algae still keep omics technologies, random mutagenesis, and transformation up to date.

References

Al-Widyan MI, Al-Shyoukh AO (2002) Experimental evaluation of the transesterification of waste palm oil into biodiesel. Bioresour Technol 85(3):253–256

Anand V, Kashyap M, Samadhiya K, Kiran B (2019) Strategies to unlock lipid production improvement in algae. Int J Environ Sci Technol 16(3):1829–1838

Anila N, Chandrashekar A, Ravishankar GA, Sarada R (2011) Establishment of Agrobacterium tumefaciens-mediated genetic transformation in *Dunaliella bardawil*. Eur J Phycol 46(1):36–44

Antolın G, Tinaut FV, Briceno Y, Castano V, Perez C, Ramırez AI (2002) Optimisation of biodiesel production by sunflower oil transesterification. Bioresour Technol 83(2):111–114

Arriola MB, Velmurugan N, Zhang Y, Plunkett MH, Hondzo H, Barney BM (2018) Genome sequences of *Chlorella sorokiniana* UTEX 1602 and Micractinium conductrix SAG 241.80: implications to maltose excretion by a green alga. Plant J 93(3):566–586

Astafyeva Y, Alawi M, Indenbirken D, Danso D, Grundhoff A, Hanelt D et al (2020) Draft genome sequence of the green alga *Scenedesmus acuminatus* SAG 38.81. Microbiol Resour Announc 9(24):e01278–e01219

Bai X, Song H, Lavoie M, Zhu K, Su Y, Ye H et al (2016) Proteomic analyses bring new insights into the effect of a dark stress on lipid biosynthesis in *Phaeodactylum tricornutum*. Sci Rep 6(1): 1–10

Bashir KMI, Kim MS, Stahl U, Cho MG (2016) Microalgae engineering toolbox: selectable and screenable markers. Biotechnol Bioprocess Eng 21(2):224–235

Bashir KMI, Moo-Sang K, Stahl U, Cho MG (2018) Agrobacterium-mediated genetic transformation of *Dictyosphaerium pulchellum* for the expression of erythropoietin. J Appl Phycol 30(6): 3503

Beacham TA, Macia VM, Rooks P, White DA, Ali ST (2015) Altered lipid accumulation in *Nannochloropsis salina* CCAP849/3 following EMS and UV induced mutagenesis. Biotechnol Rep 7:87–94

Beaulieu L (2019) Insights into the regulation of algal proteins and bioactive peptides using proteomic and transcriptomic approaches. Molecules 24(9):1708

Bengtsson A, Bengtsson H (2006) Microarray image analysis: background estimation using quantile and morphological filters. BMC Bioinformatics 7(1):1–15

Benites LF, Poulton N, Labadie K, Sieracki ME, Grimsley N, Piganeau G (2019) Single cell ecogenomics reveals mating types of individual cells and ssDNA viral infections in the smallest photosynthetic eukaryotes. Philos Trans R Soc B 374(1786):20190089

Beranek DT (1990) Distribution of methyl and ethyl adducts following alkylation with monofunctional alkylating agents. Mutat Res 231:11–30

Bhattacharya D, Price DC, Chan CX, Qiu H, Rose N, Ball S et al (2013) Genome of the red alga *Porphyridium purpureum*. Nat Commun 4(1):1–10

Bjellqvist B, Ek K, Righetti PG, Gianazza E, Görg A, Westermeier R, Postel W (1982) Isoelectric focusing in immobilized pH gradients: principle, methodology and some applications. J Biochem Biophys Methods 6(4):317–339

Blanc G, Duncan G, Agarkova I, Borodovsky M, Gurnon J, Kuo A et al (2010) The *Chlorella variabilis* NC64A genome reveals adaptation to photosymbiosis, coevolution with viruses, and cryptic sex. Plant Cell 22(9):2943–2955

Blanc G, Agarkova I, Grimwood J, Kuo A, Brueggeman A, Dunigan DD et al (2012) The genome of the polar eukaryotic microalga *Coccomyxa subellipsoidea* reveals traits of cold adaptation. Genome Biol 13(5):1–12

Blanc-Mathieu R, Verhelst B, Derelle E, Rombauts S, Bouget FY, Carré I et al (2014) An improved genome of the model marine alga *Ostreococcus tauri* unfolds by assessing Illumina de novo assemblies. BMC Genomics 15(1):1–12

Bleakley S, Hayes M (2017) Algal proteins: extraction, application, and challenges concerning production. Foods 6(5):33

Bogen C, Al-Dilaimi A, Albersmeier A, Wichmann J, Grundmann M, Rupp O et al (2013) Reconstruction of the lipid metabolism for the microalga *Monoraphidium neglectum* from its genome sequence reveals characteristics suitable for biofuel production. BMC Genomics 14(1): 1–18

Bose JL (2014) Chemical and UV mutagenesis. In: Bose JL (ed) The genetic manipulation of *Staphylococci*. Humana Press, New York, NY, pp 111–115

Bouaid A, Martinez M, Aracil J (2007) Long storage stability of biodiesel from vegetable and used frying oils. Fuel 86(16):2596–2602

Bowler C, Allen AE, Badger JH, Grimwood J, Jabbari K, Kuo A et al (2008) The *Phaeodactylum* genome reveals the evolutionary history of diatom genomes. Nature 456(7219):239–244

Bradford MM (1976) A rapid and sensitive method for the quantitation of microgram quantities of protein utilizing the principle of protein-dye binding. Anal Biochem 72(1–2):248–254

Buermans HPJ, Den Dunnen JT (2014) Next generation sequencing technology: advances and applications. Biochim Biophys Acta Mol Bas Dis 1842(10):1932–1941

de Carvalho LM, Borelli G, Camargo AP, de Assis MA, de Ferraz SMF, Fiamenghi MB et al (2019) Bioinformatics applied to biotechnology: a review towards bioenergy research. Biomass Bioenergy 123:195–224

Cazzaniga S, Dall'Osto L, Szaub J, Scibilia L, Ballottari M, Purton S, Bassi R (2014) Domestication of the green alga *Chlorella sorokiniana*: reduction of antenna size improves light-use efficiency in a photobioreactor. Biotechnol Biofuels 7(1):1–13

Cecchin M, Marcolungo L, Rossato M, Girolomoni L, Cosentino E, Cuine S et al (2019) *Chlorella vulgaris* genome assembly and annotation reveals the molecular basis for metabolic acclimation to high light conditions. Plant J 100(6):1289–1305

Cha TS, Chen CF, Yee W, Aziz A, Loh SH (2011) Cinnamic acid, coumarin and vanillin: alternative phenolic compounds for efficient *Agrobacterium*-mediated transformation of the unicellular green alga, *Nannochloropsis* sp. J Microbiol Methods 84(3):430–434

Chang WC, Zheng HQ, Chen CNN (2016) Comparative transcriptome analysis reveals a potential photosynthate partitioning mechanism between lipid and starch biosynthetic pathways in green microalgae. Algal Res 16:54–62

Chatterjee N, Walker GC (2017) Mechanisms of DNA damage, repair, and mutagenesis. Environ Mol Mutagen 58(5):235–263

Chen Y, Hu H (2019) High efficiency transformation by electroporation of the freshwater alga *Nannochloropsis limnetica*. World J Microbiol Biotechnol 35(8):1–10

Chen Z, Lee WG (2019) Electroporation for microalgal biofuels: a review. Sustain Energy Fuels 3(11):2954–2967

Chen X, Wei S, Ji Y, Guo X, Yang F (2015) Quantitative proteomics using SILAC: principles, applications, and developments. Proteomics 15(18):3175–3192

Chen CY, Kao AL, Tsai ZC, Chow TJ, Chang HY, Zhao XQ et al (2016) Expression of type 2 diacylglycerol acyltransferase gene DGTT1 from *Chlamydomonas reinhardtii* enhances lipid production in Scenedesmus obliquus. Biotechnol J 11(3):336–344

Chen C, Harst A, You W, Xu J, Ning K, Poetsch A (2019a) Proteomic study uncovers molecular principles of single-cell-level phenotypic heterogeneity in lipid storage of *Nannochloropsis oceanica*. Biotechnol Biofuels 12(1):1–14

Chen D, Yuan X, Liang L, Liu K, Ye H, Liu Z et al (2019b) Overexpression of acetyl-CoA carboxylase increases fatty acid production in the green alga *Chlamydomonas reinhardtii*. Biotechnol Lett 41(10):1133–1145

Chen HH, Xue LL, Liang MH, Jiang JG (2020) Intervention of triethylamine on *Dunaliella tertiolecta* reveals metabolic insights into triacylglycerol accumulation. Algal Res 47:101876

Chen MY, Teng WK, Zhao L, Hu CX, Zhou YK, Han BP et al (2021) Comparative genomics reveals insights into cyanobacterial evolution and habitat adaptation. ISME J 15(1):211–227

Cheng Q, Day A, Dowson-Day M, Shen GF, Dixon R (2005) The *Klebsiella pneumoniae* nitrogenase Fe protein gene (nifH) functionally substitutes for the chlL gene in *Chlamydomonas reinhardtii*. Biochem Biophys Res Commun 329(3):966–975

Cheng R, Ma R, Li K, Rong H, Lin X, Wang Z et al (2012) Agrobacterium tumefaciens mediated transformation of marine microalgae *Schizochytrium*. Microbiol Res 167(3):179–186

Chernokalskaya E, Gutierrez S, Pitt AM, Leonard JT (2004) Ultrafiltration for proteomic sample preparation. Electrophoresis 25(15):2461–2468

Chiu CW, Chen HM, Wu TT, Shih YC, Huang KK, Tsai YF et al (2015) Differential proteomics of monosodium urate crystals-induced inflammatory response in dissected murine air pouch membranes by iTRAQ technology. Proteomics 15(19):3338–3348

Choi YE, Hwang H, Kim HS, Ahn JW, Jeong WJ, Yang JW (2013) Comparative proteomics using lipid over-producing or less-producing mutants unravels lipid metabolisms in *Chlamydomonas reinhardtii*. Bioresour Technol 145:108–115

Choi JI, Yoon M, Joe M, Park H, Lee SG, Han SJ, Lee PC (2014) Development of microalga Scenedesmus dimorphus mutant with higher lipid content by radiation breeding. Bioprocess Biosyst Eng 37(12):2437–2444

Chungjatupornchai W, Fa-Aroonsawat S (2021) Enhanced triacylglycerol production in oleaginous microalga *Neochloris oleoabundans* by co-overexpression of lipogenic genes: Plastidial LPAAT1 and ER-located DGAT2. J Biosci Bioeng 131(2):124–130

Chungjatupornchai W, Kitraksa P, Fa-aroonsawat S (2016) Stable nuclear transformation of the oleaginous microalga *Neochloris oleoabundans* by electroporation. J Appl Phycol 28(1): 191–199

Cock JM, Sterck L, Rouzé P, Scornet D, Allen AE, Amoutzias G et al (2010) The Ectocarpus genome and the independent evolution of multicellularity in brown algae. Nature 465(7298): 617–621

Cong WT, Hwang SY, Jin LT, Choi JK (2008) Sensitive fluorescent staining for proteomic analysis of proteins in 1-D and 2-D SDS-PAGE and its comparison with SYPRO Ruby by PMF. Electrophoresis 29(21):4304–4315

Courchesne NMD, Parisien A, Wang B, Lan CQ (2009) Enhancement of lipid production using biochemical, genetic and transcription factor engineering approaches. J Biotechnol 141(1–2): 31–41

Crossway A, Oakes JV, Irvine JM, Ward B, Knauf VC, Shewmaker CK (1986) Integration of foreign DNA following microinjection of tobacco mesophyll protoplasts. Mol Gen Genet MGG 202(2):179–185

Cui Y, Wang J, Jiang P, Bian S, Qin S (2010) Transformation of *Platymonas* (*Tetraselmis*) *subcordiformis* (*Prasinophyceae*, *Chlorophyta*) by agitation with glass beads. World J Microbiol Biotechnol 26(9):1653–1657

Curtis BA, Tanifuji G, Burki F, Gruber A, Irimia M, Maruyama S et al (2012) Algal genomes reveal evolutionary mosaicism and the fate of nucleomorphs. Nature 492(7427):59–65

Dahlin LR, Gerritsen AT, Henard CA, Van Wychen S, Linger JG, Kunde Y et al (2019) Development of a high-productivity, halophilic, thermotolerant microalga *Picochlorum renovo*. Commun Biol 2(1):1–9

Dasgupta CN, Nayaka S, Toppo K, Singh AK, Deshpande U, Mohapatra A (2018) Draft genome sequence and detailed characterization of biofuel production by oleaginous microalga *Scenedesmus quadricauda* LWG002611. Biotechnol Biofuels 11(1):1–15

Day A, Goldschmidt-Clermont M (2011) The chloroplast transformation toolbox: selectable markers and marker removal. Plant Biotechnol J 9(5):540–553

Dementyeva P, Freudenberg RA, Baier T, Rojek K, Wobbe L, Weisshaar B, Kruse O (2021) A novel, robust and mating-competent *Chlamydomonas reinhardtii* strain with an enhanced transgene expression capacity for algal biotechnology. Biotechnol Rep 31:e00644

Desouky O, Ding N, Zhou G (2015) Targeted and non-targeted effects of ionizing radiation. J Radiat Res Appl Sci 8(2):247–254

Doan TTY, Obbard JP (2012) Enhanced intracellular lipid in *Nannochloropsis* sp. via random mutagenesis and flow cytometric cell sorting. Algal Res 1(1):17–21

Dobin A, Davis CA, Schlesinger F, Drenkow J, Zaleski C, Jha S, Batut P, Chaisson M, Gingeras TR (2013) STAR: ultrafast universal RNA-seq aligner. Bioinformatics 29:15–21

Dong Z, Chen Y (2013) Transcriptomics: advances and approaches. Sci China Life Sci 56(10): 960–967

Doron L, Segal NA, Shapira M (2016) Transgene expression in microalgae—from tools to applications. Front Plant Sci 7:505

Douchi D, Mosey M, Astling DP, Knoshaug EP, Nag A, McGowen J, Laurens LM (2021) Nuclear and chloroplast genome engineering of a productive non-model alga *Desmodesmus armatus*: insights into unusual and selective acquisition mechanisms for foreign DNA. Algal Res 53: 102152

Dunahay TG (1993) Transformation of *Chlamydomonas reinhardtii* with silicon carbide whiskers. Biotechniques 15(3):452–455

Dupree EJ, Jayathirtha M, Yorkey H, Mihasan M, Petre BA, Darie CC (2020) A critical review of bottom-up proteomics: the good, the bad, and the future of this field. Proteomes 8(3):14

Esland L, Larrea-Alvarez M, Purton S (2018) Selectable markers and reporter genes for engineering the chloroplast of *Chlamydomonas reinhardtii*. Biology 7(4):46

Fan J, Cui Y, Wan M, Wang W, Li Y (2014) Lipid accumulation and biosynthesis genes response of the oleaginous *Chlorella pyrenoidosa* under three nutrition stressors. Biotechnol Biofuels 7(1): 1–14

Fan J, Ning K, Zeng X, Luo Y, Wang D, Hu J et al (2015) Genomic foundation of starch-to-lipid switch in oleaginous *Chlorella* spp. Plant Physiol 169(4):2444–2461

Fattore N, Bellan A, Pedroletti L, Vitulo N, Morosinotto T (2021) Acclimation of photosynthesis and lipids biosynthesis to prolonged nitrogen and phosphorus limitation in *Nannochloropsis gaditana*. Algal Res 58:102368

Fayyaz M, Chew KW, Show PL, Ling TC, Ng IS, Chang JS (2020) Genetic engineering of microalgae for enhanced biorefinery capabilities. Biotechnol Adv 43:107554

Gafken PR, Lampe PD (2006) Methodologies for characterizing phosphoproteins by mass spectrometry. Cell Commun Adhes 13(5–6):249–262

Gan Q, Jiang J, Han X, Wang S, Lu Y (2018) Engineering the chloroplast genome of oleaginous marine microalga *Nannochloropsis oceanica*. Front Plant Sci 9:439

Gao C, Wang Y, Shen Y, Yan D, He X, Dai J, Wu Q (2014) Oil accumulation mechanisms of the oleaginous microalga *Chlorella protothecoides* revealed through its genome, transcriptomes, and proteomes. BMC Genomics 15(1):1–14

Garcia PA, Ge Z, Moran JL, Buie CR (2016) Microfluidic screening of electric fields for electroporation. Sci Rep 6(1):1–11

Goncalves EC, Koh J, Zhu N, Yoo MJ, Chen S, Matsuo T et al (2016) Nitrogen starvation-induced accumulation of triacylglycerol in the green algae: evidence for a role for ROC 40, a transcription factor involved in circadian rhythm. Plant J 85(6):743–757

Görg A, Weiss W, Dunn MJ (2004) Current two-dimensional electrophoresis technology for proteomics. Proteomics 4(12):3665–3685

Görg A, Drews O, Weiss W (2006) Isoelectric focusing in immobilized pH gradient strips using the IPGphor unit: sample in-gel rehydration. CSH Protoc 1:526–531

Grant GR, Farkas MH, Pizarro AD, Lahens NF, Schug J, Brunk BP, Stoeckert CJ, Hogenesch JB, Pierce EA (2011) Comparative analysis of RNA-Seq alignment algorithms and the RNA-Seq unified mapper (RUM). Bioinformatics 27:2518–2528

Griebel T, Zacher B, Ribeca P, Raineri E, Lacroix V, Guigo R, Sammeth M (2012) Modelling and simulating generic RNA-Seq experiments with the flux simulator. Nucleic Acids Res 40:10073–10083

Guarnieri MT, Pienkos PT (2015) Algal omics: unlocking bioproduct diversity in algae cell factories. Photosynth Res 123(3):255–263

Guarnieri MT, Nag A, Yang S, Pienkos PT (2013) Proteomic analysis of Chlorella vulgaris: potential targets for enhanced lipid accumulation. J Proteome 93:245–253

Guerrera IC, Kleiner O (2005) Application of mass spectrometry in proteomics. Biosci Rep 25(1–2):71–93

Guillory WX, Onyshchenko A, Ruck EC, Parks M, Nakov T, Wickett NJ, Alverson AJ (2018) Recurrent loss, horizontal transfer, and the obscure origins of mitochondrial introns in diatoms (*Bacillariophyta*). Genome Biol Evol 10(6):1504–1515

Guo SL, Zhao XQ, Tang Y, Wan C, Alam MA, Ho SH et al (2013) Establishment of an efficient genetic transformation system in *Scenedesmus obliquus*. J Biotechnol 163(1):61–68

Haslam RP, Hamilton ML, Economou CK, Smith R, Hassall KL, Napier JA, Sayanova O (2020) Overexpression of an endogenous type 2 diacylglycerol acyltransferase in the marine diatom *Phaeodactylum tricornutum* enhances lipid production and omega-3 long-chain polyunsaturated fatty acid content. Biotechnol Biofuels 13:1–17

Hatey F, Tosser-Klopp G, Clouscard-Martinato C, Mulsant P, Gasser F (1998) Expressed sequence tags for genes: a review. Genet Sel Evol 30(6):521–541

Hawkins RL, Nakamura M (1999) Expression of human growth hormone by the eukaryotic alga, *Chlorella*. Curr Microbiol 38(6):335–341

Huang W, Ye J, Zhang J, Lin Y, He M, Huang J (2016) Transcriptome analysis of *Chlorella zofingiensis* to identify genes and their expressions involved in astaxanthin and triacylglycerol biosynthesis. Algal Res 17:236–243

Hung CH, Ho MY, Kanehara K, Nakamura Y (2013) Functional study of diacylglycerol acyltransferase type 2 family in *Chlamydomonas reinhardtii*. FEBS Lett 587(15):2364–2370

Ibáñez-Salazar A, Rosales-Mendoza S, Rocha-Uribe A, Ramírez-Alonso JI, Lara-Hernández I, Hernández-Torres A et al (2014) Over-expression of Dof-type transcription factor increases lipid production in *Chlamydomonas reinhardtii*. J Biotechnol 184:27–38

Ikehata H, Ono T (2011) The mechanisms of UV mutagenesis. J Radiat Res 52(2):115–125

Issaq HJ, Veenstra TD (2008) Two-dimensional polyacrylamide gel electrophoresis (2D-PAGE): advances and perspectives. Biotechniques 44(5):697–700

Jackson HO, Berepiki A, Baylay AJ, Terry MJ, Moore CM, Bibby TS (2019) An inducible expression system in the alga *Nannochloropsis gaditana* controlled by the nitrate reductase promoter. J Appl Phycol 31(1):269–279

Jaeger D, Winkler A, Mussgnug JH, Kalinowski J, Goesmann A, Kruse O (2017) Time-resolved transcriptome analysis and lipid pathway reconstruction of the oleaginous green microalga *Monoraphidium neglectum* reveal a model for triacylglycerol and lipid hyperaccumulation. Biotechnol Biofuels 10(1):1–34

Jaiswal D, Sengupta A, Sohoni S, Sengupta S, Phadnavis AG, Pakrasi HB, Wangikar PP (2018) Genome features and biochemical characteristics of a robust, fast growing and naturally transformable cyanobacterium *Synechococcus elongatus* PCC 11801 isolated from India. Sci Rep 8(1):1–13

Jallet D, Xing D, Hughes A, Moosburner M, Simmons MP, Allen AE, Peers G (2020) Mitochondrial fatty acid β-oxidation is required for storage-lipid catabolism in a marine diatom. New Phytol 228(3):946–958

Jeon S, Koh HG, Cho JM, Kang NK, Chang YK (2021) Enhancement of lipid production in Nannochloropsis salina by overexpression of endogenous NADP-dependent malic enzyme. Algal Res 54:102218

Jinturkar KA, Rathi MN, Misra A (2011) Gene delivery using physical methods. In: Mishra A (ed) Challenges in delivery of therapeutic genomics and proteomics. Elsevier, Amsterdam, pp 83–126

Kaeppler HF, Gu W, Somers DA, Rines HW, Cockburn AF (1990) Silicon carbide fiber-mediated DNA delivery into plant cells. Plant Cell Rep 9(8):415–418

Kallio MA, Tuimala JT, Hupponen T, Klemelä P, Gentile M, Scheinin I et al (2011) Chipster: user-friendly analysis software for microarray and other high-throughput data. BMC Genomics 12(1):1–14

Kaloudas D, Pavlova N, Penchovsky R (2021) Lignocellulose, algal biomass, biofuels and biohydrogen: a review. Environ Chem Lett 19:1–16

Kang NK, Jeon S, Kwon S, Koh HG, Shin SE, Lee B et al (2015) Effects of overexpression of a bHLH transcription factor on biomass and lipid production in *Nannochloropsis salina*. Biotechnol Biofuels 8(1):1–13

Kang NK, Kim EK, Kim YU, Lee B, Jeong WJ, Jeong BR, Chang YK (2017) Increased lipid production by heterologous expression of AtWRI1 transcription factor in *Nannochloropsis salina*. Biotechnol Biofuels 10(1):1–14

Kania K, Zienkiewicz M, Drożak A (2020) Stable transformation of unicellular green alga *Coccomyxa subellipsoidea* C-169 via electroporation. Protoplasma 257(2):607–611

Kannan N, Donnellan P (2021) Algae-assisted microbial fuel cells: a practical overview. Bioresour Technol Rep 15:100747

Karthikaichamy A, Deore P, Rai V, Bulach D, Beardall J, Noronha S, Srivastava S (2017) Time for multiple extraction methods in proteomics? a comparison of three protein extraction methods in the Eustigmatophyte alga Microchloropsis gaditana CCMP526. OMICS 21(11):678–683

Kathiresan S, Chandrashekar A, Ravishankar GA, Sarada R (2009) Agrobacterium-mediated transformation in the green alga *Haematococcus pluvialis* (Chlorophyceae, Volvocales) 1. J Phycol 45(3):642–649

Kaye Y, Grundman O, Leu S, Zarka A, Zorin B, Didi-Cohen S et al (2015) Metabolic engineering toward enhanced LC-PUFA biosynthesis in *Nannochloropsis oceanica*: overexpression of endogenous Δ12 desaturase driven by stress-inducible promoter leads to enhanced deposition of polyunsaturated fatty acids in TAG. Algal Res 11:387–398

Khan S, Fu P (2020) Biotechnological perspectives on algae: a viable option for next generation biofuels. Curr Opin Biotechnol 62:146–152

Kim YH (2018) Control of the culture conditions of *Chlamydomonas reinhardtii* for efficient delivery of exogenous materials in electroporation. Algal Res 35:388–394

Kim YI, Cho JY (2019) Gel-based proteomics in disease research: is it still valuable? Biochim Biophys Acta Prot Proteom 1867(1):9–16

Kim DH, Kim YT, Cho JJ, Bae JH, Hur SB, Hwang I, Choi TJ (2002) Stable integration and functional expression of flounder growth hormone gene in transformed microalga, *Chlorella ellipsoidea*. Mar Biotechnol 4(1):63–73

Kim J, Kwak HS, Sim SJ, Jin E (2019a) Overexpression of malic enzyme isoform 2 in *Chlamydomonas reinhardtii* PTS42 increases lipid production. Bioresour Technol Rep 7: 100239

Kim YH, Kwon SG, Bae SJ, Park SJ (2019b) Optimization of the droplet electroporation method for wild type *Chlamydomonas reinhardtii* transformation. Bioelectrochemistry 126:29–37

Kindle KL (1990) High-frequency nuclear transformation of *Chlamydomonas reinhardtii*. Proc Natl Acad Sci 87(3):1228–1232

Klaitong P, Watcharawipas A, Fa-aroonsawat S, Chungjatupornchai W (2021) Nitrogen starvation-inducible promoter of microalga *Neochloris oleoabundans* lipogenic gene encoding diacylglycerol acyltransferase 2. J Appl Phycol 33(1):331–341

Knoshaug EP, Nag A, Astling DP, Douchi D, Laurens LM (2020) Draft genome sequence of the biofuel-relevant microalga *Desmodesmus armatus*. Microbiol Resour Announc 9(6):e00896–e00819

Kotnik T, Frey W, Sack M, Megli č SH, Peterka M, Miklavčič D (2015) Electroporation-based applications in biotechnology. Trends Biotechnol 33(8):480–488

Krishnan A, Likhogrud M, Cano M, Edmundson S, Melanson JB, Huesemann M et al (2021) *Picochlorum celeri* as a model system for robust outdoor algal growth in seawater. Sci Rep 11(1):1–13

Kukurba KR, Montgomery SB (2015) RNA sequencing and analysis. Cold Spring Harb Protoc 2015(11):pdb-top084970

Kumar SV, Misquitta RW, Reddy VS, Rao BJ, Rajam MV (2004) Genetic transformation of the green alga—*Chlamydomonas reinhardtii* by Agrobacterium tumefaciens. Plant Sci 166(3): 731–738

Kurtzman AL (1991) Direct gene transfer and transient gene expression in a marine red alga using the biolistic method. J Phycol 27:42

Laemmli UK (1970) Cleavage of structural proteins during the assembly of the head of bacteriophage T4. Nature 227:680–685

Lane CE, van den Heuvel K, Kozera C, Curtis BA, Parsons BJ, Bowman S, Archibald JM (2007) Nucleomorph genome of *Hemiselmis andersenii* reveals complete intron loss and compaction as a driver of protein structure and function. Proc Natl Acad Sci 104(50):19908–19913

Larsson L, Frisén J, Lundeberg J (2021) Spatially resolved transcriptomics adds a new dimension to genomics. Nat Methods 18(1):15–18

Lawrence CW (1991) [18] Classical mutagenesis techniques. Methods Enzymol 194:273–281

Lee B, Choi GG, Choi YE, Sung M, Park MS, Yang JW (2014) Enhancement of lipid productivity by ethyl methane sulfonate-mediated random mutagenesis and proteomic analysis in *Chlamydomonas reinhardtii*. Korean J Chem Eng 31(6):1036–1042

Lee JW, Lee MW, Ha JS, Kim DS, Jin E, Lee HG, Oh HM (2020a) Development of a species-specific transformation system using the novel endogenous promoter calreticulin from oleaginous microalgae Ettlia sp. Sci Rep 10(1):1–12

Lee BS, Koo KM, Ryu J, Hong MJ, Kim SH, Kwon SJ et al (2020b) Overexpression of fructose-1, 6-bisphosphate aldolase 1 enhances accumulation of fatty acids in *Chlamydomonas reinhardtii*. Algal Res 47:101825

Li Y, Mu J, Chen D, Han F, Xu H, Kong F et al (2013a) Production of biomass and lipid by the microalgae Chlorella protothecoides with heterotrophic-Cu (II) stressed (HCuS) coupling cultivation. Bioresour Technol 148:283–292

Li Y, Yuan Z, Mu J, Chen D, Feng B (2013b) Proteomic analysis of lipid accumulation in Chlorella protothecoides cells by heterotrophic N deprivation coupling cultivation. Energy Fuel 27(7): 4031–4040

Li Y, Han F, Xu H, Mu J, Chen D, Feng B, Zeng H (2014) Potential lipid accumulation and growth characteristic of the green alga Chlorella with combination cultivation mode of nitrogen (N) and phosphorus (P). Bioresour Technol 174:24–32

Li Y, Mu J, Chen D, Xu H, Han F, Feng B, Zeng H (2015) Proteomics analysis for enhanced lipid accumulation in oleaginous Chlorella vulgaris under a heterotrophic-Na+ induction two-step regime. Biotechnol Lett 37(5):1021–1030

Li DW, Cen SY, Liu YH, Balamurugan S, Zheng XY, Alimujiang A et al (2016) A type 2 diacylglycerol acyltransferase accelerates the triacylglycerol biosynthesis in heterokont oleaginous microalga Nannochloropsis oceanica. J Biotechnol 229:65–71

Li H, Han J, Pan J, Liu T, Parker CE, Borchers CH (2017) Current trends in quantitative proteomics–an update. J Mass Spectrom 52(5):319–341

Li Z, Meng T, Ling X, Li J, Zheng C, Shi Y et al (2018a) Overexpression of malonyl-CoA: ACP transacylase in Schizochytrium sp. to improve polyunsaturated fatty acid production. J Agric Food Chem 66(21):5382–5391

Li DW, Xie WH, Hao TB, Cai JX, Zhou TB, Balamurugan S et al (2018b) Constitutive and chloroplast targeted expression of acetyl-CoA carboxylase in oleaginous microalgae elevates fatty acid biosynthesis. Mar Biotechnol 20(5):566–572

Liang C, Cao S, Zhang X, Zhu B, Su Z, Xu D et al (2013) De novo sequencing and global transcriptome analysis of Nannochloropsis sp. (Eustigmatophyceae) following nitrogen starvation. Bioenergy Res 6(2):494–505

Liu J, Sun Z, Gerken H, Huang J, Jiang Y, Chen F (2014) Genetic engineering of the green alga Chlorella zofingiensis: a modified norflurazon-resistant phytoene desaturase gene as a dominant selectable marker. Appl Microbiol Biotechnol 98(11):5069–5079

Liu J, Pei G, Diao J, Chen Z, Liu L, Chen L, Zhang W (2017) Screening and transcriptomic analysis of Crypthecodinium cohnii mutants with high growth and lipid content using the acetyl-CoA carboxylase inhibitor sethoxydim. Appl Microbiol Biotechnol 101(15):6179–6191

Liu X, Zhang D, Zhang J, Chen Y, Liu X, Fan C et al (2021) Overexpression of the Transcription Factor AtLEC1 Significantly Improved the Lipid Content of Chlorella ellipsoidea. Front Bioeng Biotechnol 9:113

Loechler E, Green CL, Essigmann JM (1984) In vivo mutagenesis by O6-methylguanine built into a unique site in a viral genome. Proc Natl Acad Sci U S A 81:6271–6275

Longworth J, Wu D, Huete-Ortega M, Wright PC, Vaidyanathan S (2016) Proteome response of Phaeodactylum tricornutum, during lipid accumulation induced by nitrogen depletion. Algal Res 18:213–224

Lowe R, Shirley N, Bleackley M, Dolan S, Shafee T (2017) Transcriptomics technologies. PLoS Comput Biol 13(5):e1005457

Lowry OH, Rosebrough NJ, Farr AL, Randall RJ (1951) Protein measurement with the Folin phenol reagent. J Biol Chem 193:265–275

Lu Y, Zhang X, Gu X, Lin H, Melis A (2021) Engineering microalgae: transition from empirical design to programmable cells. Crit Rev Biotechnol 41:1–24

Lumbreras V, Stevens DR, Purton S (1998) Efficient foreign gene expression in Chlamydomonas reinhardtii mediated by an endogenous intron. Plant J 14(4):441–447

Ma Q, Wang J, Lu S, Lv Y, Yuan Y (2013a) Quantitative proteomic profiling reveals photosynthesis responsible for inoculum size dependent variation in Chlorella sorokiniana. Biotechnol Bioeng 110(3):773–784

Ma Y, Wang Z, Zhu M, Yu C, Cao Y, Zhang D, Zhou G (2013b) Increased lipid productivity and TAG content in Nannochloropsis by heavy-ion irradiation mutagenesis. Bioresour Technol 136: 360–367

Mackintosh JA, Choi HY, Bae SH, Veal DA, Bell PJ, Ferrari BC et al (2003) A fluorescent natural product for ultra sensitive detection of proteins in one-dimensional and two-dimensional gel electrophoresis. Proteomics 3(12):2273–2288

Mahan KM, Polle JEW, McKie-Krisberg Z, Lipzen A, Kuo A, Grigoriev IV, Lane TW (2021) Davis AK annotated genome sequence of the high-biomass-producing yellow-green alga *Tribonema minus*. Microbiol Resour Announc 10(24):e0032721. https://doi.org/10.1128/MRA.00327-2

Manandhar-Shrestha K, Hildebrand M (2013) Development of flow cytometric procedures for the efficient isolation of improved lipid accumulation mutants in a Chlorella sp. microalga. J Appl Phycol 25(6):1643–1651

Mao X, Lao Y, Sun H, Li X, Yu J, Chen F (2020) Time-resolved transcriptome analysis during transitions of sulfur nutritional status provides insight into triacylglycerol (TAG) and astaxanthin accumulation in the green alga *Chromochloris zofingiensis*. Biotechnol Biofuels 13(1):1–18

Mehtani J, Arora N, Patel A, Jain P, Pruthi PA, Poluri KM, Pruthi V (2017) Augmented lipid accumulation in ethyl methyl sulphonate mutants of oleaginous microalga for biodiesel production. Bioresour Technol 242:121–127

Meleady P (2018) Two-dimensional gel electrophoresis and 2D-DIGE. In: Ohlendieck K (ed) Difference gel electrophoresis. Humana Press, New York, NY, pp 3–14

Merchant SS, Prochnik SE, Vallon O, Harris EH, Karpowicz SJ, Witman GB et al (2007) The Chlamydomonas genome reveals the evolution of key animal and plant functions. Science 318(5848):245–250

Mezlini AM, Smith EJM, Fiume M, Buske O, Savich GL, Shah S, Aparicio S, Chiang DY, Goldenberg A, Brudno M (2013) iReckon: simultaneous isoform discovery and abundance estimation from RNA-seq data. Genome Res 23:519–529. PubMed: 23204306

Michelet L, Lefebvre-Legendre L, Burr SE, Rochaix JD, Goldschmidt-Clermont M (2011) Enhanced chloroplast transgene expression in a nuclear mutant of *Chlamydomonas*. Plant Biotechnol J 9(5):565–574

Mock T, Otillar RP, Strauss J, McMullan M, Paajanen P, Schmutz J et al (2017) Evolutionary genomics of the cold-adapted diatom *Fragilariopsis cylindrus*. Nature 541(7638):536–540

Moha-León JD, Pérez-Legaspi IA, Ortega-Clemente LA, Rubio-Franchini I, Ríos-Leal E (2019) Improving the lipid content of *Nannochloropsis oculata* by a mutation-selection program using UV radiation and quizalofop. J Appl Phycol 31(1):191–199

Moore CE, Curtis B, Mills T, Tanifuji G, Archibald JM (2012) Nucleomorph genome sequence of the cryptophyte alga *Chroomonas mesostigmatica* CCMP1168 reveals lineage-specific gene loss and genome complexity. Genome Biol Evol 4(11):1162–1175

Mortz E, Krogh TN, Vorum H, Görg A (2001) Improved silver staining protocols for high sensitivity protein identification using matrix-assisted laser desorption/ionization-time of flight analysis. Proteomics 1(11):1359–1363

Mosey M, Douchi D, Knoshaug EP, Laurens LM (2021) Methodological review of genetic engineering approaches for non-model algae. Algal Res 54:102221

Muñoz CF, de Jaeger L, Sturme MH, Lip KY, Olijslager JW, Springer J et al (2018) Improved DNA/protein delivery in microalgae–a simple and reliable method for the prediction of optimal electroporation settings. Algal Res 33:448–455

Muñoz CF, Weusthuis RA, D'Adamo S, Wijffels RH (2019) Effect of single and combined expression of lysophosphatidic acid acyltransferase, glycerol-3-phosphate acyltransferase, and diacylglycerol acyltransferase on lipid accumulation and composition in *Neochloris oleoabundans*. Front Plant Sci 10:1573

Myers SA, Klaeger S, Satpathy S, Viner R, Choi J, Rogers J et al (2018) Evaluation of advanced precursor determination for tandem mass tag (TMT)-based quantitative proteomics across instrument platforms. J Proteome Res 18(1):542–547

Ndimba BK, Ndimba RJ, Johnson TS, Waditee-Sirisattha R, Baba M, Sirisattha S et al (2013) Biofuels as a sustainable energy source: an update of the applications of proteomics in bioenergy crops and algae. J Proteome 93:234–244

Ng IS, Tan SI, Kao PH, Chang YK, Chang JS (2017) Recent developments on genetic engineering of microalgae for biofuels and bio-based chemicals. Biotechnol J 12(10):1600644

Ngan CY, Wong CH, Choi C, Yoshinaga Y, Louie K, Jia J et al (2015) Lineage-specific chromatin signatures reveal a regulator of lipid metabolism in microalgae. Nat Plants 1(8):1–12

Niu YF, Zhang MH, Li DW, Yang WD, Liu JS, Bai WB, Li HY (2013) Improvement of neutral lipid and polyunsaturated fatty acid biosynthesis by overexpressing a type 2 diacylglycerol acyltransferase in marine diatom *Phaeodactylum tricornutum*. Mar Drugs 11(11):4558–4569

Noochanong W, Jirakranwong P, Chanprame S (2018) EMS-induced mutation followed by quizalofop-screening increased lipid productivity in Chlorella sp. Bioprocess Biosyst Eng 41(5):613–619

Noor Z, Ahn SB, Baker MS, Ranganathan S, Mohamedali A (2021) Mass spectrometry–based protein identification in proteomics—a review. Brief Bioinform 22(2):1620–1638

Norzagaray-Valenzuela CD, Germán-Báez LJ, Valdez-Flores MA, Hernández-Verdugo S, Shelton LM, Valdez-Ortiz A (2018) Establishment of an efficient genetic transformation method in Dunaliella tertiolecta mediated by Agrobacterium tumefaciens. J Microbiol Methods 150:9–17

Ogura A, Akizuki Y, Imoda H, Mineta K, Gojobori T, Nagai S (2018) Comparative genome and transcriptome analysis of diatom, *Skeletonema costatum*, reveals evolution of genes for harmful algal bloom. BMC Genomics 19(1):1–12

Ortiz-Matamoros MF, Islas-Flores T, Voigt B, Menzel D, Baluška F, Villanueva MA (2015) Heterologous DNA uptake in cultured *Symbiodinium* spp. aided by Agrobacterium tumefaciens. PLoS One 10(7):e0132693

Ortiz-Matamoros MF, Villanueva MA, Islas-Flores T (2018) Genetic transformation of cell-walled plant and algae cells: delivering DNA through the cell wall. Brief Funct Genom 17(1):26–33

Palenik B, Grimwood J, Aerts A, Rouzé P, Salamov A, Putnam N et al (2007) The tiny eukaryote *Ostreococcus* provides genomic insights into the paradox of plankton speciation. Proc Natl Acad Sci 104(18):7705–7710

Pasaribu B, Jiang PL (2021) Agrobacterium tumefaciens Mediated Transformation of *Symbiodinium* spp. Turk J Fish Aquat Sci 21(6):291–298

Patel VK, Maji D, Pandey SS, Rout PK, Sundaram S, Kalra A (2016) Rapid budding EMS mutants of *Synechocystis* PCC 6803 producing carbohydrate or lipid enriched biomass. Algal Res 16: 36–45

Peng H, Wei D, Chen G, Chen F (2016) Transcriptome analysis reveals global regulation in response to CO 2 supplementation in oleaginous microalga *Coccomyxa subellipsoidea* C-169. Biotechnol Biofuels 9(1):1–17

Perin G, Bellan A, Segalla A, Meneghesso A, Alboresi A, Morosinotto T (2015) Generation of random mutants to improve light-use efficiency of Nannochloropsis gaditana cultures for biofuel production. Biotechnol Biofuels 8(1):1–13

Peterson CL, Reece DL, Thompson JC, Beck SM, Chase C (1996) Ethyl ester of rapeseed used as a biodiesel fuel—a case study. Biomass Bioenergy 10(5–6):331–336

Pivato M, Perozeni F, Licausi F, Cazzaniga S, Ballottari M (2021) Heterologous expression of cyanobacterial Orange Carotenoid Protein (OCP2) as a soluble carrier of ketocarotenoids in *Chlamydomonas reinhardtii*. Algal Res 55:102255

Polle JE, Barry K, Cushman J, Schmutz J, Tran D, Hathwaik LT et al (2017) Draft nuclear genome sequence of the halophilic and beta-carotene-accumulating green alga *Dunaliella salina* strain CCAP19/18. Genome Announc 5(43):e01105–e01117

Poulsen N, Chesley PM, Kröger N (2006) Molecular genetic manipulation of the diatom *Thalassiosira pseudonana* (Bacillariophyceae) 1. J Phycol 42(5):1059–1065

Prasad B, Vadakedath N, Jeong HJ, General T, Cho MG, Lein W (2014) Agrobacterium tumefaciens-mediated genetic transformation of haptophytes (*Isochrysis* species). Appl Microbiol Biotechnol 98(20):8629–8639

Prasad B, Lein W, Lindenberger CP, Buchholz R, Vadakedath N (2019) Stable nuclear transformation of rhodophyte species Porphyridium purpureum: advanced molecular tools and an optimized method. Photosynth Res 140(2):173–188

Pratheesh PT, Vineetha M, Kurup GM (2014) An efficient protocol for the Agrobacterium-mediated genetic transformation of microalga *Chlamydomonas reinhardtii*. Mol Biotechnol 56(6):507–515

Prochnik SE, Umen J, Nedelcu AM, Hallmann A, Miller SM, Nishii I et al (2010) Genomic analysis of organismal complexity in the multicellular green alga *Volvox carteri*. Science 329(5988): 223–226

Qin S, Lin H, Jiang P (2012) Advances in genetic engineering of marine algae. Biotechnol Adv 30(6):1602–1613

Qu D, Miao X (2021) Carbon flow conversion induces alkali resistance and lipid accumulation under alkaline conditions based on transcriptome analysis in *Chlorella* sp. BLD. Chemosphere 265:129046

Radakovits R, Jinkerson RE, Fuerstenberg SI, Tae H, Settlage RE, Boore JL, Posewitz MC (2012) Draft genome sequence and genetic transformation of the oleaginous alga *Nannochloropsis gaditana*. Nat Commun 3(1):1–11

Rai V, Karthikaichamy A, Das D, Noronha S, Wangikar PP, Srivastava S (2016) Multi-omics frontiers in algal research: techniques and progress to explore biofuels in the postgenomics world. OMICS 20(7):387–399

Rai V, Muthuraj M, Gandhi MN, Das D, Srivastava S (2017) Real-time iTRAQ-based proteome profiling revealed the central metabolism involved in nitrogen starvation induced lipid accumulation in microalgae. Sci Rep 7(1):1–16

Rasala BA, Chao SS, Pier M, Barrera DJ, Mayfield SP (2014) Enhanced genetic tools for engineering multigene traits into green algae. PLoS One 9(4):e94028

Rathod JP, Prakash G, Pandit R, Lali AM (2013) Agrobacterium-mediated transformation of promising oil-bearing marine algae *Parachlorella kessleri*. Photosynth Res 118(1):141–146

Razeghifard R (2013) Algal biofuels. Photosynth Res 117(1):207–219

Read BA, Kegel J, Klute MJ, Kuo A, Lefebvre SC, Maumus F et al (2013) Pan genome of the phytoplankton *Emiliania underpins* its global distribution. Nature 499(7457):209–213

Reijnders MJ, van Heck RG, Lam CM, Scaife MA, dos Santos VAM, Smith AG, Schaap PJ (2014) Green genes: bioinformatics and systems-biology innovations drive algal biotechnology. Trends Biotechnol 32(12):617–626

Remacle C, Cline S, Boutaffala L, Gabilly S, Larosa V, Barbieri MR et al (2009) The ARG9 gene encodes the plastid-resident N-acetyl ornithine aminotransferase in the green alga *Chlamydomonas reinhardtii*. Eukaryot Cell 8(9):1460–1463

Remmers IM, D'Adamo S, Martens DE, de Vos RC, Mumm R, America AH et al (2018) Orchestration of transcriptome, proteome and metabolome in the diatom *Phaeodactylum tricornutum* during nitrogen limitation. Algal Res 35:33–49

Rossoni AW, Price DC, Seger M, Lyska D, Lammers P, Bhattacharya D, Weber AP (2019) The genomes of polyextremophilic cyanidiales contain 1% horizontally transferred genes with diverse adaptive functions. elife 8:e45017

Ruecker O, Zillner K, Groebner-Ferreira R, Heitzer M (2008) Gaussia-luciferase as a sensitive reporter gene for monitoring promoter activity in the nucleus of the green alga *Chlamydomonas reinhardtii*. Mol Gen Genomics 280(2):153–162

Run C, Fang L, Fan J, Fan C, Luo Y, Hu Z, Li Y (2016) Stable nuclear transformation of the industrial alga *Chlorella pyrenoidosa*. Algal Res 17:196–201

Saal LH, Troein C, Vallon-Christersson J, Gruvberger S, Borg Å, Peterson C (2002) BioArray Software Environment (BASE): a platform for comprehensive management and analysis of microarray data. Genome Biol 3(8):1–6

Sachdeva N, Gupta RP, Mathur AS, Tuli DK (2016) Enhanced lipid production in thermo-tolerant mutants of *Chlorella pyrenoidosa* NCIM 2738. Bioresour Technol 221:576–587

Saeed AI, Bhagabati NK, Braisted JC, Liang W, Sharov V, Howe EA et al (2006) [9] TM4 microarray software suite. Methods Enzymol 411:134–193

Sahoo S, Mahapatra SR, Das N, Parida BK, Rath S, Misra N, Suar M (2020) Functional elucidation of hypothetical proteins associated with lipid accumulation: prioritizing genetic engineering targets for improved algal biofuel production. Algal Res 47:101887

Salama ES, Govindwar SP, Khandare RV, Roh HS, Jeon BH, Li X (2019) Can omics approaches improve microalgal biofuels under abiotic stress? Trends Plant Sci 24(7):611–624

San Cha T, Yee W, Aziz A (2012) Assessment of factors affecting Agrobacterium-mediated genetic transformation of the unicellular green alga, *Chlorella vulgaris*. World J Microbiol Biotechnol 28(4):1771–1779

Sanitha M, Radha S, Fatima AA, Devi SG, Ramya M (2014) *Agrobacterium*-mediated transformation of three freshwater microalgal strains. Pol J Microbiol 63(4):387–382

Sarayloo E, Simsek S, Unlu YS, Cevahir G, Erkey C, Kavakli IH (2018) Enhancement of the lipid productivity and fatty acid methyl ester profile of *Chlorella vulgaris* by two rounds of mutagenesis. Bioresour Technol 250:764–769

Schubert OT, Röst HL, Collins BC, Rosenberger G, Aebersold R (2017) Quantitative proteomics: challenges and opportunities in basic and applied research. Nat Protoc 12(7):1289–1294

Scranton MA, Ostrand JT, Fields FJ, Mayfield SP (2015) Chlamydomonas as a model for biofuels and bio-products production. Plant J 82(3):523–531

Shahar N, Landman S, Weiner I, Elman T, Dafni E, Feldman Y et al (2020) The integration of multiple nuclear-encoded transgenes in the green alga *Chlamydomonas reinhardtii* results in higher transcription levels. Front Plant Sci 10:1784

Shang C, Bi G, Yuan Z, Wang Z, Alam MA, Xie J (2016) Discovery of genes for production of biofuels through transcriptome sequencing of *Dunaliella parva*. Algal Res 13:318–326

Shang C, Zhu S, Wang Z, Qin L, Alam MA, Xie J, Yuan Z (2017) Proteome response of *Dunaliella parva* induced by nitrogen limitation. Algal Res 23:196–202

Sharma T, Chauhan RS (2016) Comparative transcriptomics reveals molecular components associated with differential lipid accumulation between microalgal sp., *Scenedesmus dimorphus* and *Scenedesmus quadricauda*. Algal Res 19:109–122

Sharma PK, Goud VV, Yamamoto Y, Sahoo L (2021) Efficient Agrobacterium tumefaciens-mediated stable genetic transformation of green microalgae, Chlorella sorokiniana. 3 Biotech 11(4):1–11

Shi H, Chen H, Gu Z, Zhang H, Chen W, Chen YQ (2016) Application of a delta-6 desaturase with α-linolenic acid preference on eicosapentaenoic acid production in *Mortierella alpina*. Microb Cell Factories 15(1):1–14

Shiio Y, Aebersold R (2006) Quantitative proteome analysis using isotope-coded affinity tags and mass spectrometry. Nat Protoc 1(1):139–145

Shoguchi E, Beedessee G, Tada I, Hisata K, Kawashima T, Takeuchi T et al (2018) Two divergent *Symbiodinium* genomes reveal conservation of a gene cluster for sunscreen biosynthesis and recently lost genes. BMC Genomics 19(1):1–11

Shrager J, Hauser C, Chang CW, Harris EH, Davies J, McDermott J et al (2003) *Chlamydomonas reinhardtii* genome project. A guide to the generation and use of the cDNA information. Plant Physiol 131(2):401–408

Sickmann A, Mreyen M, Meyer HE (2003) Mass spectrometry—a key technology in proteom research. Adv Biochem Eng Biotechnol 83:141–176

Siddiqui A, Wei Z, Boehm M, Ahmad N (2020) Engineering microalgae through chloroplast transformation to produce high-value industrial products. Biotechnol Appl Biochem 67(1): 30–40

Sierra LS, Dixon CK, Wilken LR (2017) Enzymatic cell disruption of the microalgae Chlamydomonas reinhardtii for lipid and protein extraction. Algal Res 25:149–159

Simon DP, Narayanan A, Gouda KM, Sarada R (2015) Vir gene inducers in *Dunaliella salina*; an insight in to the *Agrobacterium*-mediated genetic transformation of microalgae. Algal Res 11: 121–124

Simon DP, Anila N, Gayathri K, Sarada R (2016) Heterologous expression of β-carotene hydroxylase in *Dunaliella salina* by Agrobacterium-mediated genetic transformation. Algal Res 18: 257–265

Sirikhachornkit A, Suttangkakul A, Vuttipongchaikij S, Juntawong P (2018) De novo transcriptome analysis and gene expression profiling of an oleaginous microalga *Scenedesmus acutus* TISTR8540 during nitrogen deprivation-induced lipid accumulation. Sci Rep 8(1):1–12

Sivakumar G, Xu J, Thompson RW, Yang Y, Randol-Smith P, Weathers PJ (2012) Integrated green algal technology for bioremediation and biofuel. Bioresour Technol 107:1–9

Smalley T, Fields FJ, Berndt AJ, Ostrand JT, Heredia V, Mayfield SP (2020) Improving biomass and lipid yields of Desmodesmus armatus and Chlorella vulgaris through mutagenesis and high-throughput screening. Biomass Bioenergy 142:105755

Smita P, Gunjan P, Lali AM (2020) Reduced chlorophyll antenna mutants of Chlorella saccharophila for higher photosynthetic efficiency and biomass productivity under high light intensities. J Appl Phycol 32(3):1559–1567

Song P, Li L, Liu J (2013) Proteomic analysis in nitrogen-deprived *Isochrysis galbana* during lipid accumulation. PLoS One 8(12):e82188

Srinivasan R, Gothandam KM (2016) Synergistic action of D-glucose and acetosyringone on *Agrobacterium* strains for efficient *Dunaliella* transformation. PLoS One 11(6):e0158322

Stephenson PG, Moore CM, Terry MJ, Zubkov MV, Bibby TS (2011) Improving photosynthesis for algal biofuels: toward a green revolution. Trends Biotechnol 29(12):615–623

Sturme MH, Gong Y, Heinrich JM, Klok AJ, Eggink G, Wang D et al (2018) Transcriptome analysis reveals the genetic foundation for the dynamics of starch and lipid production in *Ettlia oleoabundans*. Algal Res 33:142–155

Suzuki S, Yamaguchi H, Nakajima N, Kawachi M (2018) *Raphidocelis subcapitata* (= *Pseudokirchneriella subcapitata*) provides an insight into genome evolution and environmental adaptations in the Sphaeropleales. Sci Rep 8(1):1–13

Takahashi H, Tanaka S, Hayashi S, Miyaki S, Takahashi A, Onai S et al (2018) Draft genome sequence of *Trebouxiophyceae* sp. strain KSI-1, isolated from an island hot spring. Microbiol Resour Announc 7(16):e01185–e01118

Tan KWM, Lee YK (2017) Expression of the heterologous *Dunaliella tertiolecta* fatty acyl-ACP thioesterase leads to increased lipid production in *Chlamydomonas reinhardtii*. J Biotechnol 247:60–67

Tan C, Qin S, Zhang Q, Jiang P, Zhao F (2005) Establishment of a micro-particle bombardment transformation system for *Dunaliella salina*. J Microbiol 43(4):361–365

Teh KY, Afifudeen CW, Aziz A, Wong LL, Loh SH, San Cha T (2019) De novo whole genome sequencing data of two mangrove-isolated microalgae from Terengganu coastal waters. Data Brief 27:104680

Teng C, Qin S, Liu J, Yu D, Liang C, Tseng C (2002) Transient expression of lacZ in bombarded unicellular green alga *Haematococcus pluvialis*. J Appl Phycol 14(6):497–500

Tokunaga S, Sanda S, Uraguchi Y, Nakagawa S, Sawayama S (2019) Overexpression of the DOF-type transcription factor enhances lipid synthesis in *Chlorella vulgaris*. Appl Biochem Biotechnol 189(1):116–128

Toyomizu M, Suzuki K, Kawata Y, Kojima H, Akiba Y (2001) Effective transformation of the cyanobacterium *Spirulina platensis* using electroporation. J Appl Phycol 13(3):209–214

Tran NAT, Padula MP, Evenhuis CR, Commault AS, Ralph PJ, Tamburic B (2016) Proteomic and biophysical analyses reveal a metabolic shift in nitrogen deprived *Nannochloropsis oculata*. Algal Res 19:1–11

Trapnell C, Pachter L, Salzberg SL (2009) TopHat: discovering splice junctions with RNA-Seq. Bioinformatics 25:1105–1111

Trapnell C, Williams BA, Pertea G, Mortazavi A, Kwan G, van Baren MJ, Salzberg SL, Wold BJ, Pachter L (2010) Transcript assembly and quantification by RNA-Seq reveals unannotated transcripts and isoform switching during cell differentiation. Nat Biotechnol 28:511–515

Tyers M, Mann M (2003) From genomics to proteomics. Nature 422(6928):193–197

Ünlü M, Morgan ME, Minden JS (1997) Difference gel electrophoresis. A single gel method for detecting changes in protein extracts. Electrophoresis 18(11):2071–2077

Valenzuela J, Mazurie A, Carlson RP, Gerlach R, Cooksey KE, Peyton BM, Fields MW (2012) Potential role of multiple carbon fixation pathways during lipid accumulation in *Phaeodactylum tricornutum*. Biotechnol Biofuels 5(1):1–17

Vigeolas H, Duby F, Kaymak E, Niessen G, Motte P, Franck F, Remacle C (2012) Isolation and partial characterization of mutants with elevated lipid content in Chlorella sorokiniana and *Scenedesmus obliquus*. J Biotechnol 162(1):3–12

Vignard J, Mirey G, Salles B (2013) Ionizing-radiation induced DNA double-strand breaks: a direct and indirect lighting up. Radiother Oncol 108(3):362–369

Walker TL, Collet C, Purton S (2005) Algal transgenics in the genomic era 1. J Phycol 41(6): 1077–1093

Wan M, Jin X, Xia J, Rosenberg JN, Yu G, Nie Z et al (2014) The effect of iron on growth, lipid accumulation, and gene expression profile of the freshwater microalga *Chlorella sorokiniana*. Appl Microbiol Biotechnol 98(22):9473–9481

Wang C, Wang Y, Su Q, Gao X (2007) Transient expression of the GUS gene in a unicellular marine green alga, *Chlorella* sp. MACC/C95, via electroporation. Biotechnol Bioprocess Eng 12(2):180–183

Wang K, Singh D, Zeng Z, Coleman SJ, Huang Y, Savich GL, He X, Miecz-kowski P, Grimm SA, Perou CM et al (2010) MapSplice: accurate mapping of RNA-seq reads for splice junction discovery. Nucleic Acids Res 38:e178

Wang D, Ning K, Li J, Hu J, Han D, Wang H et al (2014) *Nannochloropsis* genomes reveal evolution of microalgal oleaginous traits. PLoS Genet 10(1):e1004094

Wang X, Liu YH, Hu DX, Balamurugan S, Lu Y, Yang WD et al (2015) Identification of a putative patatin-like phospholipase domain-containing protein 3 (PNPLA3) ortholog involved in lipid metabolism in microalga Phaeodactylum tricornutum. Algal Res 12:274–279

Wang H, Gao L, Shao H, Zhou W, Liu T (2017) Lipid accumulation and metabolic analysis based on transcriptome sequencing of filamentous oleaginous microalgae Tribonema minus at different growth phases. Bioprocess Biosyst Eng 40(9):1327–1335

Wang C, Li Y, Lu J, Deng X, Li H, Hu Z (2018) Effect of overexpression of LPAAT and GPD1 on lipid synthesis and composition in green microalga *Chlamydomonas reinhardtii*. J Appl Phycol 30(3):1711–1719

Wannathong T, Waterhouse JC, Young RE, Economou CK, Purton S (2016) New tools for chloroplast genetic engineering allow the synthesis of human growth hormone in the green alga *Chlamydomonas reinhardtii*. Appl Microbiol Biotechnol 100(12):5467–5477

Wardman P (2009) The importance of radiation chemistry to radiation and free radical biology (The 2008 Silvanus Thompson Memorial Lecture). Br J Radiol 82(974):89–104

Wase N, Black PN, Stanley BA, DiRusso CC (2014) Integrated quantitative analysis of nitrogen stress response in *Chlamydomonas reinhardtii* using metabolite and protein profiling. J Proteome Res 13(3):1373–1396

Weaver JC, Chizmadzhev YA (1996) Theory of electroporation: a review. Bioelectrochem Bioenerg 41(2):135–160

Wei H, Shi Y, Ma X, Pan Y, Hu H, Li Y et al (2017) A type-I diacylglycerol acyltransferase modulates triacylglycerol biosynthesis and fatty acid composition in the oleaginous microalga, *Nannochloropsis oceanica*. Biotechnol Biofuels 10(1):1–18

Williams RS, Johnston SA, Riedy M, DeVit MJ, McElligott SG, Sanford JC (1991) Introduction of foreign genes into tissues of living mice by DNA-coated microprojectiles. Proc Natl Acad Sci 88(7):2726–2730

Worden AZ, Lee JH, Mock T, Rouzé P, Simmons MP, Aerts AL et al (2009) Green evolution and dynamic adaptations revealed by genomes of the marine picoeukaryotes Micromonas. Science 324(5924):268–272

Wu TD, Nacu S (2010) Fast and SNP-tolerant detection of complex variants and splicing in short reads. Bioinformatics 26:873–881

Wu H, Kerr MK, Cui X, Churchill GA (2003) MAANOVA: a software package for the analysis of spotted cDNA microarray experiments. In: Parmigiani G et al (eds) The analysis of gene expression data. Springer, New York, NY, pp 313–341

Xie WH, Zhu CC, Zhang NS, Li DW, Yang WD, Liu JS et al (2014) Construction of novel chloroplast expression vector and development of an efficient transformation system for the diatom *Phaeodactylum tricornutum*. Mar Biotechnol 16(5):538–546

Xing G, Li J, Li W, Lam SM, Yuan H, Shui G, Yang J (2021a) AP2/ERF and R2R3-MYB family transcription factors: potential associations between temperature stress and lipid metabolism in *Auxenochlorella protothecoides*. Biotechnol Biofuels 14(1):1–16

Xing W, Zhang R, Shao Q, Meng C, Wang X, Wei Z et al (2021b) Effects of Laser Mutagenesis on Microalgae Production and Lipid Accumulation in Two Economically Important Fresh *Chlorella* Strains under Heterotrophic Conditions. Agronomy 11(5):961

Xue J, Niu YF, Huang T, Yang WD, Liu JS, Li HY (2015) Genetic improvement of the microalga *Phaeodactylum tricornutum* for boosting neutral lipid accumulation. Metab Eng 27:1–9

Yan N, Fan C, Chen Y, Hu Z (2016) The potential for microalgae as bioreactors to produce pharmaceuticals. Int J Mol Sci 17(6):962

Yang ZK, Ma YH, Zheng JW, Yang WD, Liu JS, Li HY (2014) Proteomics to reveal metabolic network shifts towards lipid accumulation following nitrogen deprivation in the diatom *Phaeodactylum tricornutum*. J Appl Phycol 26(1):73–82

Yang F, Xiang W, Li T, Long L (2018) Transcriptome analysis for phosphorus starvation-induced lipid accumulation in *Scenedesmus* sp. Sci Rep 8(1):1–11

Yao Y, Lu Y, Peng KT, Huang T, Niu YF, Xie WH et al (2014) Glycerol and neutral lipid production in the oleaginous marine diatom *Phaeodactylum tricornutum* promoted by overexpression of glycerol-3-phosphate dehydrogenase. Biotechnol Biofuels 7(1):1–9

Ye N, Zhang X, Miao M, Fan X, Zheng Y, Xu D et al (2015) *Saccharina* genomes provide novel insight into kelp biology. Nat Commun 6(1):1–11

Yi Z, Su Y, Xu M, Bergmann A, Ingthorsson S, Rolfsson O et al (2018) Chemical mutagenesis and fluorescence-based high-throughput screening for enhanced accumulation of carotenoids in a model marine diatom *Phaeodactylum tricornutum*. Mar Drugs 16(8):272

Yin W, Hu H (2021) High-efficiency transformation of a centric diatom *Chaetoceros muelleri* by electroporation with a variety of selectable markers. Algal Res 55:102274

You W, Wei L, Gong Y, El Hajjami M, Xu J, Poetsch A (2020) Integration of proteome and transcriptome refines key molecular processes underlying oil production in *Nannochloropsis oceanica*. Biotechnol Biofuels 13(1):1–19

Young RE, Purton S (2016) Codon reassignment to facilitate genetic engineering and biocontainment in the chloroplast of *Chlamydomonas reinhardtii*. Plant Biotechnol J 14(5):1251–1260

Zhang Y, Fonslow BR, Shan B, Baek MC, Yates JR III (2013) Protein analysis by shotgun/bottom-up proteomics. Chem Rev 113(4):2343–2394

Zhang K, Tang C, Liang X, Zhao Q, Zhang J (2018) Isobaric tags for relative and absolute quantification (iTRAQ)-based untargeted quantitative proteomic approach to identify change of the plasma proteins by salbutamol abuse in beef cattle. J Agric Food Chem 66(1):378–386

Zhang Y, Wang H, Yang R, Wang L, Yang G, Liu T (2020a) Genetic transformation of *Tribonema minus*, a eukaryotic filamentous oleaginous yellow-green alga. Int J Mol Sci 21(6):2106

Zhang Z, Qu C, Zhang K, He Y, Zhao X, Yang L et al (2020b) Adaptation to extreme Antarctic environments revealed by the genome of a sea ice green alga. Curr Biol 30(17):3330–3341

Zhang YT, Wei W, Wang Y, Ni BJ (2021a) Enhancing methane production from algae anaerobic digestion using diatomite. J Clean Prod 315:128138

Zhang Y, Pan Y, Ding W, Hu H, Liu J (2021b) Lipid production is more than doubled by manipulating a diacylglycerol acyltransferase in algae. GCB Bioenergy 13(1):185–200

Zhang Y, Ye Y, Bai F, Liu J (2021c) The oleaginous astaxanthin-producing alga *Chromochloris zofingiensis*: potential from production to an emerging model for studying lipid metabolism and carotenogenesis. Biotechnol Biofuels 14(1):1–37

Zhao D, Huang D, Li Y, Wu M, Zhong W, Cheng Q et al (2016) A flow-through cell electroporation device for rapidly and efficiently transfecting massive amounts of cells in vitro and ex vivo. Sci Rep 6(1):1–9

Zhou H, Wilkens A, Hanelt D, von Schwartzenberg K (2021) Expanding the molecular toolbox for Zygnematophyceae–transient genetic transformation of the desmid *Micrasterias radians var.* evoluta. Eur J Phycol 56(1):51–60

Zienkiewicz K, Zienkiewicz A, Poliner E, Du ZY, Vollheyde K, Herrfurth C et al (2017) *Nannochloropsis*, a rich source of diacylglycerol acyltransferases for engineering of triacylglycerol content in different hosts. Biotechnol Biofuels 10(1):1–20

Zou LG, Chen JW, Zheng DL, Balamurugan S, Li DW, Yang WD et al (2018) High-efficiency promoter-driven coordinated regulation of multiple metabolic nodes elevates lipid accumulation in the model microalga *Phaeodactylum tricornutum*. Microb Cell Factories 17(1):1–8

Chapter 3
Algal Butanol Production: Recent Developments

Ritika, Aparna Agarwal, Rizwana, and Nidhi Jaiswal

Abstract The world's energy catastrophe, including a restricted reservoir of petrol-based fuels, has refocused global attention on developing a sustainable solution enabling the generation of alternative fuels. As a result of the availability of plentiful renewable biomass, it is possible to build a cost-effective biofuel manufacturing method at a large scale to influence sustainability. Rising fossil fuel usage, notably in the transport industry, is being blamed for the degradation of the environment, global warming, and an upsurge in the health concerns among humans associated/linked with pollution. Energy resources can be replaced by biofuels, which are also recognized as large emitters of greenhouse gases and are therefore seen to lessen the issues mentioned above. This chapter aims to evaluate the function and relevance of algae as catalysts for the long-term generation of biofuel-biobutanol. Multiple elements of biobutanol production are extensively addressed and thoroughly analyzed, including origins, synthesizing chemically, biosynthesis, generation microbially (covering utilization of natural including biologically modified organisms), as well as current advancements in biobutanol production. In recent years, this has sparked a surge in research into butanol production from renewable biomass.

Keywords Butanol · Algal bloom · Butanol production · Algal butanol production · Recent developments

Ritika
Department of Home Science, University of Delhi, Delhi, India

A. Agarwal (✉) · N. Jaiswal
Department of Food & Nutrition and Food Technology, Lady Irwin College, University of Delhi, Delhi, India
e-mail: aparna.gupta@lic.du.ac.in

Rizwana
Department of Food Technology, Bhaskaracharya College of Applied Sciences, University of Delhi, Delhi, India

N. Srivastava, P. K. Mishra (eds.), *Technological Advancement in Algal Biofuels Production*, Clean Energy Production Technologies,
https://doi.org/10.1007/978-981-19-6806-8_3

3.1 Introduction

Climate change and increase in temperatures is growing, as has the expense and oil reserves, as well as legal restrictions upon non-renewable energy resources, chemical components, and ignites generated by natural sources, have piqued awareness globally (Ranjan and Moholkar 2012). Food, water, minerals, and energy, among other elements, are in limited supply in developing nations. Fuel is the most essential of these resources since it is widely utilized in everyday transportation, energy production, and other applications. The ever-increasing global population has increased the need for petroleum products and other forms of energy (Bharathiraja et al. 2017; Demirbas 2008; Galbe and Zacchi 2007). Because fossil fuels seem to be the significant primary source of power until 2030, the rising depletion of oil deposits may be addressed mainly by using sustainable energy alternatives to supplement supply, improve the environment, and raise the country's economic worth (Festel 2008).

Butanol is an essential commercial chemical that, being a highly robust liquified fuel, can substitute gasoline. The ability to efficiently manufacture butanol from clean and sustainable carbon source materials has been steadily improving. New fermentation methods for converting large biomass to butanol are being explored based on biotechnology and process engineering improvements.

Marine biomass, particularly algal biomass, has lately attracted interest due to several significant benefits, including richness, fast growth, high glucose content, absence of lignin, and the fact that it does not struggle for land. The biofuels they produce are referred to as third-generation bioenergy. With algal biomass's renewable and sustainable qualities, enhanced biofuels, organic acids, biohydrogen, biomaterials, or other high-valuated goods are emerging from its transformation into diverse valuation biochemicals and biomaterials.

3.2 Butanol

An OH group is attached to one of the carbons at the end of the four-carbon chain in 1-butanol, also known as butyl alcohol. Indeed, it is exploited as a paint thinning agent and a solvent; it might be employed as a biofuel in the future.

$CH_3CH_2CH_2CH_2OH$ =

Butyl alcohol is a primary (1°) alcohol that oxidizes quickly.

The additional structural isomers of 1-butanol are 2-butanol (sec-butyl alcohol), 2-methyl-1-propanol (isobutyl alcohol), and 2-methyl-2-propanol (tert-butyl alcohol).

In 2-butanol, sometimes called sec-butanol, sec-butyl alcohol, or s-butyl alcohol, it is on the second carbon that an OH group is found. There are two additional carbons connected to the alcoholic carbon, making it secondary. It is employed in the production of other chemicals as both a solvent and a catalyst.

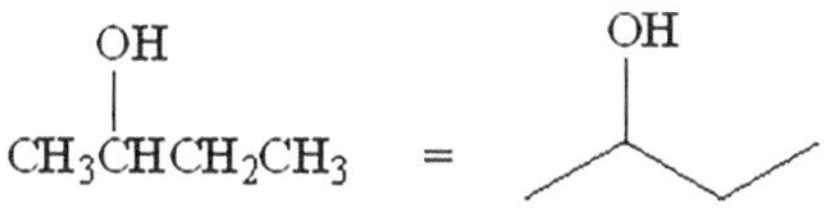

Secondary (2°) alcohol sec-butyl alcohol gets quickly oxidized.

2-Butanol is an enantiomeric chiral compound with two forms: (R)-2-butanol and (S)-2-butanol.

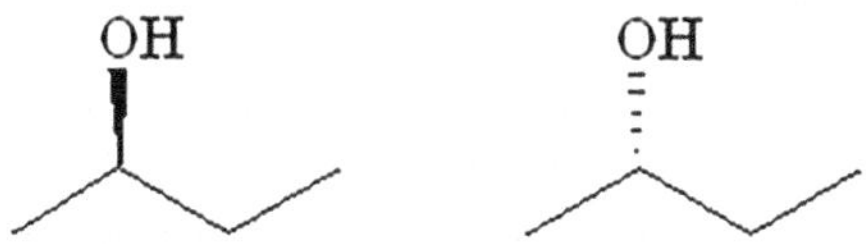

(*R*)-2-butanol (*S*)-2-butanol

The middle carbon of the three-carbon chain has a methyl group and has an OH group at the end of it, making it isobutanol, commonly known as 2-methyl-1-propanol or isobutyl alcohol. There are several applications for this solvent, including paints, inks, etc.

Isobutyl alcohol is a kind of primary (1°) alcohol prone to oxidation.

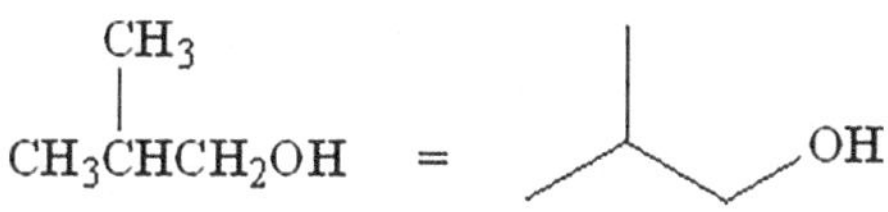

It is also known as 2-methyl-2-propanol, tert-butanol, tert-butyl alcohol, or *t*-butyl alcohol because it has three carbons, one of which has an OH group and the other of which has a methyl group. Because it is connected to three other carbons, the alcoholic carbon is referred to as tertiary. It is utilized as a solvent, ethanol denaturant, gasoline octane enhancer, and certain pain relievers.

Tertiary (3°) alcohol that is not prone to oxidization is tert-butyl alcohol. *t*-Butoxide anion, produced by the reaction between sodium or strong bases like sodium hydride and, in organic synthesis, tert-butanol, is employed to remove the moderately acidic hydrogen out of the OH group and charge oxygen negatively. Although *t*-butoxide is a strong base, it is incapable of nucleophilic substitution reactions due to its steric mass, increasing the likelihood of elimination processes in a given reaction.

Table 3.1 Butanol's characteristics compared to those of different fuels (Bankar et al. 2013; Jin et al. 2011)

Characteristics	*n*-Butanol (C_4H_9OH)	Ethanol (C_2H_5OH)	Gasoline (C_4C_{12})
Molecular weight	74.11	46.07	111.19
Cetane number	25	5–8	0–10
Octane number	96	85	80–99
Density (g/mL) at 20 °C	0.808	0.789	0.72–0.78
Flashpoint (°C)	35	14	−45 to −38
Boiling point (°C)	117.7	78.5	25–215
Flammability (%vol.)	1.4–11.2	3.3–19	0.6–8
Saturation pressure (kPa) at 38 °C	2.27	0.78522	31.01
Viscosity (mm^2/s)	2.63	1.2–1.074	0.4–0.8
Energy density (MJ/L)	29.2	25	32

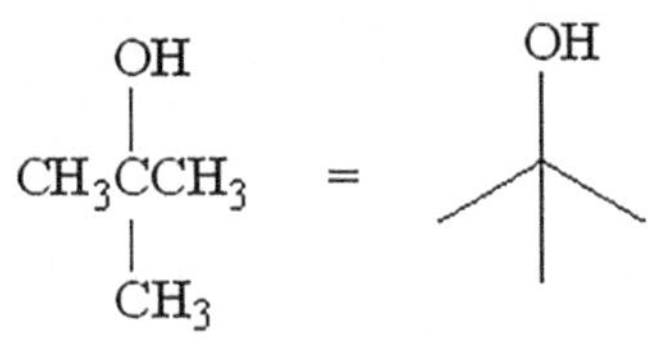

Butanol offers numerous benefits over ethanol (Shapovalov and Ashkinazi 2008). Butanol contains 25% higher energy than ethanol, making it a more valuable energy source. Being less volatile and safer as it evaporates at a pace six times slower than ethanol and 13.5 times slower than gasoline, because butanol can be delivered via operating gasoline pipelines, it is considerably less invasive than ethanol, whereas ethanol requires rail transportation. It may be mixed with more enormous proportions with gasoline. It is utilized rather than gasoline, whereas ethanol may only be used as a supplement. It has the potential to alleviate infrastructural issues in the hydrogen supply chain.

Compared to ethanol, butanol's energy output is higher (10 W h/g). As a result of burning butanol, it does not create sulfur or nitrogen oxides. Butanol, unlike ethanol and methanol, has properties that are more akin to diesel (Rakopoulos et al. 2010). As a result, biobutanol overcomes the world's concerns with rapid biofuel consumption. It is highly flexible when mixed with gasoline due to its reduced vapor pressure and minimal water reactivity (Table 3.1).

3.3 Algae

Microalgae and macroalgae are the two types of algae. Microalgae, the most common eukaryotes in phytoplankton, are unicellular plants in marine, freshwater, and terrestrial settings as single cells, aggregates, or filaments, as stated by Metting

Table 3.2 Cultivation methods for microalgae (Sanjukta Banerjee et al. 2020)

Growth pattern	Source of energy	Source of carbon	Need for light	Cost of upkeep
Photoautotrophs	Natural/artificial light	Inorganic carbon (CO_2)	Required	Cost-effective and appropriate to outdoor cultivation
Heterotrophs	Organic carbon	Organic and inorganic carbon	Not required	Massive price and PBR production is favored
Mixotrophs	Organic carbon or natural light	Organic and inorganic carbon	Not required	The price is reasonable, and it is well-suited to outdoor cultivation

(1996). They produce over 50% of the atmospheric oxygen and conduct over 50% of the chief photosynthetic output on the planet (Chisti 2007). Various aspects of microalgae are discussed in Table 3.2.

Certain organisms can withstand higher CO_2 levels than others and serve a significant role in CO_2 collection from flue gases. Because it can thrive under 40% CO_2, the green alga in the sea *Chlorococcum littorale* is a plant that can withstand severe CO_2 stress (Kodama et al. 1993). In addition to producing critical omega-3 and omega-6 fatty acids, microalgae serve as an essential part of the food chain and are gathered by higher-level organisms, such as "fat fish."

Microalgae harvesting is a costly procedure that is made marginally easier if the species flocculates. Microalgae carbohydrate concentration varies substantially with species and strains, as well as containing a moderate content. Exopolysaccharides (EPS) are released by several organisms due to their physiology or in response to stress (Delattre et al. 2016). Although cellulose and starch are the most common polysaccharides (PS) in microalgae, numerous others have been discovered with essential characteristics (Chew et al. 2017).

Multicellular photosynthetic organisms which live in water are known as macroalgae, sometimes referred to as seaweed. They are widespread in seas, especially around coastlines, where they can exist as unrestricted creatures or cling to stones and various hard planes. Macroalgae come in three varieties, green, brown, and red algae, with their colors driving from natural colorants, including chlorophylls. There are pigments present in green algae (Chlorophyta), brown algae (Phaeophyta), and red algae (Rhodophyta) that reflect or absorb certain wavelengths of light. Chemically, macroalgae are very different from land plants. Glucose polysaccharides are found in seaweed. When it comes to green and red algae, cellulose and starch make up 25–50% of the dry weight, respectively, whereas brown algae contain cellulose as well as laminarin (30–50% dry weight) (Huesemann et al. 2012), as well as additional complex polysaccharides.

Green seaweeds Green seaweeds may be found in shallow waters such as bays and estuaries. They are occasionally referred to as opportunistic seaweeds. Green seaweed biomass, such as *Ulva* sp., *Codium* sp., *Enteromorpha* sp., *Monostroma* sp., *Chlorella* sp., and others, has been studied extensively for its significant carbohydrate composition (e.g., glucose, rhamnose, xylose, and glucuronic acid) since the first report (Santhakumaran et al. 2019). Although monosaccharides account up the majority of algal biomass, with glucose being the greatest, green seaweed has a greater glucose concentration than other algal species because primarily, starch serves as a source of energy (Kim et al. 2011).

Brown seaweed Unlike green seaweed, brown seaweed grows deeper in water than red seaweed (Prazukin et al. 2018). Algin, alginate, and cellulose are among the macromolecules found in high concentrations. Protein, mannitol, minerals, and other inorganic elements, like sulfate, are also present in trace amounts. Fucoxanthin, a yellow-brown pigment covering chlorophyll's green color, gives brown macroalgae their distinctive olive-green to dark brown hue. The kelp (*Laminaria*) is the most enormous and most complex macroalga in this category, attaining lengths over 100 m. *Saccharina latissima*, *Alaria* sp., *Fucus* sp., *Macrocystis* sp., and *Sargassum* sp. are all brown seaweeds, as is the genus *Laminaria*. Water-soluble polymer *laminarin* contains glucose (20–25 glucose molecules), which is the primary storage polysaccharide in brown macroalgae. In laminarin, mannitol, the glucose unit's reducing terminus, is bonded to the sugar alcohol (Percival 1979). Laminarin accounts up to almost 35% of brown algae's dry weight (Mautner 1954). As a structural polysaccharide, alginate accounts for up to 40% of the dry weight of brown seaweeds (Draget et al. 2005). Mannuronic and guluronic acids are the two kinds of uronic acids that make up alginate. These are either chains of alternating mannuronic and guluronic units or continuous blocks of exclusively mannuronic or just guluronic acid. Guluronic-richer alginate differs from alginic acid, with higher mannuronic content in its characteristics (Percival 1979). In addition to 1,2-linked-L-fucose-4-sulfate units, fucoidan contains minimal D-xylose, D-galactose, and glucuronic acid. Fucoidan seems to be a densely branched and heterogeneous polysaccharide.

Red macroalgae Red seaweed is found chiefly along tropical and subtropical beaches, with about 3700 species; *Chlorella* sp., *Asparagus schoberioides*, *Pyropia* sp., *Gelidium amansii*, and *Gloiopeltis* sp. are the most prevalent (Zanolla et al. 2018). The red hue of red algae is due to phycoerythrin, which is found in higher concentrations in red algae than in other algae. Cellulose and long-chain carrageenan or agar, both of which include -D-galactose and its derivatives, make up the red algal cell walls' structural polysaccharides (Brawley et al. 2017), while starch is the power reserve polysaccharide, with other carbohydrates such as xylan, mannan, and fructan (Park et al. 2020).

3.4 Algal Bloom and Cultivation

In recent years, considerable interest and emphasis on the cultivation of algae have boosted the synthesis of biofuels and the replacement of first- and second-generation biomass. Algal growth and cultivation can be achieved with minimal and cost-effective resources, as they can proliferate in wide and favorable conditions leading to the generation of biomass with a greater production rate. They have the advantage over other land crops in that they may be cultivated on non-agricultural terrain but have the ability to receive nutrients from seawater. Other distinctive features include the following: they do not require a large area for growth and have high carbon uptake and CO_2 absorption, thereby mitigating its level and reducing the greenhouse effect (Das 2015; Demirbas 2009; Ganesana et al. 2020).

Algae utilize only sunlight, CO_2, and nutrients for their growth, and microalgal cells retain around 15–20 times more high-energy lipids (like triacylglycerol) than crops, making them a valuable source of high-energy lipids. When transesterified, these triglycerides generate fatty acid and methyl esters. Triglycerides serve as a building block for the production of biofuels (such as biodiesel). Therefore, there is a need to create a substantial amount of lipids; extensive algae cultivation is required to meet the demand for biofuels globally. Apart from lipids, they are rich in carbohydrates, protein, and various pigments like carotenoids. However, the number of macronutrients and micronutrients and the accessibility and quantity of sunlight are all factors that influence algal development and determine the concentration of metabolic products (Walker 1954). Therefore, proper techniques, approaches, and conditions are required for algae cultivation (Adeniyi et al. 2018; Mata et al. 2010; Banerjee et al. 2020; Widjaja et al. 2009).

One of the essential requirement which influences the photosynthesis is light, and the growth of algae as different wavelengths and intensity results in drastic change affects the microalgal cell lipid accumulation. Photoautotrophs (have to be powered by sunlight), photoheterotrophs, and mixotrophs are the algal species that are categorized on the basis of sunlight and carbon sources required for their production. It is vital to consider the temperature since it affects algae development, nutritional needs, and biochemical makeup, determining protein content, starch, and pigments. The appropriate regions are temperate and subtropical, increasing the growth. Devastating algal development occurs at temperatures over 35 °C. In addition, pH, illumination, and medium culture are the other physical parameters influencing their growth (Mata et al. 2010; Banerjee et al. 2020; Ganesana et al. 2020; Widjaja et al. 2009; Vitova et al. 2015).

There are three methods of cultivation system—open, closed, and hybrid system. The open system is a cost-effective, most common method and involves outdoor cultivation setups like large lakes, canals, or the existence of adequate light and atmospheric carbon dioxide readily available for mitigation; large-scale algal production can be achieved. However, there are certain limitations such as higher demand of the land area, difficulty in monitoring, and a greater chance of contamination. On the other hand, a closed system involves solar panels, a biomass unit, a gas exchange column, and a pump that make up the flexible systems of photobioreactors, wherein physical and chemical parameters can be adjusted and controlled in greater biomass yield. However, the cost of a closed cultivation system is two and a half times higher, and therefore, innovative approaches are needed to be followed to control the cost and increase biomass production. To get beyond the drawbacks of both open and closed systems, hybridization resulted in hybrid systems. In this system, the initial operating cost is low, and the chances of contamination are less. Algae are initially cultivated in photobioreactors to generate a high biomass density before being moved to an open pond system, enhancing the process' economic compatibility (Banerjee et al. 2020; Brennan and Owende 2010; Ganesana et al. 2020; Gustavo et al. 2013; Muraza 2014; Schenk et al. 2008; Shaikh et al. 2017; Slade and Bauen 2013).

3.5 Butanol Production

Butanol has historically been manufactured employing acetone-butanol-ethanol (ABE) fermentation utilizing carbon feedstocks including potatoes, corn, tapioca, sugar beets, sugarcane, and millet (Jones and Woods 1986; Qureshi et al. 2008). In 1861, a French scientist named Louis Pasteur invented acetone-butanol-ethanol (ABE) fermentation. In World War I, large quantities of acetone were necessary (Fernbach and Strange 1912), a patent was submitted for a tested method in the United Kingdom, and research in butanol was revived. Nevertheless, it was necessary to find a creature capable of increasing the synthesis of acetone.

Strange and Chaim Weizmann harvested garden soil's *Clostridium acetobutylicumin* for ABE fermentation, generating significant quantities of acetone (Weismann and Alliston 1922). Solventogenic *Clostridium* is most recognized for its ability to generate butanol naturally (Lee et al. 2008; Ujor et al. 2016; Xin et al. 2018). Noteworthy is the generation of acetone, butanol, and ethanol while utilizing these strains. The best exploited solventogenic *Clostridium* species include *Clostridium acetobutylicum* ATCC 824 and *Clostridium beijerinckii NCIMB 8052*. Much attention has been given to *Clostridium saccharobutylicum* and *Clostridium saccharoperbutylacetonium* recently because they produce a good amount of butanol during ABE fermentation (Dürre 2007). A petrochemical method for the synthesis of *n*-butanol was discovered in the 1950s. The majority of the process included acetaldehyde aldol condensation, followed by dehydration and crotonaldehyde hydrogenation. Fermentation techniques were abolished due to this rapidly expanding Oxo synthesis, another term for industrial discovery (Uyttebroek et al. 2015).

3.5.1 Route of Biochemical Production

In the presence of numerous *Clostridiaceae* bacteria, sugar, glycerol, or lignocellulose feed is fermented, making it the most appealing approach for producing biobutanol (Chen et al. 2014; Yadav et al. 2014; Yang et al. 2014). Acetone-butanol-ethanol (ABE) fermentation has numerous benefits since it is based mainly on affordable and plentiful raw resources. Because it employs waste products from biodiesel manufacturing, as an alternative approach, producing biobutanol by fermenting glycerol produced by biodiesel is being studied. Due to the increased oil prices, butanol production has been resurgent, emphasizing its utility chemical and a viable biofuel (Kumar and Gayen 2011; Lee et al. 2008).

When making biobutanol or *n*-butanol from food crops that are first-generation feedstock, it is better to use a second-generation feedstock that is not meant for human food. For ABE fermentation, the second-generation feedstock is favored, as it is non-food and will not compete for natural assets with crop production. It is believed that these plant-based innovative technologies employing a more significant percentage of it will have long-term economic advantages as a result. Third-generation feedstock research is also underway, but there is not much literature on the subject.

First-generation bioenergy substrates When food crops (such as maize and sugarcane) were initially utilized as feedstock to generate bioethanol, biodiesel, and other types of fuels, they were dubbed the first generation of bioenergy pioneers because of their prominence (Buckeridge et al. 2019). First-generation biobutanol is made using a straightforward method involving primarily hexose sugar fermentation. Starchy cereals, including maize, wheat, rice, and cassava, are hydrolyzed to provide these sugars.

Second-generation bioenergy substrates Second-generation biofuels are usually referred to as such because they are made from unsuitable to be eaten leftovers from fodder or inedible plant materials, including grasses, trees, and many more (Abdullah et al. 2019). These new biofuels are not in competition with the food and feed supply chain and are available in a wide range of quantities which is a crucial advantage.

Third-generation bioenergy substrates Marine biomass, particularly algal biomass, has recently acquired appeal due to numerous significant benefits, including availability, fast proliferation, significant carbohydrate composition, lignin deficit, and lack of land competition (Gaurav et al. 2017). Third-generation bioenergy is the term for the biofuels they generate.

Clostridium spp. bacteria may ferment a wide range of carbohydrates to generate organic chemicals, acids, alcohols, or various dissolvents. Nevertheless, only a few saccharolytic and mesophilic species that can generate butyrate can also make fermentation yields of butanol relatively high: *Clostridium acetobutylicum*, *Clostridium aurantibutylicum*, *Clostridium beijerinckii*, and *Clostridium tetanomorphum* are *Clostridium* species (George and Chen 1983; George et al. 1983; Gottwald et al. 1984; Jones and Woods 1986).

Acidogenesis (turning sugar into organic acids) and solventogenesis (generating solvents) are the two primary fermentation processes used in batch butanol production (Jones and Woods 1986). In the initial fermentation phase, the bacteria multiply at an exponential pace, generating acids (predominantly acetate and butyrate). Consequently, the pH is reduced to 4.5. In reaction to the low pH, bacterial cells switch from acidogenesis to solventogenesis throughout their metabolic activities; the production rate decreases toward the end of acidogenesis (Johnson et al. 1931;

Bahl et al. 1982). The solventogenic enzymes required for the second step are induced by the acid products (Ballongue et al. 1985).

During this phase, acetone and butanol are produced using acetate and butyrate as feedstock, with no bacterial growth visible. Solventogenesis requires organic acids as co-substrates (Fond et al. 1985; Kell et al. 1973). Butanol, along with acetone/ethanol combination, is the main by-product of the second phase. The fluidity and cell membranes' functionality is affected by the presence of solvents (Moreira et al. 1981; Vollherbst-Schneck et al. 1984). Butanol, including various other compounds, accumulates to the point that bacterial activity is halted at solventogenesis (Bowles and Ellefson 1985).

3.5.1.1 Metabolism

The Embden-Meyerhof-Parnas (EMP) route degrades carbohydrates to pyruvate, initiating the acidogenesis and solventogenesis metabolic pathways (Rogers 1986). With two units of pyruvate, a single hexose molecule generates two adenosine triphosphate (ATP) copies with two parts of nicotinamide adenine dinucleotide (NADH) (Doelle 1975; Gottschalk 1986; Thauer et al. 1977). The phosphogluconate route, which directly corresponds to glycolysis, can be used by *Clostridia*-producing solvents to utilize pentoses (Cynkin and Delwiche 1958; Ounine et al. 1983). In all, three pentose molecules are converted to pyruvate, yielding five molecules, including both ATP and NADH.

Pyruvate is a crucial component in *Clostridium* spp. metabolism, as this route description shows (Kotze 1969). Three enzymes are responsible for the efficacy of fermentation: [FeFe] hydrogenase, NADH-ferredoxin reductase, and NADPH-ferredoxin reductase. Fermentation requires ferredoxin, a reduced electron transporter. Ferredoxin-NAD (P) reductase may transfer electrons to pyridine nucleotides or transfer electrons to hydrogen via hydrogenase. Pyruvate-ferredoxin oxidoreductase converts pyruvate to acetyl-CoA, which might subsequently be converted into acetone, acetate, or CO_2 (oxidized products) or butanol, ethanol, or butyrate (reduced products) (Fig. 3.1).

3.5.2 Chemistry-Based Production of n*-Butanol with Ethanol as the Starting Material*

To increase the productivity and conversion rate of ethanol to *n*-butanol, a chemistry-based approach usually consists of a few primary stages performed in the presence of catalysts. The chemical approach has the benefit of just requiring one step for converting *n*-butanol from ethanol; in contrast, the biochemical process may need many stages. The chemical steps for *n*-butanol are many and extensively established (Yadav et al. 2014; Uyttebroek et al. 2015).

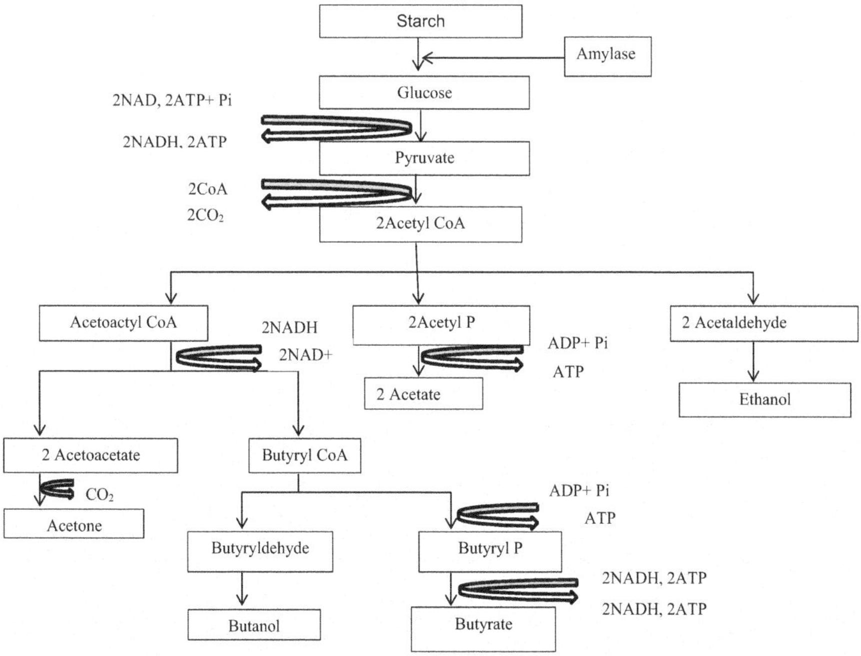

Fig. 3.1 Biochemical pathway of ABE fermentation. (Source: P. Dürre 2007)

3.5.2.1 Utilization of Ethanol to Produce *n*-Butanol Follows a Reaction Process

To make *n*-butanol from ethanol, follow three simple procedures. Dehydrogenation of ethanol to generate acetaldehydes is the initial liquid phase reaction (Santacesaria et al. 2012), proceeded by aldol condensation of acetaldehyde (Juben et al. 2007) and hydrogenation to *n*-butanol. A technique for converting ethanol to butanol in the gas phase utilizing a zeolite catalyst was proposed by Ndou et al. and Ueda et al., reviewed by Ndaba et al. (2015). The production of intermediates, including acetaldehydes and crotonaldehydes, is not said to occur in the gas phase process. There have been few publications on the gaseous-phase synthesis of *n*-butanol and its mechanism. A typical industrial approach for raising the carbon count of alcohols is to convert ethanol to *n*-butanol by joining two molecules together. The Guerbet reaction was named after Marcel Guerbet, who first studied aldol condensation during the 1890s (Uyttebroek et al. 2015). Even though this approach is rarely used currently, it may resurface in the future (Swodenk 1983).

The three primary chemical methods for butanol synthesis are crotonaldehyde hydrogenation, oxo synthesis (hydroformylation), and Reppe synthesis. Before the 1950s, acetaldehyde was converted to butanol by crotonaldehyde hydrogenation, the most common industrial technique. The crotonaldehyde hydrogenation method includes aldol condensation of acetaldehyde with the involvement of alkaline catalysts at ambient temperatures and pressures; further dehydration occurs as a result of acidification with amino acids or H_3PO_4. A copper (Cu) catalyst is used in distillation and hydrogenation to produce crotonaldehyde in the gaseous or liquid phases. Using Rh or Ru catalyst, the reaction between petroleum-derived olefins including C_2H_4 and propylene (C_3H_6) and carbon monoxide (CO) produces intermediate molecules (aldehydes) containing a formyl group (CHO) in the primary step of chemical butanol production using the oxo method. Compared to the olefin, the starting molecule, the aldehydes generated have one additional carbon atom. Considering the catalyst employed, this reaction can use much energy, operating at temperatures between 80 and 200 °C and pressures between 20 and 30 MPa. After the initial step of synthesizing the aldehyde mixture (1-butanal and 2-methyl propanal), the aldehydes are hydrogenated in the liquid or vapor phase with Cu, Ni, or a blend of the two to create the respective alcohol types in the second stage (25% 2-methyl-1-propanol or isobutanol and 75% 1-butanol).

Additionally, C_3H_6 is carbonylated with CO and H_2O using a catalyst (tertiary ammonium salt or polynuclear iron carbonyl hydrides) at 0.5–2 MPa and 100 °C in the Reppe process, which was developed in 1942 (Cotton et al. 1999; Uyttebroek et al. 2015). 1-Butanol and isobutanol are formed in a 43:7 ratio during the procedure. Butanol is produced in more significant quantities using the Reppe technique; despite the milder reaction conditions, it has not been commercially deployed since it is more expensive (Fig. 3.2).

3.6 The Utilization of Algae as a Sustainable Biofuel Producing Source

Consequently, fossil fuels now prevail in the energy market at lower costs than renewable fuels since technological improvements. Unfortunately, uncontrolled use of fossil fuels will not only deplete finite supplies shortly but will also cause a slew of ecological issues, including an increase in CO_2, NO_x, SO_x, and other pollutants in the air, as well as GHG emissions. In response, solutions must be developed to improve energy conversion efficiency, while CO_2 emissions will be decreased, and renewable energy sources will be used more frequently.

Creating biofuel from earthly biomass can be beneficial if it is carried out sustainably that does not harm the ecosystem or biodiversity; otherwise, the food vs. fuel argument will resurface. Terrestrial plant-based fuels are divisive because they need the cultivation of land that could otherwise be utilized to raise food, feed, and fiber for the world's rising population. The use of readily accessible algal biomass for sustainable biofuel production is one viable option. Algal biomass has several benefits over terrestrial biomass, including eliminating a need for cultivation land and the capacity to grow in seawater and wastewater; and in

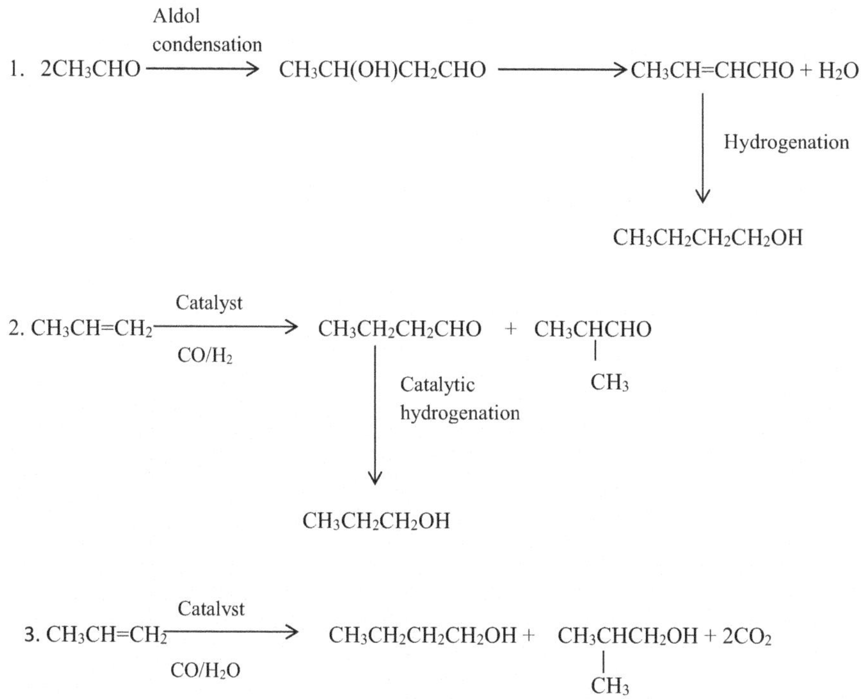

Fig. 3.2 Butanol is synthesized chemically using three different methods: (1) crotonaldehyde, (2) oxo synthesis, and (3) Reppe synthesis. (Source: B. Ndaba et al. 2015)

comparison to plants on land, they have a faster rate of growth and a higher yield per unit area.

Algae are polyphyletic creatures that can create biomass from sunlight, water, and CO_2 (Clark et al. 2008). With their versatility, they can grow and flourish anywhere they are planted, which means they may be gathered all year round. Even without interfering with agricultural commodities, it might grow in non-farm soil or brackish/marine water. It grew and cultivated quickly, having the capacity to twice biomass in a matter of hours. Microalgae and macroalgae are two different kinds of algae. Microalgae are made up of unicellular organisms that are tiny in size, whereas macroalgae have numerous cells and a plant-like structure with roots, stems, and leaves. Depending on their pigmentation, they are classified as red, green, or brown. Compared to microalgae, macroalgae has a lower protein and fat content but more excellent carbohydrate content (Monlau et al. 2014).

Carbohydrates, lipids, and pigments, including carotenoids, phycocyanin, and phycobiliprotein, which have economic and aesthetic value, are abundant in algae. In addition, algae-derived biofuel produces less NOx than fossil fuel, and they can resist and buffer excessive CO_2 levels in the gas stream (Adeniyi et al. 2018). As a result, algal biomass may be utilized for CO_2 sequestration and wastewater treatment in addition to biofuel generation. As a result of their dependence on sunshine and carbon, algae may be grown in three different ways: autotrophic, mixotrophic, and heterotrophic. It may be grown in either an open or closed setting. Massive-scale algal cultivation is generally done in an external culture setup with enough sunlight for algae growth and CO_2 from the atmosphere as a source of carbon, as suggested by Banerjee et al. (2020) in their paper. Photobioreactors (PBR) that replicate environmental conditions are employed in closed culture systems to produce high biomass production with little contamination and greater control over physical and chemical characteristics (Mata et al. 2010). Nevertheless, compared to an open system, the expenses of these systems with airflow aided by artificial lights and mechanical pumps (Nguyen et al. 2009) are about 2.5 times higher (Davis et al. 2011).

The two most common liquid fuels made from microalgae are biodiesel and bioethanol. Biohydrogen (Bowles and Ellefson 1985), long-chain hydrocarbons similar to crude oil (Sharma and Singh 2009), and biogas (Sheehan et al. 1998) are some of the other fuels that algae may create. Algal biodiesel is a lucrative feedstock for exploitation in the near future on the grounds that it is environmentally friendly and renewable and the fact that it emits 78% less CO_2 than petroleum (Chen et al. 2012). Several bacteria, yeasts, and fungi may use microalgal carbohydrates as a carbon source to create bioethanol, biohydrogen, and biobutanol through fermentation (Clark et al. 2008). Since algal cell walls contain less lignin and hemicelluloses, there is no need for pretreatment, rendering it a more efficient feed than lignocellulosic matter (Harun et al. 2010; Shaishav et al. 2013). The CO_2 created as a result of the method can be used to grow algae (Chen et al. 2012). Biohydrogen (H_2) may also be produced by algae through photofermentation or biophotolysis (Park et al. 2020). Consequently, as inputs, H_2O, CO_2, and solar energy are used to generate fermentation outputs. It is known as "photofermentation"; as photosynthetic efficiency improves, biofuel generation becomes more cost-effective.

In the vicinity of sunlight, microalgae break down water into H_2 and O_2 all through direct biophotolysis. Cyanobacteria (blue-green algae) perform indirect photolysis via ferredoxin in photosystem I. Algae that break down organic substrates into H_2 and CO_2 in the presence of sunshine are capable of photofermentation (Chen et al. 2013). Anaerobic digestion, which generates methane (CH_4), CO_2, and hydrogen sulfide, can employ algae biomass as a substrate in addition to other gases (Cantrell et al. 2008). Organic molecules are hydrolyzed into soluble sugars, which are subsequently fermented for generating gases. This method has the advantage of using wet biomass with a moisture level of 80–90% (Jones and Mayfield 2012). Methanogens later transform this combination into CH_4 (60–70%) and CO_2 (30–40%) (Yen and Brune 2007; Demirbas 2001). At high temperatures (800–1000 °C), partially oxidizing algae biomass can produce flammable gases, including carbon monoxide (CO), CO_2, CH_4, H_2, and N_2 (McKendry 2002; Walker 1954). It may be efficiently combusted primarily for fuel in its gaseous phase to be used in turbines and motors and does have a calorific value between 4 and 6 MJ/m^3 (Juneja et al. 2013). Despite several advantages, the advancement of reaching a commercial degree of algae technology has been hampered by many obstacles. The low biomass content in water is the most significant constraint. Photosynthesis is impeded in open settings because of low light penetration, limiting biomass output (Adeniyi et al. 2018).

Furthermore, algae's tiny size and density make separation from water challenging, resulting in high harvesting costs and a negative energy balance in the process. As a result, species selection is critical so that, in addition to biofuel generation, biomass may be utilized to extract value-added by-products (Suganya et al. 2016). These disadvantages may be addressed with appropriate strain selection and optimization of operational conditions for optimal utilization, thanks to rising environmental awareness and technical advancement.

3.7 Role of Algae in Production of Butanol

Algal biomass could thrive in sea, saline-alkali soils, and even low-quality water settings due to its rapid growth, flexibility, and high consumption rate and is occasionally used as sewage filters (Voloshin et al. 2016). Furthermore, because of its reduced hemicellulose concentration and paucity of lignin, it has higher hydrolytic efficiency and lower pretreatment expenses than lignocelluloses, rendering it an even more commercially viable substrate for valuation good transformation (Cesário et al. 2018). Due to cell membrane obstruction, microbes are unable to break down the amorphous polysaccharides found in algal biomass and similar biomass (Cui et al. 2018). As a result, developing an effective hydrolysis technique must come first that can effectively break polysaccharides down into fermentable sugars including glucose, galactose, mannose, xylose, and others, along with various organic compounds such as sodium alginate, for example (Rengasamy et al. 2020). Sugar concentration and hydrolytic efficacy of hydrolysates are influenced mainly by the conditions employed, ranging substantially based on temperature, processing length, additives, and biomass input (van der Wal et al. 2013). Furthermore, algal hydrolysates can comprise complicated macromolecules, including lipids, proteins, and peptides, interfering with subsequent microbial metabolism (Zang et al. 2020). Additionally, throughout hydrolysis, the application of different chemicals (e.g., H_2SO_4) can produce poisonous chemicals like furfural, which will inhibit microbial development (Lin et al. 2016) (Fig. 3.3).

Clostridia species are essential for completing the ABE fermentation process via acidogenic and solventogenic routes. At 35–37 °C, *C. acetobutylicum* may ferment microalgal biomass (Wang et al. 2017). During carbohydrate metabolism, they are converted to acids like butyric acid and acetic acid and gases including carbon dioxide and hydrogen, via the acidogenic pathway. The solventogenic technique used to make organic compounds like acetone, butanol, and ethanol is used after the growth curve has reached its stationary phase and is triggered through acid reduction (Ibrahim et al. 2018). The pH of the fermentation medium starts at 5.5–6.0 and rises to 7 throughout fermentation. By sparging N_2 gas into the fermenter, anaerobic conditions are maintained. Based on the culture environment and stains employed, a

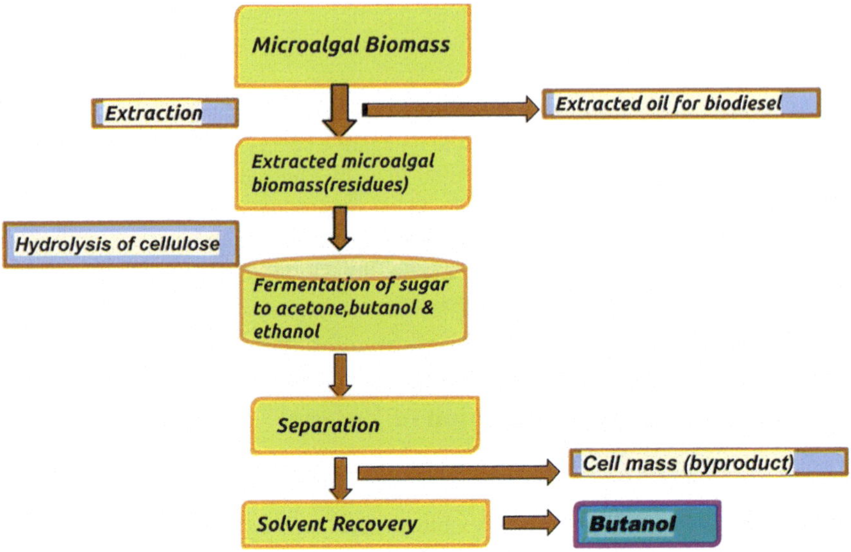

Fig. 3.3 Converting microalgal biomass to biobutanol. (Source: Ajay Kumar 2020)

72-h fermentation period is usually necessary. 3:6:1 is reported as the mass ratio of acetone, butanol, and ethanol generated by ABE fermentation.

$$\underset{\text{Glucose}}{10C_6H_{12}O_6} \xrightarrow{\text{CLOSTRIDIA}} \underset{\text{Butanol}}{5C_4H_{10}O} + \underset{\text{Acetone}}{3C_3H_6O} + \underset{\text{Ethanol}}{C_2H_6O} + \underset{\text{Acetone}}{C_2H_4O_2} + \underset{\text{Water}}{H_2O} + \underset{\text{Carbon-dioxide}}{23CO_2} + \underset{\text{Hydrogen}}{16H_2}$$

C. acetobutylicum, *C. beijerinckii*, *Clostridium saccaroperbutylacetonicum*, *Clostridium saccharoacetobutylicum*, *Clostridium aurantibutyricum*, *Clostridium pasteurianum*, *Clostridium sporogenes*, *Clostridium cadaveris*, and *Clostridium tetanomorphum* are among *Clostridia* species being looked for butanol synthesis (Yeong et al. 2018). A UV-visible spectrophotometer is used to determine optical density at 600 nm for the clostridial development profile in a batch, fed-batch, continuous, adsorbed surface fermentation, including biofilm bioreactor. HPLC (high-performance liquid chromatography) and the dinitrosalicylic acid (DNSA) technique determine the amount of residual sugar. According to various researchers, the entire dissolvents (acetone, butanol, and ethanol), as well as acids (acetic and butyric acid), are examined using gas chromatography (Birgen et al. 2019a). The utilization of microalgae for butanol synthesis has received little attention in the literature as a result of the poor ABE fermentation product yield (Birgen et al. 2019b). Among the drawbacks of ABE fermentation include high substrate overhead expenses (microalgal biomass) and substrate inhibition (high CO_2 concentration), sluggish clostridial growth with low cell concentration, poor butanol level (20 g/L),

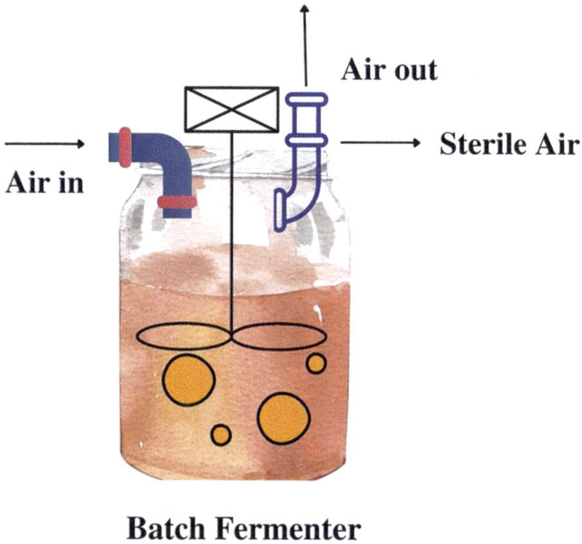

Fig. 3.4 Butanol synthesis via batch fermentation. (Source: A. Pugazhendhi et al. 2019)

output (0.33 g/L), and efficiency (0.5 g/L h), unfavorable butanol preference, and increased power usage (Li et al. 2020).

Butanol is made using various fermentation techniques, including feed-batch fermentation, two-stage continuous fermentation, continuous fermentation, and alternating fermentation. In a typical fermentation plant, continuous fermentation yields higher butanol compared to batch fermentation (Kujawska et al. 2015). Fermentation using fed-batch methods involves supplying the fermentor with several nutrients or feedstock and retrieving the finished product (Fig. 3.4). Biobutanol is made via fed-batch fermentation employing a variety of biological substrates. The overall cell count is frequently maintained at a consistent level in batch fermentation via measuring cell density at frequent durations. The cloudiness of the culture determines the rate of nutrient supply and culture removal. Batch fermentation produces a low yield while squandering substrate and medium. The remaining culture and substrate are discarded after the procedure (Mariano et al. 2011; Qureshi 2015).

As long as the feed is continuously given and the product is continuously collected, the fermentation is **continuous** (As shown in Fig. 3.5). Continuous fermentation involves draining or emptying 50% of the medium and transferring it to a culture vessel. In the fermentation containers, the inoculum is recycled. Fermentation yields are increased when bacterial cells are recycled. A fresh medium is then introduced to the fermentation tank to ensure continuous output. This approach yields a larger yield with less waste. Continuous fermentation in the laboratory is both valuable and straightforward. Furthermore, due to its temporal efficiency, continuous fermentation is the best choice for massive butanol synthesis, according

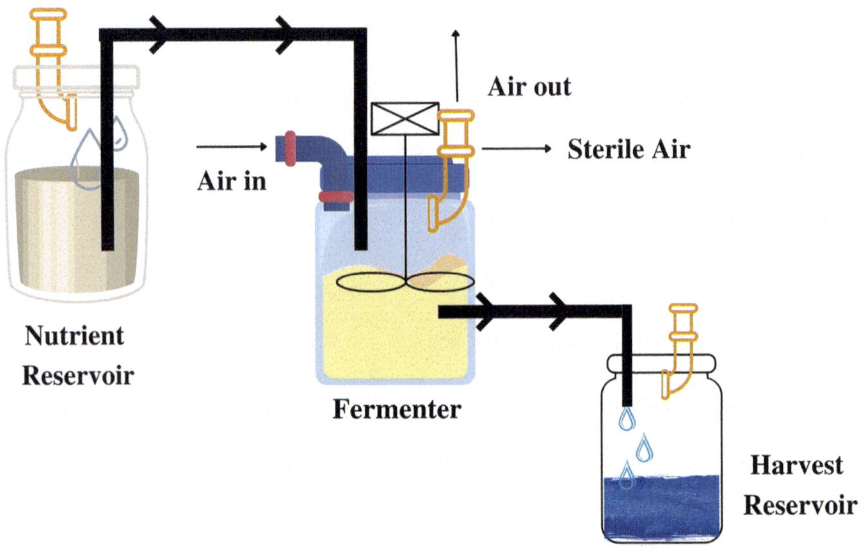

Fig. 3.5 Butanol synthesis via continuous fermentation. (Source: A. Pugazhendhi et al. 2019)

to a cost-benefit study of the batch, fed-batch, including continuous fermentation frameworks (Ranjan and Moholkar 2012).

The initial phase of a two-phased continuous fermentation process is the acidogenic turbidostatic phase, which is comprised of growth of cells at a maximal rate and consequent transmission of the growth medium to the next phase. When it comes to fermenters, stage I fermenters are for growing, whereas stage II or later fermenters are for synthesizing. Such a situation often occurs in fermenters where the microorganism's multiplication and synthesis operations are not co-occurring (Fig. 3.6).

After the fermentation is accomplished, the product is retrieved utilizing several unit processes for separation and purification. In terms of energy efficiency, butanol is challenging to restore because of its little content (20 g/L) and significant boiling point (117 °C). Several procedures are often used to separate fermentation products, including (a) liquid-liquid extraction, (b) pervaporation, (c) adsorption, (d) membrane-based approaches, and (e) gas stripping.

The collection of butanol from the fermentation broth takes a long period and requires much energy. Distillation is among the most popular and widely utilized procedures for recovering butanol because of the dilution of butanol during fermentation (Kujawska et al. 2015). Butanol's total manufacturing cost may rise due to its more remarkable boiling point and increased energy usage during distillation. As a result, distillation is uneconomical and inefficient in terms of energy production. Researchers have devised various recovery strategies for the cost-effective and the most often utilized processes which are adsorption, gas stripping, and pervaporation for butanol separation (Ezeji et al. 2004; Qureshi et al. 1992, 2001).

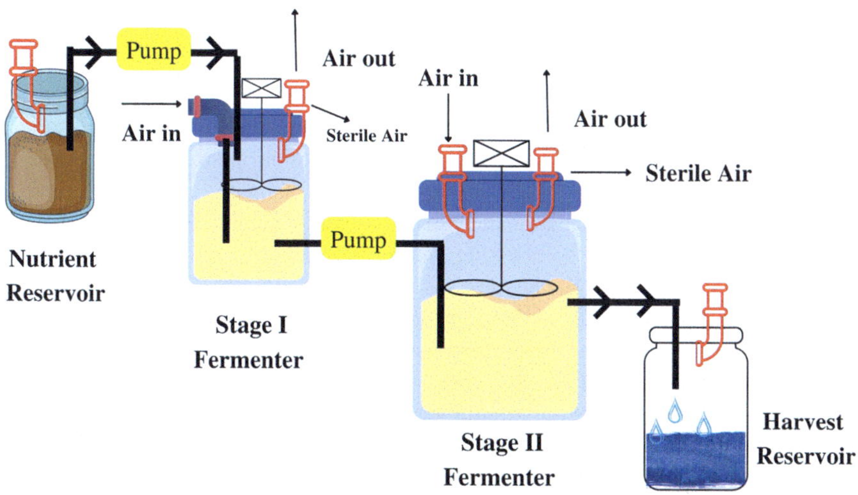

Fig. 3.6 Butanol synthesis via two-stage fermenter. (Source: A. Pugazhendhi et al. 2019)

When adsorbing butanol on the adsorbed, while desorbing when the temperature is increased otherwise, the displacer is employed in a minimal energy process known as **adsorption** (Abdehagh et al. 2017). Although it is a straightforward approach, the yield achieved by this method is lower than that produced by other methods (Martin-Calvo et al. 2018; Qureshi et al. 2005, 2001). Various types of adsorption materials have been utilized, including silicalite, resins, charcoal, polyvinyl pyridine, and so on (Qureshi et al. 2005). However, although adsorbents are easy to utilize, extremely specific, and feasible means of butanol retrieval, concerns regarding compatibility and contamination hazards persist.

Gas stripping is a feasible method for extracting butanol from an ABE fermentation bioreactor. This approach involves injecting the solution into a column in the opposite direction as the gas to separate the selected component or product (Kujawska et al. 2015). This method is straightforward, non-fouling, and simple to use, but it has two main drawbacks: high foam generation and butanol selectivity (Xue et al. 2017). Nevertheless, gas stripping has shown to be the most effective integrated recovery approach in biobutanol fermentation so far.

During ABE fermentation with *C. acetobutylicum*, **pervaporation** is a critical approach for removing hazardous compounds. It is a method that relies on the membrane's molecular interactions with the feed elements. The introduction of vacuum causes selective diffusion of elements across the membrane in this approach (Van Hecke et al. 2012). For successful product recovery, selecting the right pervaporation membrane is crucial. The substantial cost of producing pressure drop by the low-pressure edge of the membrane is a significant drawback of this technique (Xue et al. 2017).

3.8 Improvements in Algal Butanol Synthesis in Current Years

A new generation of alcoholic propellant derived from natural biomass is far more cost-effective and environmentally benign than traditional petroleum-based fuels. However, biomass-derived fuels do not contribute to planetary climatic disruption and are carbon-neutral because of their high oxygen content. They also appear to have an excellent value for combusting and a prime rate of heat transfer (Serrano-Ruiz and Dumesic 2011). Nonetheless, there is currently research being done on algal biomass conversion to alcohols in a laboratory environment, necessitating technology development for large-scale commercial purposes (Ashokkumar et al. 2017; Brockmann et al. 2015).

For its low engine corrosivity, significant calorific value, and excellent hydrophobicity, biobutanol is indeed a preferred alternative in transit (Li et al. 2020). It is synthesized by solventogenic *Clostridia*, such as *C. acetobutylicum* or *C. beijerinckii*, utilizing the conventional ABE fermentation method (Salaeh et al. 2019). Because of its enormous stocks of inclusion carbohydrates, seaweed biomass seems to be a highly enticing feedstock for the butanol industries. Hou et al. (2017) found that the hydrolysate of the brown algae *Laminaria* digitate yielded 0.42 g/g of butanol when fermented anaerobically with *C. beijerinckii* DSM-6422. Moreover, Hong et al. (2020) using a novel isolate that is not affected by pH, *Clostridium* sp. *WK* improved H_2SO_4-based hydrolysis on *Gelidium amansii* (red algal biomass) to reach a peak hydrolytic rate at 80.95% and produced the hydrolysate yielded 0.2 g/g biobutanol. Nevertheless, numerous variables impact the transformation of biobutanol from algal biomass, such as the seaweed morphology, the amount of generated reducing sugars, and the efficiency of hydrolysis. Even though hydrolyzing brown algae *Sargassum fulvellum* generated an excellent yield of reducing sugar (0.425 g/g), owing to their poor catabolism efficiency on the algae framework comprising galactose as well as mannitol, however, according to Sunwoo et al. (2018), *C. acetobutylicum* and *C. tyrobutyricum* produce butanol at a rate of around 0.087 g/g (0.156 g of total ABE). As a result, strategies for boosting microbial strains' consumption of complex carbohydrates via seaweed biomass must be addressed.

Algal cells have a low resistance to the produced alcohol, making producing alcoholic fuel using algae challenging. Advances in systemic biology with metabolic programming, on the other hand, have discovered new genes and mechanisms within *Synechocystis* sp. PCC6803 that could be switched off as a substitute to improve biofuel yields (Jiang et al. 2014). Increased algal biofuel generation is now possible because of the advent of genomic technologies for DNA construction for heterologous expressions of metabolic processes (Olson and Carter 2016). Researchers were able to uncover cellular pathways and genes that have been altered during alcohol generation in *C. reinhardtii* under diverse stresses employing an iTRAQ-LC-MS/MS-dependent quantitative proteomics approach. Various mechanisms that aid in the defense against the toxicity of organic solvents have also been identified. Butanol overexpression was shown to be related to the Cre.770 protein (Wang et al. 2017).

The WT sta7 gene was used for developing corresponding sta7-10 (c5) and sta7-10 (c19) variants from starch-free sta6 and sta7 variants; improved starch buildup in *C. reinhardtii* was found (Jang 2013). Through fermentation methods, these complementing strains with enhanced starch concentration can be utilized to produce bioethanol or biobutanol. *Thermotoga neapolitana*, a hyperthermophilic bacterium, was used to modify *C. reinhardtii* by inserting the amylase gene from it, to enhance and harvest fermentable carbohydrate levels in algae (Patel et al. 2019). Altered algae expressing a starch hydrolyzing enzyme can decrease the need for further pretreatment, making the biomass more appropriate as a feedstock for bioethanol or biobutanol synthesis.

3.9 Conclusion and Future Perspectives

The rising need for fuel has piqued academics' interest in finding alternate, green energy alternatives that are environmentally favorable. For the entire globe, the utilization of sustainable feedstock offers a cost-effective but also ecologically beneficial source of energy. A potent biofuel, biobutanol, can replace non-renewable energy sources in the coming generations.

Algae are unquestionably an underutilized resource. As per numerous writers, improving the various components in a cascade biorefinery determines the economic viability of employing micro- as well as macroalgae as sustainable and ecologically beneficial substrates.

Algae's competitiveness might be boosted through cascading biorefineries by harvesting all accessible high-value elements (proteins, lipids, pigments, and ashes as fertilizer). More study is needed to create novel technologies that might allow the recovery of distinct algae constituents while also lowering operational expenses.

Notwithstanding technical obstacles and constraints, the third-generation biorefinery (algae) displays substantial benefits and pragmatic possibilities and may play a key role in building a renewable and sustainable marine economy. Problems encountered and appropriate alternatives relating to biomass usage and product transformation are suggested based on current developments to offer perspectives for balancing sustainability and economic profits. Feedstock selection, procedure efficiency, and metabolic flux management should all be considered in future advancements.

References

Abdehagh N, Dai B, Thibault J, Handan Tezel F (2017) Biobutanol separation from ABE model solutions and fermentation broths using a combined adsorption–gas stripping process. J Chem Technol Biotechnol 92:245–251

Abdullah B, Muhammad, Syed SAFA, Shokravi Z, Ismail S, Kassim KA, Mahmood AN, Aziz MMA (2019) Fourth generation biofuel: a review on risks and mitigation strategies. Renew Sust Energ Rev 107:37–50

Adeniyi OM, Azimov U, Burluka A (2018) Algae biofuel: current status and future applications. Renew Sust Energ Rev 90:316–335
Ashokkumar V, Salim MR, Salam Z, Sivakumar P, Chong CT, Elumalai S, Suresh V, Ani FN (2017) Production of liquid biofuels (biodiesel and bioethanol) from brown marine macroalgae Padina tetrastromatica. Energy Convers Manag 135:351–361
Bahl H, Andersch W, Braun K, Gottschalk G (1982) Effect on pH and butyrate concentration on the production of acetone and butanol by Clostridium acetobutylicum growing in a continuous culture. Eur J Appl Microbiol Biotechnol 14:17–20
Ballongue J, Amine J, Masion E, Petitdemange H, Gay R (1985) Induction of acetoacetate decarboxylase in Clostridium acetobutylicum. FEMS Microbiol Lett 29:273–277
Banerjee S, Banerjee S, Ghosh AK, Das D (2020) Maneuvering the genetic and metabolic pathway for improving biofuel production in algae: present status and future prospective. Renew Sust Energ Rev 133:110155
Bankar SB, Survase SA, Ojamo H, Granström T (2013) Biobutanol: the outlook of an academic and industrialist. RSC Adv 3:24734–24757
Bharathiraja B, Jayamuthunagai J, Sudharsanaa T, Bharghavi A, Praveenkumar R, Chakravarthy M et al (2017) Biobutanol – an impending biofuel for future: a review on upstream and downstream processing techniques. Renew Sust Energ Rev 2017(68):788–807
Birgen C, Berglihn OT, Preisig HA, Wentzel A (2019a) Kinetic study of butanol production from mixtures of glucose and xylose and investigation of different pre-growth strategies. Biochem Eng J 147:110–117
Birgen C, Dürre P, Preisig HA, Wentzel A (2019b) Butanol production from lignocellulosic biomass: revisiting fermentation performance indicators with exploratory data analysis. Biotechnol Biofuels 12(1):167
Bowles LK, Ellefson WL (1985) Effects of butanol on Clostridium acetobutylicum. Appl Environ Microbiol 50:1165–1170
Brawley SH, Blouin NA, Ficko-Blean E, Wheeler GL, Lohr M, Goodson HV, Jenkins JW, Blaby-Haas CE, Helliwell KE, Chan CX, Marriage TN, Bhattacharya D, Klein AS, Badis Y, Brodie J, Cao Y, Collen J, Dittami SM, Gachon CMM, Green BR, Karpowicz SJ, Kim JW, Kudahl UJ, Lin S, Michel G, Mittag M, Olson B, Pangilinan JL, Peng Y, Qiu H, Shu S, Singer JT, Smith AG, Sprecher BN, Wagner V, Wang W, Wang ZY, Yan J, Yarish C, Zauner-Riek S, Zhuang Y, Zou Y, Lindquist EA, Grimwood J, Barry KW, Rokhsar DS, Schmutz J, Stiller JW, Grossman AR, Prochnik SE (2017) Insights into the red algae and eukaryotic evolution from the genome of Porphyra umbilicalis (Bangiophyceae, Rhodophyta). Proc Natl Acad Sci U S A 114(31):E6361–E6370
Brennan L, Owende P (2010) Biofuels from microalgae—a review of the technologies for production processing and extractions of biofuels and co-products. Renew Sust Energ Rev 14(557):577
Brockmann D, Pradinaud C, Champenois J, Benoit M, Helias A (2015) Environmental assessment of bioethanol from onshore grown green seaweed. Biofuels Bioprod Biorefin 9(6):696–708
Buckeridge MS, Grandis A, Tavares EQP (2019) Disassembling the glycomic code of sugarcane cell walls to improve second-generation bioethanol production. In: Ray RC, Ramachandran S (eds) Bioethanol production from food crops, pp 31–43
Cantrell KB, Ducey T, Ro KS, Hunt PG (2008) Livestock waste-to-bioenergy generation opportunities. Bioresour Technol 99:7941–7953
Cesário MT, da Fonseca MMR, Marques MM, de Almeida MCMD (2018) Marine algal carbohydrates as carbon sources for the production of biochemicals and biomaterials. Biotechnol Adv 36:798–817
Chen R, Yue Z, Deitz L, Liu Y, Mulbry W, Liao W (2012) Use of an algal hydrolysate to improve enzymatic hydrolysis of lignocelluloses. Bioresour Technol 108:149–154
Chen CY, Zhao XQ, Yen HW, Ho SH, Cheng CL, Lee DJ et al (2013) Microalgae-based carbohydrates for biofuel production. Biochem Eng J 78:1–10
Chen C, Wang L, Xiao G, Liu Y, Xiao Z, Deng Q, Yao P (2014) Continuous acetone–butanol–ethanol (ABE) fermentation and gas production under slight pressure in a membrane bioreactor. Bioresour Technol 163:6–11

Chew KW, Yap JY, Show PL, Suan NH, Juan JC, Ling TC, Lee D-J, Chang J-S (2017) Microalgae biorefinery: high value products perspectives. Bioresour Technol 229:53–62
Chisti Y (2007) Biodiesel from microalgae. Biotechnol Adv 25:294–306
Clark JH, Deswarte F, Stevens CV (2008) Introduction to chemicals from biomass. Wiley series in renewable resources. Wiley, New York, NY
Cotton AF, Wilkinson G, Bochmann M, Murillo CA (1999) Advance inorganic chemistry. John Wiley and Sons, New York, NY
Cui Y, Liu X, Li S, Hao L, Du J, Gao D, Kang Q, Lu J (2018) Extraction, characterization and biological activity of sulfated polysaccharides from seaweed Dictyopteris divaricata. Int J Biol Macromol 117:256–263
Cynkin MA, Delwiche EA (1958) Metabolism of pentoses by clostridia. I. Enzymes of ribose dissimilation in extracts of Clostridium perfringens. J Bacteriol 75:331–334
Das D (2015) Algal biorefinery: an integrated approach. Springer, Cham
Davis R, Aden A, Pienkos PT (2011) Techno-economic analysis of autotrophic microalgae for fuel production. Appl Energy 88:3524–3531
Delattre C, Pierre G, Laroche C, Michaud P (2016) Production, extraction and characterization of microalgal and cyanobacterial exopolysaccharides. Biotechnol Adv 34:1159–1179
Demirbas A (2001) Biomass resource facilities and biomass conversion processing for fuels and chemicals. Energy Convers Manag 42:1357–1378
Demirbas A (2008) Biofuels sources, biofuel policy, biofuel economy and global biofuel projections. Energy Convers Manag 49:2106–2116
Demirbas MF (2009) Biorefineries for biofuel upgrading: a critical review. Appl Energy 86:151–161
Doelle HW (1975) Bacterial metabolism. Academic Press, New York, NY
Draget KI, Smidsrød O, Skjåk-Bræk G (2005) Alginates from algae. Biopolymers online. Wiley-VCH Verlag GmbH & Co. KGaA, Weinheim
Dürre P (2007) Biobutanol: an attractive biofuel. Biotechnol J 2:1525–1534
Ezeji T, Qureshi N, Blaschek H (2004) Acetone butanol ethanol (ABE) production from concentrated substrate: reduction in substrate inhibition by fed-batch technique and product inhibition by gas stripping. Appl Microbiol Biotechnol 63:653–658
Fernbach A, Strange H (1912) Fermentation process for the production of acetone and higher alcohols from starch, sugars, and other carbohydrate material
Festel GW (2008) Biofuels-economic aspects. Chem Eng Technol 31:715–720
Fond O, Matta-Ammouri G, Petitdemange H, Engasser JM (1985) The role of acids on the production of acetone and butanol by Clostridium acetobutylicum. Appl Microbiol Biotechnol 22:195–200
Galbe M, Zacchi G (2007) Pretreatment of lignocellulosic materials for efficient bioethanol production. In: Olsson L (ed) Biofuels. Springer, New York, NY, pp 41–65
Ganesana R, Manigandanb S, Samuelc MS, Shanmuganathand R, Brindhadevie K, Chie NTL, Ducf PA, Pugazhendhie A (2020) A review on prospective production of biofuel from microalgae. Biotechnol Rep 27:1–13
Gaurav N, Sivasankari S, Kiran GS, Ninawe A, Selvin J (2017) Utilization of bioresources for sustainable biofuels: a review. Renew Sust Energ Rev 73:205–214
George HA, Chen JS (1983) Acidic conditions are not obligatory for onset of butanol formation by Clostridium beijerinckii (synonym. C. butylicum). Appl Environ Microbiol 46:321–327
George HA, Johnson JL, Moore WEC, Holdeman LV, Chen JS (1983) Acetone, isopropanol, and butanol production by Clostridium beijerinckii (syn. Clostridium butylicum) and Clostridium aurantibutylicum. Appl Environ Microbiol 45:1160–1163
Gottschalk G (1986) Bacterial metabolism. Springer, New York, NY
Gottwald M, Hippe H, Gottschalk G (1984) Formation of n-butanol from D-glucose by strains of "Clostridium tetanomorphum" group. Appl Environ Microbiol 48:573–576
Gustavo BL, Abdelaziz AEM, Hallenbeck PC (2013) Algal biofuels: challenges and opportunities. Bioresour Technol 145(134):141

Harun R, Danquah MK, Forde GM (2010) Microalgal biomass as a fermentation feedstock for bioethanol production. J Chem Technol Biotechnol 85:199–203

Hong Y, Chen C, Wu YR (2020) Biobutanol production from sulfuric acid-pretreated red algal biomass by a newly isolated Clostridium sp. strain WK. Biotechnol Appl Biochem 67(5): 738–743

Hou X, From N, Angelidaki I, Huijgen WJJ, Bjerre AB (2017) Butanol fermentation of the brown seaweed Laminaria digitata by Clostridium beijerinckii DSM-6422. Bioresour Technol 238:16–21

Huesemann MH, Kuo L-J, Urquhart L, Gill GA, Roesijadi G (2012) Acetone-butanol fermentation of marine macroalgae. Bioresour Technol 108:305–309

Ibrahim MF, Kim SW, Abd-Aziz S (2018) Advanced bioprocessing strategies for biobutanol production from biomass. Renew Sust Energ Rev 91:1192–1204

Jang S-S (2013) Potential use of Gelidium amansii acid hydrolysate for lactic acid production by Lactobacillus rhamnosus. Food Technol Biotechnol 51:131–136

Jiang W, Brueggeman AJ, Horken KM, Plucinak TM, Weeks DP (2014) Successful transient expression of Cas9 and single guide RNA genes in Chlamydomonas reinhardtii. Eukaryot Cell 3(11):1465–1469

Jin C, Yao M, Liu H, Chia-fon FL, Ji J (2011) Progress in the production and application of n-butanol as a biofuel. Renew Sust Energ Rev 15:4080–4106

Johnson MJ, Peterson WH, Fred EB (1931) Oxidation and reduction relations between substrate and products in the acetone-butyl alcohol fermentation. J Biol Chem 91:569–591

Jones CS, Mayfield SP (2012) Algae biofuels: versatility for the future of bioenergy. Curr Opin Biotechnol 23:346–351

Jones DT, Woods DR (1986) Acetone-butanol fermentation revisited. Microbiol Rev 50:484–524

Juben N, James C, Dumesi A (2007) An overview of dehydration, aldol-condensation and hydrogenation processes for production of liquid alkanes from biomass-derived carbohydrates. Catal Today 123:59–70

Juneja A, Ceballos R, Murthy G (2013) Effects of environmental factors and nutrient availability on the biochemical composition of algae for biofuels production: a review. Energies 6:4607–4638

Kell DB, Peck MW, Rodger G, Morris JG (1973) On the permeability to weak acids and bases of the cytoplasmic membrane of Clostridium pasteurianum. Biochem Biophys Res Commun 99: 81–88

Kim NJ, Li H, Jung K, Chang HN, Lee PC (2011) Ethanol production from marine algal hydrolysates using Escherichia coli KO11. Bioresour Technol 102(16):7466–7469

Kodama M, Ikemoto H, Miyachi S (1993) A new species of highly CO2-tolerant fast-growing marine microalga suitable for high-density culture. J Mar Biotechnol 1:21–25

Kotze JP (1969) Glycolytic and related enzymes in clostridial classification. Appl Microbiol 18: 744–747

Kujawska A, Kujawski J, Bryjak M, Kujawski W (2015) ABE fermentation products recovery methods—a review. Renew Sust Energ Rev 48:648–661

Kumar A (2020) Current and future perspective of microalgae for simultaneous wastewater treatment and feedstock for biofuels production

Kumar M, Gayen K (2011) Developments in biobutanol production: new insights. Appl Energy 88: 1999–2012

Lee SY, Park JH, Jang SH, Nielsen LK, Kim J, Jung KS (2008) Fermentative butanol production by Clostridia. Biotechnol Bioeng 101:209–228

Li S, Huang L, Ke C, Pang Z, Liu L (2020) Pathway dissection, regulation, engineering and application: lessons learned from biobutanol production by solventogenic clostridia. Biotechnol Biofuels 13(1):39

Lin N, Liu T, Lin L, Lin S, Zang Q, He J, Abliz Z, Li C, Wang A, Jin H (2016) Comparison of in vivo immunomodulatory effects of 5-hydroxymethylfurfural and 5,5′-oxydimethylenebis (2-furfural). Regul Toxicol Pharmacol 81:500–511

Mariano AP, Qureshi N, Filho RM, Ezeji TC (2011) Bioproduction of butanol in bioreactors: new insights from simultaneous in situ butanol recovery to eliminate product toxicity. Biotechnol Bioeng 108:1757–1765

Martin-Calvo A, Van der Perre S, Claessens B, Calero S, Denayer JF (2018) Unravelling the influence of carbon dioxide on the adsorptive recovery of butanol from fermentation broth using ITQ-29 and ZIF-8. Phys Chem Chem Phys 20:9957–9964

Mata TM, Martins AA, Caetano NS (2010) Microalgae for biodiesel production and other applications: a review. Renew Sust Energ Rev 14:217–232

Mautner HG (1954) The chemistry of brown algae. Econ Bot 8:174–192

McKendry P (2002) Energy production from biomass (part 3): gasification technologies. Bioresour Technol 83:55–63

Metting FB (1996) Biodiversity and application of microalgae. J Ind Microbiol 17:477–489

Monlau F, Sambusiti C, Barakat A, Quéméneur M, Trably E, Steyer JP, Carrère H (2014) Do furanic and phenolic compounds of lignocellulosic and algae biomass hydrolyzate inhibit anaerobic mixed cultures? A comprehensive review. Biotechnol Adv 32:934–951

Moreira AR, Ulmer DC, Linden JC (1981) Butanol toxicity in the butylic fermentation. Biotechnol Bioeng Symp 11:567–579

Muraza O (2014) Biodiesel production from algae by using heterogeneous catalysts: a critical review. Energy 78:72–83

Ndaba B, Chiyanzu I, Marx S (2015) n-Butanol derived from biochemical and chemical routes: a review. Biotechnol Rep 8:1–9

Nguyen MT, Choi SP, Lee J, Lee JH, Sim SJ (2009) Hydrothermal acid pretreatment of Chlamydomonas reinhardtii biomass for ethanol production. J Microbiol Biotechnol 19:161–166

Olson AC, Carter CJ (2016) The involvement of hybrid cluster protein 4, HCP4, in Anaerobic Metabolism in Chlamydomonas reinhardtii. PLoS One 11:e0149816

Ounine K, Petitdemange H, Raval G, Gay R (1983) Acetone-butanol production from pentoses by Clostridium acetobutylicum. Biotechnol Lett 5:605–610

Park SH, Lee CR, Hong SK (2020) Implications of agar and agarose in industrial applications of sustainable marine biomass. Appl Microbiol Biotechnol 104(7):2815–2832

Patel VK, Soni N, Prasad V, Sapre A, Dasgupta S, Bhadra B (2019) CRISPR–Cas9 system for genome engineering of photosynthetic microalgae. Mol Biotechnol 61:1–21

Percival E (1979) The polysaccharides of green, red and brown seaweeds: their basic structure, biosynthesis and function. Br Phycol J 14:103–117

Prazukin AV, Firsov YK, Kamenir Y (2018) The vertical structure of the vegetative canopy of the brown algae Cystoseira (Black Sea). J Oceanol Limnol 38(1):124–132

Pugazhendhi A et al (2019) Biobutanol as a promising liquid fuel for the future - recent updates and perspectives. Fuel 253:637–646

Qureshi N (2015) Butanol production by fermentation: efficient bioreactors. In: Snyder SW (ed) Commercializing biobased products: opportunities, challenges, benefits, and risks. RSC, London, p 48

Qureshi N, Maddox IS, Friedl A (1992) Application of continuous substrate feeding to the ABE fermentation: relief of product inhibition using extraction, perstraction, stripping, and pervaporation. Biotechnol Prog 8:382–390

Qureshi N, Meagher M, Huang J, Hutkins R (2001) Acetone butanol ethanol (ABE) recovery by pervaporation using silicalite–silicone composite membrane from fed-batch reactor of Clostridium acetobutylicum. J Membr Sci 187:93–102

Qureshi N, Hughes S, Maddox I, Cotta M (2005) Energy-efficient recovery of butanol from model solutions and fermentation broth by adsorption. Bioprocess Biosyst Eng 27:215–222

Qureshi N, Ezeji TC, Ebener J, Dien BS, Cotta MA, Blaschek HP (2008) Butanol production by Clostridium beijerinckii. Part I: use of acid and enzyme hydrolyzed corn fiber. Bioresour Technol 99:5915–5922

Rakopoulos D, Rakopoulos C, Giakoumis E, Dimaratos A, Kyritsis D (2010) Effects of butanol–diesel fuel blends on the performance and emissions of a high-speed DI diesel engine. Energy Convers Manag 51:1989–1997
Ranjan A, Moholkar VS (2012) Biobutanol: science, engineering, and economics. Int J Energy Res 36:277–323
Rengasamy KR, Mahomoodally MF, Aumeeruddy MZ, Zengin G, Xiao J, Kim DH (2020) Bioactive compounds in seaweeds: an overview of their biological properties and safety. Food Chem Toxicol 135:111013
Rogers P (1986) Genetics and biochemistry of Clostridium relevant to development of fermentation processes. Adv Appl Microbiol 31:1–60
Salaeh S, Kongjan P, Panphon S, Hemmanee S, Reungsang A, Jariyaboon R (2019) Feasibility of ABE fermentation from Rhizoclonium spp. hydrolysate with low nutrient supplementation. Biomass Bioenergy 127:105269
Santacesaria E, Carotenuto G, Tesser R, Di Serio M (2012) Ethanol dehydrogenation to ethyl acetate by using copper and copper chromite catalysts. Chem Eng J 179:209–220
Santhakumaran P, Kookal SK, Mathew L, Ray JG (2019) Bioprospecting of three rapid growing freshwater green algae, promising biomass for biodiesel production. BioEnergy Res 12(3): 680–693
Schenk PM, Thomas-Hall SR, Stephens E, Marx UC, Mussgnug JH, Posten C et al (2008) Second-generation biofuels: high-efficiency microalgae for biodiesel production. BioEnergy Res 1:20–43
Serrano-Ruiz JC, Dumesic JA (2011) Catalytic routes for the conversion of biomass into liquid hydrocarbon transportation fuels. Energy Environ Sci 4(1):83–99
Shaikh AR, Alia SAM, Hossaina MM, deLasa H (2017) Biological CO2 fixation with production of microalgae in wastewater—a review. Renew Sust Energ Rev 76:379
Shaishav S, Singh RN, Satyendra T (2013) Biohydrogen from algae: fuel of the future. Int Res J Environ Sci 2:44–47
Shapovalov O, Ashkinazi L (2008) Biobutanol: biofuel of second generation. Russ J Appl Chem 81:2232–2236
Sharma YC, Singh B (2009) Development of biodiesel: current scenario. Renew Sust Energ Rev 13: 1646–1651
Sheehan J, Camobreco V, Duffield J, Graboski M, Shapouri H (1998) Life cycle inventory of biodiesel and petroleum diesel for use in an urban bus. National Renewable Energy Lab, Golden, CO. Final report (No. NREL/SR-580-24089)
Slade R, Bauen A (2013) Micro-algae cultivation for biofuels: cost, energy balance, environmental impacts and future prospects. Biomass Bioenergy 53:29–38
Suganya T, Varman M, Masjuki HH, Renganathan S (2016) Macroalgae and microalgae as a potential source for commercial applications along with biofuels production: a biorefinery approach. Renew Sust Energ Rev 55:909–941
Sunwoo IY, Hau NT, Ra CH, Jeong GT, Kim SK (2018) Acetone-butanol-ethanol production from waste seaweed collected from Gwangalli Beach, Busan, Korea, based on pH-controlled and sequential fermentation using two strains. Appl Biochem Biotechnol 185(4):1075–1087
Swodenk W (1983) Ethanol als Rohstoff für die chemische Industrie. Chem In Tech 55:683–688
Thauer RK, Jungermann K, Dekker K (1977) Energy conservation in chemotropic anaerobic bacteria. Bacteriol Rev 41:100–180
Ujor V, Okonkwo C, Ezeji TC (2016) Unorthodox methods for enhancing solvent production in solventogenic Clostridium species. Appl Microbiol Biotechnol 100(3):1089–1099
Uyttebroek M, Van Hecke W, Vanbroekhoven K (2015) Sustainability metrics of 1- butanol. Catal Today 239:7–10
Van Hecke W, Vandezande P, Claes S, Vangeel S, Beckers H, Diels L et al (2012) Integrated bioprocess for long-term continuous cultivation of Clostridium acetobutylicum coupled to pervaporation with PDMS composite membranes. Bioresour Technol 111:368–377

Vitova M et al (2015) Accumulation of energy reserves in algae: from cell cycles to biotechnological applications. Biotechnol Adv 33(6):1204–1218

Vollherbst-Schneck KJA, Sands A, Montencourt BS (1984) Effect of butanol on lipid composition and fluidity of Clostridium acetobutylicum ATCC 824. Appl Environ Microbiol 47:193–194

Voloshin RA, Rodionova MV, Zharmukhamedov SK, Nejat Veziroglu T, Allakhverdiev SI (2016) Review: biofuel production from plant and algal biomass. Int J Hydrog Energy 41(39): 17257–17273

van der Wal H, Sperber BL, Houweling-Tan B, Bakker RR, Brandenburg W, Lopez-Contreras AM (2013) Production of acetone, butanol, and ethanol from biomass of the green seaweed Ulva lactuca. Bioresour Technol 28:431–437

Walker JB (1954) Inorganic micronutrient requirements of Chlorella II Quantitative requirements for iron, manganese, and zinc. Arch Biochem Biophys 53:1–8

Wang Y, Ho SH, Yen HW, Nagarajan D, Ren NQ, Li S, Chang JS (2017) Current advances on fermentative biobutanol production using third-generation feedstock. Biotechnol Adv 35(8): 1049–1059

Weismann C, Alliston L (1922) Production of secondary butyl alcohol

Widjaja A, Chien C-C, Ju Y-H (2009) Study of increasing lipid production from fresh water microalgae Chlorella Vulgaris. J Taiwan Inst Chem Eng 40(1):13–20

Xin F, Yan W, Zhou J, Wu H, Dong W, Ma J, Zhang W, Jiang M (2018) Exploitation of novel wild type solventogenic strains for butanol production. Biotechnol Biofuels 11(1):252

Xue C, Zhao J, Chen L, Yang S-T, Bai F (2017) Recent advances and state-of-the-art strategies in strain and process engineering for biobutanol production by Clostridium acetobutylicum. Biotechnol Adv 35:310–322

Yadav S, Rawat G, Tripathi P, Saxena RK (2014) A novel approach for biobutanol production by Clostridium acetobutylicum using glycerol: a low cost substrate. Renew Energy 71:37–42

Yang MK, Keinänen S, Vepsäläinen M, Romar J, Tynjälä H, Lassi PU, Pappinen A (2014) The use of (green field) biomass pretreatment liquor for fermentative butanol production and the catalytic oxidation of biobutanol. Chem Eng Res Des 92:1531–1538

Yen HW, Brune DE (2007) Anaerobic co-digestion of algal sludge and waste paper to produce methane. Bioresour Technol 98:130–134

Yeong TK, Jiao K, Zeng X, Lin L, Pan S, Danquah MK (2018) Microalgae for biobutanol production—technology evaluation and value proposition. Algal Res 31:367–376

Zang G, Shah A, Wan C (2020) Techno-economic analysis of an integrated biorefinery strategy based on one-pot biomass fractionation and furfural production. J Clean Prod 260:120837

Zanolla M, Altamirano M, Carmona R, De la Rosa J, Souza-Egipsy V, Sherwood A, Tsiamis K, Barbosa AM, Muñoz AR, Andreakis N (2018) Assessing global range expansion in a cryptic species complex: insights from the red seaweed genus Asparagopsis (Florideophyceae). J Phycol 54(1):12–24

Chapter 4
Algal Synthesis of Gold Nanoparticles: Applications in Bioenergy

Shilpi Srivastava, Francisco Fuentes, and Atul Bhargava

Abstract Gold nanoparticles (AuNPs) are small particles of gold (Au) having a diameter of 1–100 nm and are usually preferred for the inherent optoelectronic and electrochemical properties that find applications in electronics, nanotechnology, as catalysts, sensory probes, and biomedicine. Algae have recently evoked interest among researchers as potential organisms for the biological synthesis of nanomaterials due to ease in cultivation and presence of different functional groups and enzymes in their cell wall which act as reducing and capping agents in the nanofabrication process at ambient conditions. Different classes of algae have been reported to produce AuNPs both extracellularly and intracellularly. Several unique properties of AuNPs have made them quite attractive as catalytic agents in the production process of biofuels. This article throws light on the biofabrication of AuNPs using algae and their possible applications in biofuel development.

Keywords Gold nanoparticles · Green synthesis · Algae · Bioenergy · Biofuels

4.1 Introduction

Gold nanoparticles (AuNPs) are small particles of gold (Au) having a diameter of 1–100 nm. Gold nanoparticles can be variously colored depending on the size, shape, and degree of aggregation and nature of the protecting organic shells on their surface. Thus, the AuNPs may appear red, orange, brown, or blue in aqueous solutions and emit bright resonance light of different wavelengths falling in the

S. Srivastava
Amity Institute of Biotechnology, Amity University Uttar Pradesh, Lucknow, India

F. Fuentes
Facultad de Agronomía e Ingeniería Forestal, Pontificia Universidad Católica de Chile, Santiago, Chile

A. Bhargava (✉)
Department of Botany, Mahatma Gandhi Central University, Motihari, Bihar, India
e-mail: atulbhargava@mgcub.ac.in

N. Srivastava, P. K. Mishra (eds.), *Technological Advancement in Algal Biofuels Production*, Clean Energy Production Technologies,
https://doi.org/10.1007/978-981-19-6806-8_4

visible region (Kumar et al. 2013). The AuNPs exhibit different shapes like spherical, octahedral, cuboctahedral, icosahedral, decahedral, multiple twined, tetrahedral, nanotriangles, hexagonal platelets, nanostars, nanorods, nanocages, nanoprisms, and nanoshells or may be irregularly shaped (Taufikurohmah et al. 2011; Yapar and Inal 2012; Nafisi and Maibach 2017). The AuNPs also exhibit an absorption peak from 500 to 550 nm in the UV-vis spectra which is a measure of the size, shape, solvent, surface ligand, core charge, temperature, and agglomeration of the nanoparticles (Srivastava et al. 2005; Das et al. 2011; Jain et al. 2006). Other properties that find use in different industries include chemical and mechanochemical stability, easy surface functionalization, catalytic activity, biocompatibility, resistance to oxidation, crystallizability, and good size dispersion properties (Daniel and Astruc 2004; Shukla et al. 2005; Henz et al. 2008; Kannan and Subbalaxmi 2011).

4.2 Applications of Gold Nanoparticles

AuNPs are usually preferred for the inherent optoelectronic and electrochemical properties that find applications in electronics, nanotechnology, as catalysts, sensory probes, and biomedicine. In fact, AuNPs were an ingredient of the "Swarna Bhasma," used for its therapeutic benefits in the ancient traditional Indian Ayurvedic system as far back as 2500 BC (Kashani et al. 2018). Recent studies have hinted toward their effectiveness and utilization in biodiagnostics, imaging drug delivery, and photothermal therapy with respect to therapeutic use (Fig. 4.1) (Dreaden et al. 2011; Dykman and Khlebtsov 2012; Yang et al. 2015; Bai et al. 2020). Besides this, the AuNPs due to their high specific surface area and easy modification by functional groups display potent antioxidant and antimicrobial effects without any host side effects (Gu et al. 2021). In agriculture, the antibacterial and antifungal activity of AuNPs finds use in protecting plants various against pathogens. Additionally, the AuNPs are also utilized for pesticide identification and water purification (Graily-Moradi et al. 2020).

4.3 Algae in Synthesis of Gold Nanoparticles

Algae have recently evoked interest among researchers as potential organisms for the biological synthesis of nanomaterials (Fig. 4.2). This is due to several reasons. Firstly, the algal cell wall is a rich source of functional groups and enzymes which act as reducing and capping agents in the nanofabrication process at ambient conditions (Crookes-Goodson et al. 2008; Siddiqi and Husen 2016). Secondly, algae can be conveniently cultivated with less expenditure in structures like open cultivation systems as well as closed cultivation systems like bioreactors (Narala et al. 2016). Both intracellular and extracellular production of nanoparticles has been reported in algae. Figure 4.3 depicts the steps involved in the production of

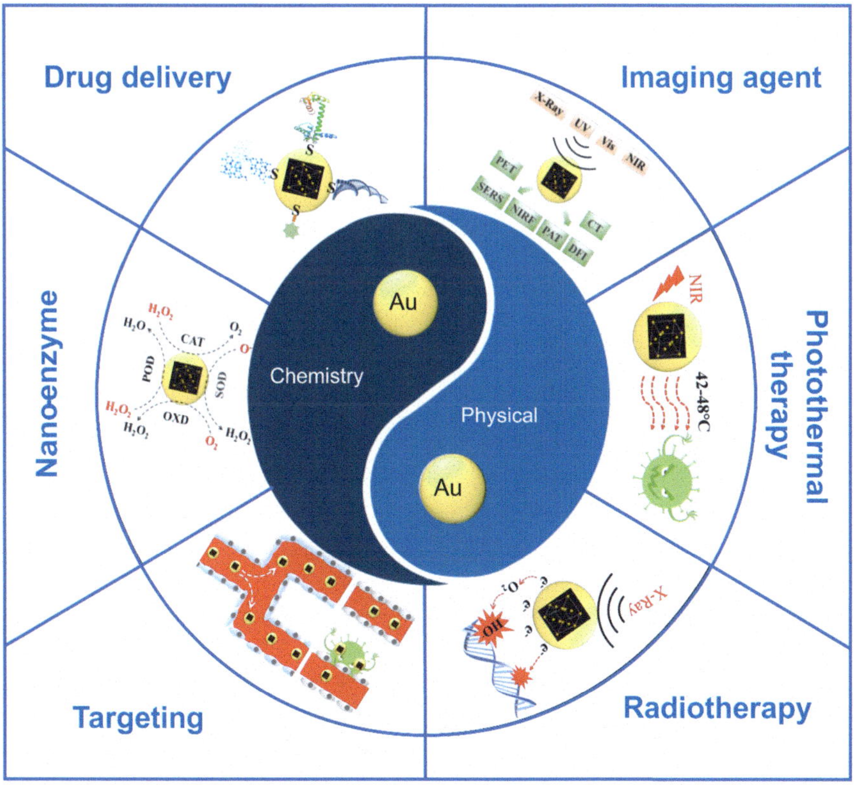

Fig. 4.1 Different applications of AuNPs in biomedical science. (Reprinted from Bai et al. 2020)

nanoparticles. Using algae for nanofabrication involves dissolution of algal extract in water or an organic solution followed by heating for a specific period of time. The filtrate is then mixed with metal solutions of desired molarity. Finally, the mixture is incubated for a specified time with or without stirring (Mukherjee et al. 2021).

The mechanism involved in AuNP synthesis is primarily mediated by the reduction of Au^{3+} ions to form AuNPs in which the enzymes and cell components are thought to play an important role leading to nanoparticle nucleation and growth (Ahmad et al. 2003; Duran et al. 2005; He et al. 2007).

4.4 Blue-Green Algae

Blue-green algae, also termed as cyanobacteria, are monophyletic microscopic photosynthetic entities that are found in fresh as well as brackish waters. Members of this primitive phylogenetically coherent group evolved 2.7–2.2 billion years ago and are thought to be responsible for the oxygenation of Earth's atmosphere during

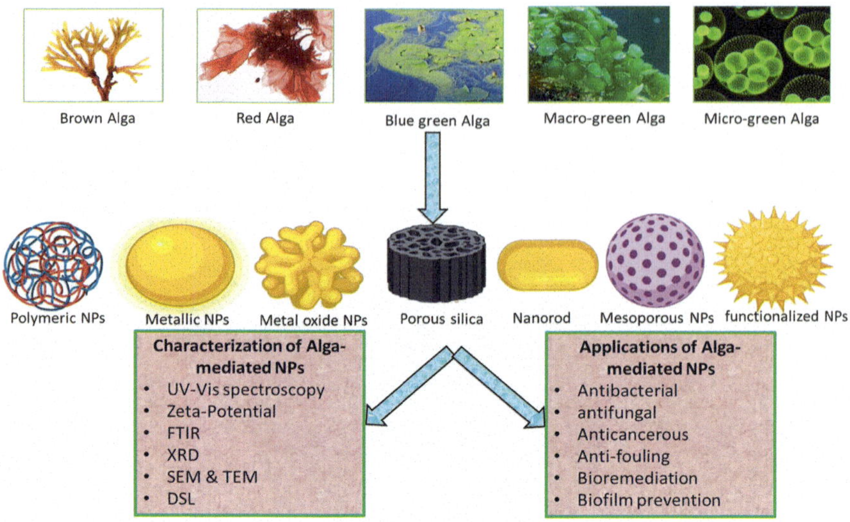

Fig. 4.2 Intracellular and extracellular synthesis of nanoparticles by microbes. (Reprinted with permission from Shedbalkar et al. 2014)

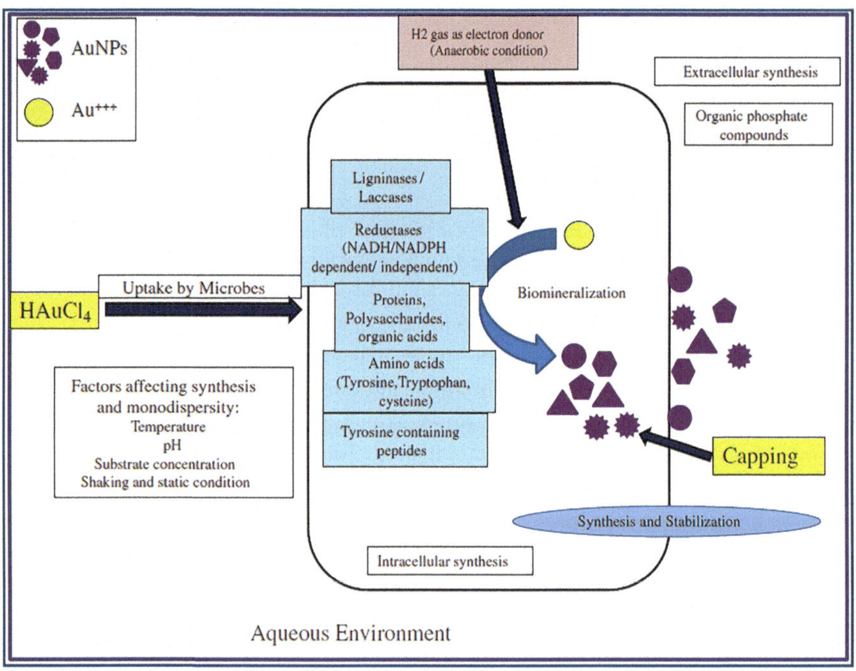

Fig. 4.3 Different types of nanoparticles synthesized by algae. (Reprinted from Chaudhary et al. 2020)

the evolutionary period (Stal 2007; Garcia-Pichel 2009). Blue-green algae cause blooms in water bodies which are of great significance in the ecology and management of many eutrophic fresh and brackish water bodies. These organisms are also known by other terms like Cyanophyceae, Cyanophyta, and Myxophyceae and range in width between 0.5 and 100 μm. This group contains specialized cells known as heterocysts, hormonia, and necridia which have diverse functions and are rarely found in other organisms. The distinguishing feature of the group is the lack of membrane-bound chloroplasts, presence of phycobiliproteins, cyanophycean starch as the storage product, and peptidoglycan walls (Srivastava and Bhargava 2021).

Cyanobacteria have become popular as promising nanoparticle-producing algae since they offer several advantages over other algal groups. These are their ubiquitous presence with the ability to survive in diverse natural habitats, easy culturing, capability of reducing metal ions in aerobic conditions, no requirement of extraneous reducing agent, and less time required for completing the reaction (Brayner et al. 2007; Flores and Herrero 2011; Dahoumane et al. 2012). Initial studies pointed out toward the role of nitrogenase enzymes and heterocysts in nanoparticle production. However, recent studies have pointed out that blue-green algae devoid of nitrogenases offer a better yield of nanoparticles (Dahoumane et al. 2012, 2014a, b). *Anabaena* has been the genus of choice for green synthesis of AuNPs. *Anabaena* is a nitrogen fixing filamentous blue-green algae with cylindrical, barrel-shaped, or spherical cells. A number of species of the genus have been successfully harnessed for biofabrication of AuNPs (Table 4.1). The dinitrogen-fixing photosynthetic blue-green alga *Anabaena laxa* was successfully utilized by Lenartowicz et al. (2017) to produce AuNPs using different concentrations of $HAuCl_4$ (0.1, 0.5, and 1.0 mM). Higher concentrations led to drastic decrease in chlorophyll content in the cells leading to high mortality. This showed high sensitivity of cyanobacterial cells toward gold ions. The formation of AuNPs was accomplished within 24 h of incubation at all three tested concentrations. The XRD pattern showed typical reflections for the crystalline gold structure for all the three concentrations. Lower concentrations led to formation of spherical AuNPs while triangular, hexagonal, and irregular AuNPs were formed at highest concentrations.

El-Sheekh et al. (2020) successfully accomplished biosynthesis of AuNPs using *Oscillatoria* sp. and *Spirulina platensis*. Octahedral, pentagonal, and triangular structures ranging in size from 15.60 to 77.13 nm were observed in the TEM images. The proteins and the polysaccharides present in the algal cell wall were thought to have played a major role in the nanofabrication process. The newly formed AuNPs were also tested for their antiviral activity against herpes simplex virus (HSV-1) using cell-line culture technique. The results show almost 90% reduction in the cytopathic effect (CPE) of the virus and proved the efficiency of AuNPs as inhibitory agents for the HSV-1 replication.

Table 4.1 Production of gold nanoparticles by blue-green algae, green algae, and diatoms

Algal group/ genus	Family	Nanoparticle size (nm)	Biological activity	Reference
Blue-green algae				
Anabaena sp.	Nostocaceae	9	nr	Rösken et al. (2014)
Anabaena cylindrica	Nostocaceae	10	nr	Rösken et al. (2016)
Anabaena flos-aquae	Nostocaceae	23–91	nr	Dahoumane et al. (2012)
Anabaena laxa	Nostocaceae	3–100	nr	Lenartowicz et al. (2017)
Anabaena sphaerica	Nostocaceae	1–50	nr	Roychoudhury et al. (2016)
Anabaena spiroides	Nostocaceae	<80	Antibacterial	Mandhata et al. (2021)
Calothrix	Rivulariaceae	5.4	nr	Brayner et al. (2007)
Cyanothece	Cyanothecaceae	80–129	Anti-myocardial infarction	Younis et al. (2019)
Leptolyngbya	Leptolyngbyaceae	100–200	Antibacterial	Zada et al. (2018)
Lyngbya majuscula	Oscillatoriaceae	<20	nr	Chakraborty et al. (2008)
		2–25	nr	Parial and Pal (2014b)
Phormidium tenue	Oscillatoriaceae	14.84	nr	Parial et al. (2012)
Phormidium valderianum	Oscillatoriaceae	7.92–24	nr	Parial et al. (2012)
Plectonema boryanum	Oscillatoriaceae	10–25	nr	Lengke et al. (2006)
Spirulina platensis	Oscillatoriaceae	~5	Antibacterial	Suganya et al. (2015)
Spirulina subsalsa	Oscillatoriaceae	5–30	nr	Parial and Pal (2014b)
Green algae				
Caulerpa racemosa	Caulerpaceae	5–25	Antibacterial	Kathiraven et al. (2015)
Chlamydomonas reinhardtii	Chlamydomonadaceae	42	Antimicrobial	Nguyen et al. (2018)
Chlorella pyrenoidosa	Oocystaceae	25–30	Nitrate reductase activity	Oza et al. (2012)
Chlorella vulgaris	Oocystaceae	2–10	Antibacterial; antifungal	Annamalai and Nallamuthu (2015)
Chlorococcum infusionum	Eriococcidae	2–52	nr	Roychoudhury et al. (2016)
Cosmarium impressulum	*Zygnematophyceae*	11–54	nr	Dahoumane et al. (2012)

(continued)

Table 4.1 (continued)

Algal group/ genus	Family	Nanoparticle size (nm)	Biological activity	Reference
Pithophora oedogonia	Cladophoraceae	25–44	Antibacterial	Sinha et al. (2015)
Rhizoclonium fontinale	Cladophoraceae	~16	nr	Parial and Pal (2014a)
Spirogyra submaxima	Zygnemataceae	2–50	nr	Roychoudhary and Pal (2014)
Tetraselmis kochinensis	Chlorophyceae	5–35	nr	Senapati et al. (2012)
Ulva intestinalis	Ulvaceae	42.39	nr	Parial et al. (2012)
Diatoms				
Aulacoseira	Aulacoseriaceae	30	Drug delivery	Briceño et al. (2021)
Diadesmis gallica	Diadesmidaceae	22	nr	Schröfel et al. (2011)
Eolimna minima	Naviculaceae	<100	nr	Feurtet-Mazel et al. (2016)
Navicula atomus	Naviculaceae	9	nr	Schröfel et al. (2011)
Nitzschia	Bacillariaceae	43	Antibacterial	Borase et al. (2017)
Stephanopyxis turris	Stephanopyxidaceae	10–30	nr	Pytlik et al. (2017)

4.5 Green Algae

Green algae, also known as Chlorophyta or Prasinophyta, is a taxonomically diverse group of autotrophs that comprise about 9000–12,000 species and are believed to be descendent of the original plastid-harboring eukaryote. Members of this group occur in freshwater, marine, as well as terrestrial environments. The group is characterized by the following (Kumar and Singh 1979; Naselli-Flores and Barone 2009):

1. Presence of chlorophyll a and b, carotenoids, and xanthophylls as major pigments
2. Cell wall made of cellulose
3. Starch as the storage product
4. Presence of whiplash, isokont flagella
5. Presence of pyrenoids

The emergence and proliferation of chlorophyta on land about 470 million years ago is considered as a remarkable moment in evolutionary history since they played a major role in the colonization and subsequent evolution of modern land plants. This led to significant changes in the atmosphere and terrestrial substrates, which facilitated the evolutionary process.

The green alga *Rhizoclonium fontinales* collected from Sundarbans (West Bengal, India) was explored for the biofabrication of AuNPs and the production monitored for diverse conditions, namely, variation in metal concentration (1, 5, 10, 15, 35, 55, 75, 95 mg/L), biomass concentration (1, 5, 15, 35, 55, 75, 95 mg/L), and pH (5, 7 and 9) (Parial and Pal 2014a). The results showed that AuNP synthesis took place at 10–55 mg/L gold solution (15 mg/L being the optimum) with higher concentrations having an inhibitory effect on the nanofabrication process. With respect to biomass, it was observed that increasing biomass ratio led to a decrease in AuNP synthesis. Acidic pH (pH 5) encouraged synthesis of AuNPs at a faster rate while neutral and basic pH levels led to decreased AuNP production. HRTEM studies confirmed the presence of AuNPs of different shapes and sizes like spherical (~5–20 nm), triangular (~15–88 nm), hexagons (~34 nm), and rod-shaped (~100 × 51.5 nm) at pH 5. Spherical nanoparticles were produced at near neutral pH, while increasing pH to 9 resulted in synthesis of monodisperse nanospheres after 3 days of salt exposure.

4.6 Diatoms

These are a monophyletic group of eukaryotic, single-celled heterokont algae belonging to Bacillariophyceae which are found in both brackish and fresh water (Sabater 2009; Mock and Medlin 2012). Diatoms have transparent, siliceous cell wall having perforation which allow the diffusion and excretion of materials. The group is characterized by chrysolaminarin as the reserve product and a perforated siliceous wall known as frustule. Diatoms form an important component of aquatic food webs and are considered as major players in vegetal primary production.

E. minima suspensions were treated with 20, 50, and 100 μg/mL $KAuCl_4$ with each reaction mixture containing 1.2×10^6 cells/mL of diatoms (Feurtet-Mazel et al. 2016). A change in color from light to red purple and SPR peak between 520 and 585 nm indicated toward successful production of AuNPs. TEM studies showed the presence of nanoparticles below 100 nm range both inside and outside of the diatom cells.

Borase et al. (2017) developed a facile method for the synthesis of AuNPs using the pennate diatom *Nitzschia* which is named after Christian Ludwig Nitzsch. A change in the color of reaction mixture to ruby red along with the absorbance peak at 529 nm indicated the formation of AuNPs. Electron microscopy studies showed the presence of irregularly shaped nanoparticles having an average size of 43 nm. The mechanistic studies pointed out the increased production of catalase and peroxidase on exposure to the gold salt which led to relieved reactive oxygen species (ROS) stress. The proteins and polysaccharides of the diatom played a major role in the reduction as well as the stabilization of nanoparticles. The AuNPs also exhibited strong antibacterial activity against several bacterial species (*Escherichia coli*,

Pseudomonas aeruginosa, and *Staphylococcus aureus*) on coupling them with antibiotics. In a more recent study, Briceño et al. (2021) synthesized *Aulacoseira* diatoms studded with AuNPs via in situ and ex situ method and used them as potential antibiotic delivery systems.

4.7 Brown Algae

Phaeophyceae or brown algae (or division Phaeophyta) are benthic multicellular macroalgae that comprise about 280 genera and 1800 species having complex morphologies. Most of the currently known species are marine with only less than 1% known to live in freshwater bodies. Brown algae have complex life cycles that exhibit sexual dimorphisms and alternation of generations (Luthringer et al. 2014). Members of this class have the following distinguishing traits (Wehr and Sheath 2003; Yoon et al. 2009):

- Presence of pigment fucoxanthin masks chlorophyll *a* and *c* giving the thallus a brownish appearance.
- Laminarin as the storage polysaccharide.
- Flagellated structures bearing a pair of laterally inserted unequal flagella, of which the anterior is larger and pantonematic, whereas the posterior one is smaller and acronematic.
- Cell wall composed of cellulose, fucinic acid, and alginic acid.

Brown algae have attracted attention for effective biosorption due to high metal uptakes in comparison to other living forms due to the presence of mucilaginous polysaccharides like alginate and sulfated fucoidans in their cell walls (Romera et al. 2007; Mata et al. 2008; Patil et al. 2018) which also confer numerous therapeutic properties to them. The biosorption potential is also aided by several functional groups, namely, carboxylic acid, amine, sulfhydryl, and sulfonates which aid in metal uptake (Raize et al. 2004; Jayakumar et al. 2021). Brown algae have also been successfully utilized for eco-friendly synthesis of AuNPs (Table 4.2).

Green synthesis of AuNPs was successfully accomplished using the marine algae, *Stoechospermum marginatum*, by Rajathi et al. (2012). This alga is found from Indian Ocean to the Pacific Ocean and is rich in sulfated fucans. The dried fronds of the alga were ground to powder and exposed to 1 mM of hydrogen tetrachloroaureate ($HAuCl_4$) solution. Within 10 min, a color change was observed from brown to red demonstrating the formation of AuNPs. It was observed in TEM analysis that the AuNPs were variously shaped ranging from spherical to hexagonal and triangular. The size of the nanoparticles was estimated to range between 18.7 and 93.7 nm in TEM and 40–85 nm in SEM images. The authors were of the view that the terpenoids and phenols present in the thallus played a possible role in the reduction of gold ions to gold nanoparticles.

Vijayan et al. (2014) developed a green method for the fabrication of AuNPs treating the aqueous extract of marine macroalgae *Turbinaria conoides* with

Table 4.2 Production of gold nanoparticles by brown and red algae

	Family	Nanoparticle size (nm)	Biological activity	Reference
Brown algae				
Cystoseira baccata	Cystoseiraceae	8.4	Anticancer	Gonzalez-Ballesteros et al. (2017)
Dictyopteris divaricata	Dictyotaceae	62.6	Neuroprotection	Young Park et al. (2019)
Ecklonia cava	Lessoniaceae	30	Antimicrobial	Venkatesan et al. (2014)
Ecklonia stolonifera	Lessoniaceae	27.9	nr	Jun et al. (2020)
Egregia sp.	Lessoniaceae	2–20	Antisenescence	Colin et al. (2018)
Fucus vesiculosus	Fucaceae	10–50	nr	Mata et al. (2009)
Laminaria japonica	Laminariaceae	15–20	nr	Ghodake and Lee (2011)
Padina gymnospora	Dictyotaceae	53–67	nr	Singh et al. (2013)
Padina pavonica	Dictyotaceae	30–100	Antibacterial	Isaac and Renitta (2015)
Padina tetrastromatica	Dictyotaceae	8–10	Antibacterial	Rajeshkumar (2017)
Sargassum muticum	Sargassaceae	5.42	nr	Namvar et al. (2015)
Sargassum myriocystum	Sargassaceae	15	nr	Stalin Dhas et al. (2012)
Sargassum swartzii	Sargassaceae	35	Anticancer	Dhas et al. (2014)
Sargassum wightii	Sargassaceae	8–12	nr	Singaravelu et al. (2007)
Stoechospermum marginatum	Dictyotaceae	18.7–93.7	nr	Rajathi et al. (2012)
Turbinaria conoides	Sargassaceae	2–19	nr	Vijayan et al. (2014)
Red algae				
Acanthophora spicifera	Rhodomelaceae	<20	Antibacterial; anti-oxidant; cytotoxic	Babu et al. (2020)
Chondrus crispus	Gigartinaceae	30–50	nr	Castro et al. (2013)
Corallina officinalis	Corallinaceae	14.6	Anticancer	El-Kassas and El-Sheekh (2014)
Galaxaura elongata	Galaxauraceae	3.85–77.13	Antibacterial	Abdel-Rouf et al. (2017)
Gelidium amansii	Gelidiaceae	4–7	Antimicrobial	Murphin Kumar et al. (2017)
Gracilaria corticata	Gracilariaceae	45–57	Antimicrobial; antioxidant	Naveen and Prakash (2013)

(continued)

Table 4.2 (continued)

	Family	Nanoparticle size (nm)	Biological activity	Reference
Iridaea cordata	Gigartinaceae	12.3	Anticancer	González-Ballesteros et al. (2021)
Kappaphycus alvarezii	Solieriaceae	10–40	nr	Rajasulochana et al. (2010)
Lemanea fluviatilis	Lemaneaceae	5–15	Antioxidant	Sharma et al. (2014)
Osmundaria obtusiloba	Rhodomelaceae	10–20	Optical sensing	Rojas-Perez et al. (2015)

$HAuCl_4 \cdot 3H_2O$. FTIR spectra showed the presence of different functional groups like O-H, C-OH, and C=O in the algal extract that played an important role in the reducing process and also served as capping agents. High-resolution transmission electron microscopy (HRTEM) revealed the presence of well dispersed, non-agglomerated small-sized triangular AuNPs ranging in size from 2 to 19 nm.

4.8 Red Algae

Rhodophyta or red algae are an anciently derived eukaryotic lineage of eukaryotic algae constituting an algal division that comprise the following characters (Garbary and Gabrielson 1990; Mayanglambam and Sahoo 2015):

- Total absence of flagella in all life stages
- Floridean starch and galactoside floridosides as the reserve food materials
- Presence of phycobiliprotein pigments like r-phycocyanin and r-phycoerythrin
- Un-stacked thylakoids
- Chloroplasts lacking an external endoplasmic reticulum

Members of this group are primarily marine in distribution and occur mostly in littoral and sublittoral areas of oceans, especially in warmer water bodies. Only about 3% of the total 6500 species occur in freshwater habitats (e.g., *Batrachospermum* and *Lemanea*) mainly as macroscopic and benthic forms.

The dried biomass of edible freshwater epilithic red alga, *Lemanea fluviatilis*, was used as both bioreductant and stabilizer for the green synthesis of gold nanoparticles using an aqueous solution of chloroauric acid solution (Sharma et al. 2014). The UV-visible spectra showed surface plasmon resonance (SPR) at ~530 nm. TEM studies concluded that the AuNPs were polydispersed and spherical and had size ranging from 5 to 15 nm, while studies using XRD confirmed the presence of face-centered cubic particles of the noble metal having average crystallite size of 5.9 nm

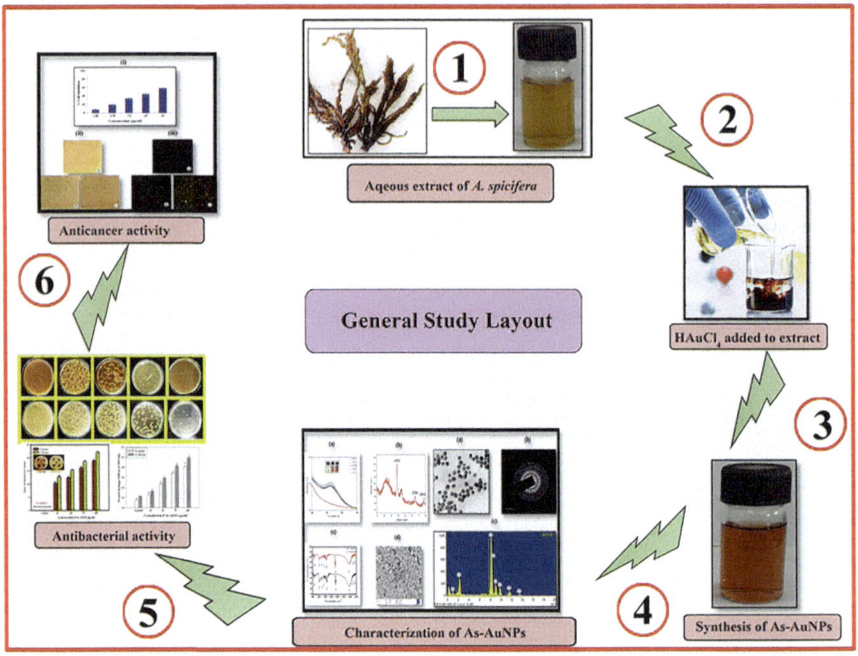

Fig. 4.4 Biofabrication of AuNPs using *Acanthophora spicifera*. (Reprinted with permission from Babu et al. 2020)

(Sharma et al. 2014). Significant antioxidant activity was exhibited by the gold nanoparticles using a modified DPPH assay.

Babu et al. (2020) used an environmentally benign process for the fabrication of pharmaceutically active AuNPs using *Acanthophora spicifera*, an erect macroalga that grows in upright clumps. The aqueous extract of the alga collected from the coastal region of Tamil Nadu (India) was treated with 1 M $HAuCl_4$ solution (Fig. 4.4), and the nanoparticles were characterized by several analytical techniques. The field-emission scanning electron microscope (FE-SEM) results confirmed that the AuNPs were spherical to oval in shape, while transmission electron microscopy estimated their average size <20 nm. The crystalline and face-centered cubic (fcc) structure was confirmed by X-ray diffraction (XRD) analysis, while their single crystallinity was confirmed by selected area electron difraction (SAED). The AuNPs exhibited high antibacterial activity against *Vibrio harveyi*, cytotoxicity against human colon adenocarcinoma (HT-29) cells (IC50 value of 21.86 μg/mL), and significant antioxidant activities which proved their broad-spectrum biological activities that can be effectively harnessed for various biomedical applications.

4.9 Applications of Gold Nanoparticles in Bioenergy Research

Biofuels have received increased attention among policy makers due to their cost-efficiency, accessibility, and eco-friendly nature (Kaparaju et al. 2009; Santoro et al. 2017; Sekoai et al. 2019). Several unique properties of nanoparticles have made them quite attractive as catalytic agents in the production process of biohydrogen, biogas, biodiesel, and bioethanol (Hussein 2015; Rai and Da Silva 2017). Gold nanoparticles can be of immense utility in generating hydrogen from water using solar energy. Gold nanoparticles find special mention here for harvesting aldehydes, esters, and carboxylic acids from different substrates. Gold nanoparticles owing to their large surface-area-to-volume-ratio induce stimulation of biohydrogen-producing processes especially the activity of biohydrogen-producing enzymes such as [Fe-Fe]- and [Ni-Fe]-hydrogenases and ferredoxins (proteins that mediate the transfer of electrons in biohydrogen producers) (Gordon and Seckbach 2012; Ramsurn and Gupta 2013). The first reports of using AuNPs for construction of direct methanol fuel cell (DMFC) were by Kuralkar et al. (2015) who used AuNPs as catalysts using two chambers (anodic chamber and cathodic chamber). DMFCs use methanol as the anode fuel to produce electricity and are popular due to their structural simplicity, rapid operation, and high energy density (Samimi and Rahimpour 2018; Alias et al. 2020). DMFCs are 15 times more efficient than a lithium-ion battery and find extensive use in portable electronic devices like laptop, mobile phone, and digital camera (Hsueh et al. 2012; Samimi and Rahimpour 2018). The use of AuNPs as catalysts had several advantages like being cheap in comparison to platinum, demonstration of high catalytic activity, and greater surface area for the attachment of methanol. The DMFC catalyzed by AuNPs showed potential in replacing lithium-ion batteries in the electronic industry since they have the ability to produce higher amount of energy in a small space. Another advantage pertains to its charging capacity since recharging is not required till there is ample supply of fuel and air to the DMFC.

Zhang and Shen (2007) were the pioneers to study the enhancement effect of nanometer-sized AuNP catalysts on hydrogen production in small batch reactors using artificial wastewater. The green synthesized AuNPs having diameters of 5 nm, 10 nm, and 20 nm enhanced production of hydrogen by about 50%, 40%, and 20%, respectively, indicating their role in enhancement of the biological activity of hydrogen-producing microbes. Naruse et al. (2011) synthesized gold nanoparticle-decorated functionalized multiwalled carbon nanotubes and explored them as anode in oxidation reactions for utilization in air-biofuel cells. The air-glucose fuel cell prototype using the hybrid nanostructured catalyst displayed an outstandingly high open circuit voltage (1.3 V) with a maximum power density. Mielby et al. (2014) successfully incorporated scores of AuNPs having diameter of 2–3 nm in zeolite silicate and reported high activity and selectivity of AuNPs for the catalytic gas phase oxidation of ethanol. The encapsulated AuNPs exhibited good stability and resulted in a 50% conversion of ethanol with 98% selectivity toward acetaldehyde at 200 °C of 220 $\mu W/cm^2$.

4.10 Conclusion

There has been increasing interest in the use of algae for gold nanoparticle production over the last few decades due to reasons like simplicity, safety, and eco-friendly approach that is devoid of use of toxic chemicals. Thus, algae can be considered as "nanobiofactories" for synthesis of AuNPs. However, their effective utilization requires the following conditions to be fulfilled:

1. In-depth analysis for better understanding of the precise mechanisms of nanoparticle formation with more emphasis on the mechanistic aspects
2. Standardization with respect to the algal species used, culture media, optimum growth stage, reaction conditions, substrate concentration, reaction time, pH, and temperature
3. Better understanding of conditions controlling the size, shape, and dispersity of the AuNPs
4. Exploring diverse uses of the biologically synthesized nanoparticles

References

Abdel-Rouf N, Al-Enazi NM, Ibraheem IBM (2017) Green biosynthesis of gold nanoparticles using *Galaxaura elongate* and characterization of their antibacterial activity. Arab J Chem 10(2): S3029–S3039

Ahmad A, Senapati S, Khan MI, Kumar R, Sastry M (2003) Extracellular biosynthesis of monodisperse gold nanoparticles by a novel extremophilic actinomycete, *Thermomonospora* sp. Langmuir 19:3550–3553

Alias MS, Kamarudin SK, Zainoodin AM, Masdar MS (2020) Active direct methanol fuel cell: an overview. Int J Hydrog Energy 45:19620–19641

Annamalai J, Nallamuthu T (2015) Characterization of biosynthesis gold nanoparticles from aqueous extract of *Chlorella vulgaris* and their antipathogenic properties. Appl Nanosci 5: 603–607

Babu B, Palanisamy S, Vinosha M, Anjali R, Kumar P, Pandi B, Tabarsa M, You S, Prabhu NM (2020) Bioengineered gold nanoparticles from marine seaweed *Acanthophora spicifera* for pharmaceutical uses: antioxidant, antibacterial, and anticancer activities. Bioprocess Biosyst Eng 43(12):2231–2242

Bai X, Wang Y, Song Z, Feng Y, Chen Y, Zhang D, Feng L (2020) The basic properties of gold nanoparticles and their applications in tumor diagnosis and treatment. Int J Mol Sci 21(7):2480

Borase HP, Patil CD, Suryawanshi RK, Koli SH, Mohite BV, Benelli G, Patil SV (2017) Mechanistic approach for fabrication of gold nanoparticles by *Nitzschia* diatom and their antibacterial activity. Bioprocess Biosyst Eng 40(10):1437–1446

Brayner R, Barberousse H, Hernadi M, Djedjat C, Yéprémian C, Coradin T, Livage J, Fiévet F, Couté A (2007) Cyanobacteria as bioreactors for the synthesis of Au, Ag, Pd, and Pt nanoparticles via an enzyme-mediated route. J Nanosci Nanotechnol 7:2696–2708

Briceño S, Chavez-Chico EA, González G (2021) Diatoms decorated with gold nanoparticles by in-situ and ex-situ methods for in vitro gentamicin release. Mater Sci Eng C 123:112018

Castro L, González F, Blázquez ML, Muñoz JA, Ballester A (2013) Biological synthesis of metallic nanoparticles using algae. IET Nanobiotechnol 7(3):109–116

Chakraborty N, Banerjee A, Lahiri S, Panda A, Ghosh AN, Pal R (2008) Biorecovery of gold using cyanobacteria and a eukaryotic alga with special reference to nanogold formation- a novel phenomenon. J Appl Phycol 21(1):145–152

Chaudhary R, Nawaz K, Khan AK, Hano C, Abbasi BH, Anjum S (2020) An overview of the algae-mediated biosynthesis of nanoparticles and their biomedical applications. *Biomolecules* 10(11): 1498

Colin JA, Pech-Pech IE, Oviedo M, Águila SA, Romo-Herrera JM, Contreras OE (2018) Gold nanoparticles synthesis assisted by marine algae extract: biomolecules shells from a green chemistry approach. Chem Phys Lett 708:210–215

Crookes-Goodson WJ, Slocik JM, Naik RR (2008) Bio-directed synthesis and assembly of nanomaterials. Chem Soc Rev 37:2403–2412

Dahoumane SA, Djediat C, Yéprémian C, Couté A, Fiévet F, Coradin T, Brayner R (2012) Species Selection for the design of gold nanobioreactor by photosynthetic organisms. J Nanopart Res 14:883–889

Dahoumane SA, Yéprémian C, Djédiat C (2014a) A global approach of the mechanism involved in the biosynthesis of gold colloids using micro-algae. J Nanopart Res 16:2607

Dahoumane SA, Wijesekera K, Filipe CD, Brennan JD (2014b) Stoichiometrically controlled production of bimetallic gold-silver alloy colloids using micro-alga cultures. J Colloid Interface Sci 416:67–72

Daniel MC, Astruc D (2004) Gold nanoparticles: assembly, supramolecular chemistry, quantum-size-related properties, and applications toward biology, catalysis, and nanotechnology. Chem Rev 104:293

Das M, Shim KH, An SSA, Yi DK (2011) Review on gold nanoparticles and their applications. Toxicol Environ Heal Sci 3(4):193–205

Dhas TS, Kumar VG, Karthick V, Govindaraju K, Shankara Narayana T (2014) Biosynthesis of gold nanoparticles using *Sargassum swartzii* and its cytotoxicity effect on HeLa cells. Spectrochim Acta A Mol Biomol Spectrosc 133:102–106

Dreaden EC, Alkilany A, Huang X, Murphy C, El-Sayed MA (2011) The golden age: gold nanoparticles for biomedicine. Chem Soc Rev 41:2740–2779

Duran N, Marcato PD, Alves OL, D'Souza G, Esposito E (2005) Mechanistic aspects of biosynthesis of silver nanoparticles by several *Fusarium oxysporum* strains. J Nanobiotechnol 3:1–7

Dykman L, Khlebtsov NG (2012) Gold nanoparticles in biomedical applications: recent advances and perspectives. Chem Soc Rev 41:2256–2282

El-Kassas HY, El-Sheekh MM (2014) Cytotoxic activity of biosynthesized gold nanoparticles with an extract of the red seaweed *Corallina officinalis* on the MCF-7 human breast cancer cell line. Asian Pac J Cancer Prev 15(10):4311–4317

El-Sheekh MM, Shabaan MT, Hassan L, Morsi HH (2020) Antiviral activity of algae biosynthesized silver and gold nanoparticles against Herps Simplex (HSV-1) virus in vitro using cell-line culture technique. Int J Environ Health Res 6:1–12

Feurtet-Mazel A, Mornet S, Charron L, Mesmer-Dudons N, Maury-Brachet R, Baudrimont M (2016) Biosynthesis of gold nanoparticles by the living freshwater diatom *Eolimna minima*, a species developed in river biofilms. Environ Sci Pollut Res Int 23(5):4334–4339

Flores E, Herrero A (2011) Compartmentalized function through cell differentiation in filamentous cyanobacteria. Nat Rev Microbiol 8:39–50

Garbary DJ, Gabrielson PW (1990) Taxonomy and evolution. In: Cole KM, Sheath RG (eds) Biology of the red algae. Cambridge University Press, Cambridge, pp 477–498

Garcia-Pichel F (2009) Cyanobacteria. In: Schaechter M (ed) Encyclopedia of microbiology. Elsevier, Amsterdam, pp 107–124

Ghodake G, Lee DS (2011) Biological synthesis of gold nanoparticles using the aqueous extract of the brown algae *Laminaria japonica*. J Nanoelectron Optoelectron 6:268–271

Gonzalez-Ballesteros N, Prado-Lopez S, Rodriguez-Gonzalez JB, Lastra M, Rodriguez-Arguelles MC (2017) Green synthesis of gold nanoparticles using brown algae *Cystoseira baccata*: its activity in colon cancer cells. Colloids Surf B: Biointerfaces 153:190–198

González-Ballesteros N, Rodríguez-Argüelles MC, Lastra-Valdor M (2021) Evaluation of the antioxidant capacities of Antarctic macroalgae and their use for nanoparticles production. Molecules 26(4):1182

Gordon R, Seckbach J (2012) The science of algal fuels. Springer, Dordrecht

Graily-Moradi F, Maadani Mallak A, Ghorbanpour M (2020) Biogenic synthesis of gold nanoparticles and their potential application in agriculture. In: Ghorbanpour M, Bhargava P, Varma A, Choudhary D (eds) Biogenic nano-particles and their use in agro-ecosystems. Springer, Singapore, pp 187–204

Gu X, Xu Z, Gu L, Han F, Chen B, Pan X (2021) Preparation and antibacterial properties of gold nanoparticles: a review. Environ Chem Lett 19:167–187

He S, Guo Z, Zhang Y, Zhang S, Wang J, Gu N (2007) Biosynthesis of gold nanoparticles using the bacteria *Rhodopseudomonas capsulate*. Mater Lett 61:3984

Henz BJ, Hawa T, Zachariah MR (2008) Mechano-chemical stability of gold nanoparticles coated with alkanethiolate SAMs. Langmuir 24:773

Hsueh KL, Tsai LD, Lai CC, Peng YM (2012) Direct methanol fuel cells. Electrochem Technol Energy Storage Convers 1–2:701–727

Hussein A (2015) Applications of nanotechnology in renewable energies-a comprehensive overview and understanding. Renew Sust Energ Rev 42:460–476

Isaac G, Renitta RE (2015) Brown algae mediated synthesis, characterization of gold nanoparticles using *Padina pavonica* and their antibacterial activity against human pathogen. Int J Pharm Tech Res 8(9):31–40

Jain PK, Lee KS, El-Sayed IH, El-Sayed MA (2006) Calculated absorption and scattering properties of gold nanoparticles of different size, shape, and composition: applications in biological imaging and biomedicine. J Phys Chem B 110(14):7238–7248

Jayakumar V, Govindaradjane S, Rajamohan N, Rajasimman M (2021) Biosorption potential of brown algae, *Sargassum polycystum*, for the removal of toxic metals, cadmium and zinc. Environ Sci Pollut Res 29:41909. https://doi.org/10.1007/s11356-021-15185-7

Jun ES, Kim YJ, Kim HH, Park SY (2020) Gold nanoparticles using *Ecklonia stolonifera* protect human dermal fibroblasts from UVA-induced senescence through inhibiting MMP-1 and MMP-3. Mar Drugs 18(9):433

Kannan N, Subbalaxmi S (2011) Biogenesis of nanoparticles - a current perspective. Rev Adv Mater Sci 27:99

Kaparaju P, Serrano M, Thomsen A, Kongjan I, Angelidaki P (2009) Bioethanol, biohydrogen and biogas production from wheat straw in a biorefinery concept. Bioresour Technol 100:2562–2568

Kashani AS, Kuruvinashetti K, Beauet D, Badilescu S, Piekny A, Packirisamy M (2018) Enhanced internalization of Indian Ayurvedic Swarna Bhasma (gold nanopowder) for effective interaction with human cells. J Nanosci Nanotechnol 18(10):6791–6798

Kathiraven T, Sundaramanickam A, Shanmugam N, Balasubramanian T (2015) Green synthesis of silver nanoparticles using marine algae *Caulerpa racemosa* and their antibacterial activity against some human pathogens. Appl Nanosci 5:499–504

Kumar HD, Singh HN (1979) Chlorophyta. A textbook on algae. Palgrave, London

Kumar D, Saini N, Jain N, Sareen R, Pandit V (2013) Gold nanoparticles: an era in bionanotechnology. Expert Opin Drug Deliv 10(3):397–409

Kuralkar M, Gaikwad S, Rai M, Gade A, Ingle A (2015) Gold nanoparticles: novel catalyst for the preparation of direct methanol fuel cell. IET Nanobiotechnol 9(2):66–70

Lenartowicz M, Marek PH, Madura ID, Lipok J (2017) Formation of variously shaped gold nanoparticles by *Anabaena laxa*. J Clust Sci 28:3035–3055

Lengke MF, Fleet ME, Southam G (2006) Morphology of gold nanoparticles synthesized by filamentous cyanobacteria from gold(I)-thiosulfate and gold(III)-chloride complexes. Langmuir 22(6):2780–2787

Luthringer R, Cormier A, Ahmed S, Peters AF, Cock JM, Coelho SM (2014) Sexual dimorphism in the brown algae. Perspect Phycol 1:11–25

Mandhata CP, Sahoo CR, Mahanta CS, Padhy RN (2021) Isolation, biosynthesis and antimicrobial activity of gold nanoparticles produced with extracts of *Anabaena spiroides*. Bioprocess Biosyst Eng 44(8):1617–1626

Mata Y, Blázquez M, Ballester A, González F, Munoz J (2008) Characterization of the biosorption of cadmium, lead and copper with the brown alga *Fucus vesiculosus*. J Hazard Mater 158(2): 316–323

Mata YN, Torres E, Blazquez ML, Ballester A, Gonzalez F, Munoz JA (2009) Gold(III) biosorption and bioreduction with the brown alga *Fucus vesiculosus*. J Hazard Mater 166:612–618

Mayanglambam A, Sahoo D (2015) Red algae. In: Sahoo D, Seckbach J (eds) The algae world. Cellular origin, life in extreme habitats and astrobiology, vol 26. Springer, Dordrecht

Mielby J, Abildstrøm JO, Wang F, Kasama T, Weidenthaler C, Kegnæs S (2014) Oxidation of bioethanol using Zeolite-encapsulated gold nanoparticles. Angew Chem Int Ed 126(46): 12721–12712

Mock T, Medlin LK (2012) Genomics and genetics of diatoms. In: Piganeau G (ed) Advances in botanical research. Elsevier, Amsterdam, pp 245–284

Mukherjee A, Sarkar D, Sasmal S (2021) A review of green synthesis of metal nanoparticles using algae. Front Microbiol 12:693899

Murphin Kumar PS, Mubarak Ali D, Saratale RG, Saratale GD, Pugazhendhi A, Gopalakrishnan K, Thajuddin N (2017) Synthesis of nano-cuboidal gold particles for effective antimicrobial property against clinical human pathogens. Microb Pathog 113:68–73

Nafisi S, Maibach HI (2017) Nanotechnology in cosmetics. In: Sakamoto K, Lochhead RY, Maibach HI, Yamashita Y (eds) Cosmetic science and technology: theoretical principles and applications. Elsevier, Amsterdam, pp 337–369

Namvar F, Azizi S, Ahmad MB, Shameli K, Mohamad R, Mahdavi M, Tahir PM (2015) Green synthesis and characterization of gold nanoparticles using the marine macroalgae Sargassum muticum. Res Chem Intermed 41:5723–5730

Narala RR, Garg S, Sharma KK, Thomas-Hall SR, Deme M, Li Y, Schenk PM (2016) Comparison of microalgae cultivation in photobioreactor, open raceway pond, and a two-stage hybrid system. Front Energy Res 4:29

Naruse J, Hoa LQ, Sugano Y, Ikeuchi T, Yoshikawa H, Saito M, Tamiya E (2011) Development of biofuel cells based on gold nanoparticle decorated multi-walled carbon nanotubes. Biosens Bioelectron 30(1):204–210

Naselli-Flores L, Barone R (2009) Green algae. In: Likens GE (ed) Encyclopedia of inland waters, vol 1. Elsevier, Oxford, pp 166–173

Naveen BE, Prakash S (2013) Biological synthesis of gold nanoparticles using algae *Gracilaria corticata* and its application as a potent antimicrobial and antioxidant agent. Asian J Pharm Clin Res 6(2):179–182

Nguyen NHA, Padil VVT, Slaveykova VI, Černík M, Ševců A (2018) Green synthesis of metal and metal oxide nanoparticles and their effect on the unicellular alga *Chlamydomonas reinhardtii*. Nanoscale Res Lett 13(1):159

Oza G, Pandey S, Mewada A, Kalita G, Sharon M (2012) Facile biosynthesis of gold nanoparticles exploiting optimum pH and temperature of fresh water algae *Chlorella pyrenoidusa*. Adv Appl Sci Res 3(3):1405–1412

Parial D, Pal R (2014a) Biosynthesis of monodisperse gold nanoparticles by green alga *Rhizoclonium* and associated biochemical changes. J Appl Phycol 27(2):975–984

Parial D, Pal R (2014b) Green synthesis of gold nanoparticles using Cyanobacteria and their characterization. Indian J Appl Res 4:69–72

Parial D, Patra HK, Dasgupta AK, Pal R (2012) Screening of different algae for green synthesis of gold nanoparticle. Eur J Phycol 47:22–29

Patil NP, Le V, Sligar AD, Mei L, Chavarria D, Yang EY, Baker AB (2018) Algal polysaccharides as therapeutic agents for atherosclerosis. Front Cardiovasc Med 5:153

Pytlik N, Kaden J, Finger M, Naumann J, Wanke S, Machill S, Brunner E (2017) Biological synthesis of gold nanoparticles by the diatom *Stephanopyxis turris* and in vivo SERS analyses. Algal Res 28:9–15
Rai M, Da Silva SS (2017) Nanotechnology for bioenergy and biofuel production. Springer, Dordrecht
Raize O, Argaman Y, Yannai S (2004) Mechanisms of biosorption of different heavy metals by brown marine algae. Biotechnol Bioeng 87(4):451–458
Rajasulochana P, Dhamotharan R, Murugakoothan P, Murugesan S, Krishnamoorthy P (2010) Biosynthesis and characterization of gold nanoparticles using the alga *Kappaphycus alvarezii*. Int J Nanosci 9:511–516
Rajathi FAA, Parthiban C, Ganesh Kumar V, Anantharaman P (2012) Biosynthesis of antibacterial gold nanoparticles using brown alga, *Stoechospermum marginatum* (kützing). Spectrochim Acta A Mol Biomol Spectrosc 99:166–173
Rajeshkumar S (2017) Phytochemical constituents of fucoidan (*Padina tetrastromatica*) and its assisted AgNPs for enhanced antibacterial activity. IET Nanobiotechnol 11(3):292–299
Ramsurn H, Gupta R (2013) Nanotechnology in solar and biofuels. ACS Sustain Chem Eng 1:779–797
Rojas-Perez A, Adorno L, Cordero M, Ruiz A, Mercado-Diaz Z, Rodriguez A, Betancourt L, Velez C, Feliciano I, Cabrea C, Diaz Vazquez LM (2015) Biosynthesis of gold nanoparticles using *Osmudaria obtusiloba* extract and their potential use in optical sensing application. Austin J Biosens Bioelectron 1(5):1–9
Romera E, González F, Ballester A, Blázquez M, Munoz J (2007) Comparative study of biosorption of heavy metals using different types of algae. Bioresour Technol 98(17):3344–3353
Rösken LM, Cappel F, Körsten S, Fischer CB, Schönleber A, van Smaalen S, Geimer S, Beresko C, Ankerhold G, Wehner S (2014) Time-dependent growth of crystalline Au^0-nanoparticles in cyanobacteria as self-reproducing bioreactors: 1. *Anabaena* sp. J Nanopart Res 16:2370
Rösken LM, Cappel F, Körsten S, Fischer CB, Schönleber A, van Smaalen S, Geimer S, Beresko C, Ankerhold G, Wehner S (2016) Time-dependent growth of crystalline Au^0-nanoparticles in cyanobacteria as self-reproducing bioreactors: 2. *Anabaena cylindrica*. Beilstein J Nanotechnol 7:312–327
Roychoudhary P, Pal R (2014) *Spirogya submaxima*-a green Alga for nanogold production. J Algal Biomass Utln 5(1):15–19
Roychoudhury P, Bhattacharya A, Dasgupta A, Pal R (2016) Biogenic synthesis of gold nanoparticle using fractioned cellular components from eukaryotic algae and cyanobacteria. Phycol Res 64(3):133–140
Sabater S (2009) Diatome. In: Likens GE (ed) Encyclopedia of inland waters. Elsevier, Amsterdam, pp 149–156
Samimi F, Rahimpour MR (2018) Direct methanol fuel cell. In: Basile A, Dalena F (eds) Methanol science and engineering. Elsevier, Amsterdam, pp 381–397
Santoro C, Arbizzani C, Erable B, Ieropoulos I (2017) Microbial fuel cells: from fundamentals to applications. A review. J Power Sources 356:225–244
Schröfel A, Kratosova G, Bohunicka M, Dobrocka E, Vavra I (2011) Biosynthesis of gold nanoparticles using diatoms-silica-gold and EPS-gold bionanocomposite formation. J Nanopart Res 13(8):3207–3216
Sekoai PT, Ouma CNM, du Preez SP, Modisha P, Engelbrecht N, Bessarabov DG, Ghimire A (2019) Application of nanoparticles in biofuels: an overview. Fuel 237:380–397
Senapati S, Syed A, Moeez S, Kumar A, Ahmad A (2012) Intracellular synthesis of gold nanoparticles using alga *Tetraselmis kochinensis*. Mater Lett 79:116–118
Sharma B, Purkayastha DD, Hazra S, Thajamanbi M, Bhattacharjee CR, Ghosh NN, Rout J (2014) Biosynthesis of fluorescent gold nanoparticles using an edible freshwater red alga, *Lemanea fluviatilis* (L.) C.Ag. and antioxidant activity of biomatrix loaded nanoparticles. Bioprocess Biosyst Eng 37(12):2559–2565

Shedbalkar U, Singh R, Wadhwani S, Gaidhani S, Chopade BA (2014) Microbial synthesis of gold nanoparticles: current status and future prospects. Adv Colloid Interf Sci 209:40–48

Shukla R, Bansal V, Chaudhary M, Basu A, Bhonde RR, Sastry M (2005) Biocompatibility of gold nanoparticles and their endocytotic fate inside the cellular compartment: a microscopic overview. Langmuir 21:10644

Siddiqi KS, Husen A (2016) Fabrication of metal and metal oxide nanoparticles by algae and their toxic effects. Nanoscale Res Lett 11(1):363

Singaravelu G, Arockiamary JS, Kumar VG, Govindaraju K (2007) A novel extracellular synthesis of monodisperse gold nanoparticles using marine alga, *Sargassum wightii* Greville. Colloids Surf B: Biointerfaces 57(1):97–101

Singh M, Kalaivani R, Manikandan S, Saneetha N, Kumaraguru AK (2013) Facile green synthesis of variable metallic gold nanoparticles using *Padina gymnospora*, a brown marine macroalga. Appl Nanosci 3:145–151

Sinha SN, Paul D, Halder N, Sengupta D, Patra SK (2015) Green synthesis of silver nanoparticles using fresh water green alga *Pithophora oedogonia* (Mont.) Wittrock and evaluation of their antibacterial activity. Appl Nanosci 5:703–709

Srivastava S, Bhargava A (2021) Green nanoparticles: the future of nanobiotechnology. Springer Nature, Cham. ISBN: 978-981-16-7105-0

Srivastava S, Frankamp BL, Rotello VM (2005) Controlled plasmon resonance of gold nanoparticles self-assembled with PAMAM dendrimers. Chem Mater 17:487–490

Stal LJ (2007) Cyanobacteria. In: Seckbach J (ed) Algae and cyanobacteria in extreme environments. Cellular origin, life in extreme habitats and astrobiology, vol 11. Springer, Dordrecht

Stalin Dhas T, Ganesh Kumar V, Stanley Abraham L, Karthick V, Govindaraju K (2012) *Sargassum myriocystum* mediated biosynthesis of gold nanoparticles. Spectrochim Acta Part A Mol Biomol Spectrosc 99:97–101

Suganya KS, Govindaraju K, Kumar VG, Dhas TS, Karthick V, Singaravelu G, Elanchezhiyan M (2015) Blue green alga mediated synthesis of gold nanoparticles and its antibacterial efficacy against Gram positive organisms. Mater Sci Eng C Mater Biol Appl 47:351–356

Taufikurohmah T, Sanjaya IGM, Syahrani A (2011) Nanogold synthesis using matrix mono glyceryl stearate as antiaging compounds in modern cosmetics. J Mater Sci Eng A 1:857–864

Venkatesan J, Manivasagan P, Kim SK, Kirthi AV, Marimuthu S, Rahuman AA (2014) Marine algae-mediated synthesis of gold nanoparticles using a novel *Ecklonia cava*. Bioprocess Biosyst Eng 37(8):1591–1597

Vijayan SR, Santhiyagu P, Singamuthu M, Ahila NK, Jayaraman R, Ethiraj K (2014) Synthesis and characterization of silver and gold nanoparticles using aqueous extract of seaweed, *Turbinaria conoides*, and their antimicrofouling activity. Sci World J 2014:938272

Wehr JD, Sheath RG (2003) Introduction to freshwater algae. In: Wehr JD, Sheath RG (eds) Freshwater algae of North America: ecology and classification. Elsevier, Amsterdam, pp 1–9

Yang X, Yang M, Pang B, Vara M, Xia Y (2015) Gold nanomaterials at work in biomedicine. Chem Rev 115:10410–10488

Yapar EA, Inal O (2012) Nanomaterials and cosmetics. J Fac Pharm Istanbul Univ 42:43–70

Yoon HS, Andersen RA, Boo SM, Bhattacharya D (2009) Stramenophiles. In: Schaecter M (ed) Encyclopedia of microbiology. Elsevier, Amsterdam, pp 721–731

Young Park S, Jin Kim Y, Park G, Kim HH (2019) Neuroprotective effect of *Dictyopteris divaricata* extract-capped gold nanoparticles against oxygen and glucose deprivation/reoxygenation. Colloids Surf B: Biointerfaces 179:421–428

Younis NS, Bakir EM, Mohamed ME, El Semary NA (2019) Cyanobacteria as nanogold factories II: chemical reactivity and anti-myocardial infraction properties of customized gold nanoparticles biosynthesized by *Cyanothece* sp. Mar Drugs 17(7):402

Zada S, Ahmad A, Khan S, Iqbal A, Ahmad S, Ali H, Fu P (2018) Biofabrication of gold nanoparticles by *Lyptolyngbya* JSC-1 extract as super reducing and stabilizing agents: synthesis, characterization and antibacterial activity. Microb Pathog 114:116–123

Zhang Y, Shen J (2007) Enhancement effect of gold nanoparticles on biohydrogen production from artificial wastewater. Int J Hydrog Energy 32(1):17–23

Chapter 5
Challenges Assessment in Economic Algal Biofuel Production

S. M. Bhatt

Abstract Microalgal biofuels are known to be the best alternative to fossil fuels to enhance the required demand worldwide to cut down the cost. Lipid quality decides biodiesel production, and other factors like conversion technology, process efficiency, and cost-effectiveness are also a decisive factor. Economically production, suitable bioreactor may make it more efficient for production of it more feasible.

Biodiesel production from bio-oil is a major challenge, but challenges can be solved by specific approaches (Khan et al. 2017); oil is basically present up to 20% and can be increased under specific condition, up to 80%. Some species like *Ulothrix*, *Tribonema*, and *Euglena* have good potential for biodiesel production. Genetic engineering approaches have more opportunity for enhanced expression of gene.

For Biodiesel production, various optimization strategies have been used like adaptive neuro-fuzzy inference system (ANFIS) (Kumar 2020) and various photobioreactor like airlift, bubble column, and flat panel (Banerjee et al. 2019).

Algae have around 80% energy as compared to petroleum. Microalgae have around 30–40% more energy as compared to palm oil. *Botryococcus braunii* is supposed to have 40% more hydrocarbon. Production of algae is fast, convenient, and easy.

S. M. Bhatt (✉)
Department of Agriculture, AGC, Amritsar, Punjab, India

N. Srivastava, P. K. Mishra (eds.), *Technological Advancement in Algal Biofuels Production*, Clean Energy Production Technologies,
https://doi.org/10.1007/978-981-19-6806-8_5

5.1 Introduction

Algae are found ubiquitously in pond lakes, rivers, oceans, and wastewater released (González-Camejo et al. 2019). They are growing in variety of condition such as temperature, pH value, saline condition, and different light intensities. Microalgae have been classified on the basis of size, since they are microscopic single cells that may be similar to cyanobacteria (chloroxybacteria) or eukaryotic, similar to green algae (*Chlorophyta*) (Figs. 5.1, 5.2, 5.3, and 5.4; Tables 5.1 and 5.2).

List of microalgae

1. *Botryococcus braunii*
2. *Nannochloropsis* sp.
3. *Dunaliella primolecta*
4. *Chlorella* sp.
5. *Crypthecodinium cohnii*

5.1.1 Biofuel Production Process (Fig. 5.5)

With traditional methods, biomass is obtained in very small quantities, making biofuel at an economically feasible proposition (Fig. 5.6).

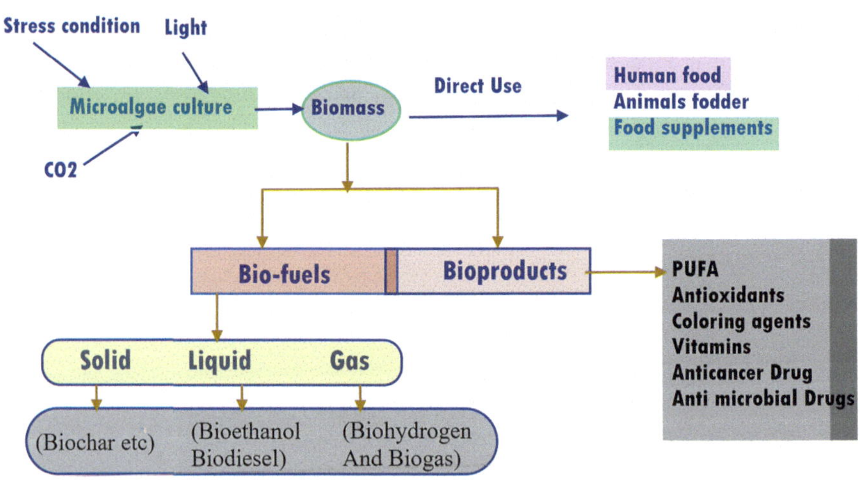

Fig. 5.1 Depicting various use of microalgae. (Adapted from M. I. Khan et al. 2018)

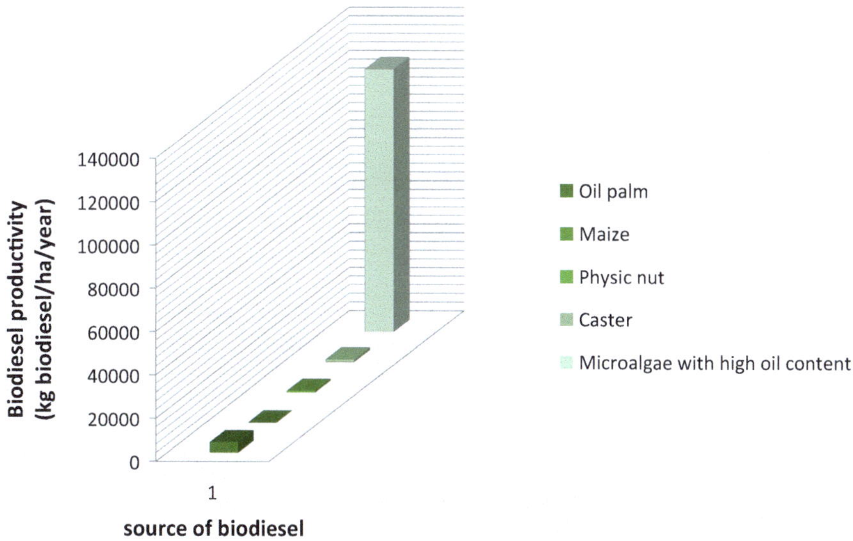

Fig. 5.2 Source of biodiesel production (Medipally et al. 2015)

5.2 Strain Improvement

Major bottleneck in commercialization of microalgae-based biodiesel production is the low productivity of bio-oil obtained from microalgae. Therefore, many scientist focus on numerous level of improvements such as (1) photosynthetic conversion economy in mass cultivation and (2) metabolic pathway modification (Kumar et al. 2020). Mainly various gene editing tools also have been used to improve the strain.

Some worker has worked on high CO_2 fixation (Cheng et al. 2019), while some prefer random mutagenesis (Arora and Philippidis 2021).

5.2.1 Cultivation Systems

5.2.1.1 Biofuel Production Process

1. Culturing of algae
2. Harvesting of algae or dewatering of algae
3. Oil extraction from algae
4. Purification of oil
5. Processing of oil into biofuels

A comparative table has been presented in Table 5.3; open pond (OP) has high water requirement, high CO_2, and high physical footprint; pH/light/temperature control is

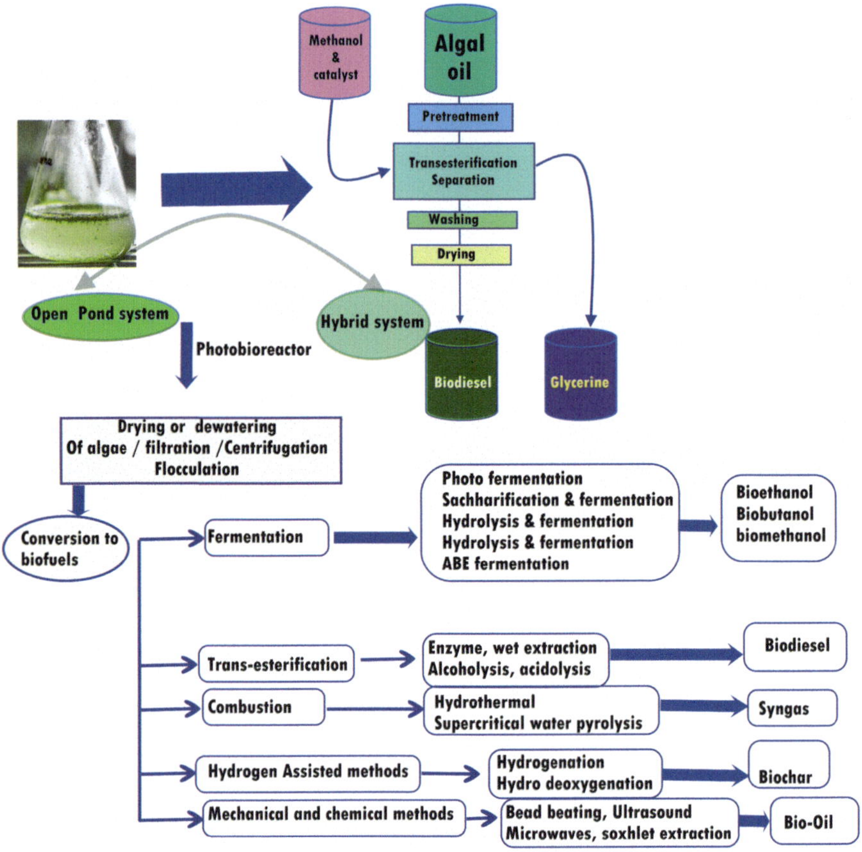

Fig. 5.3 Conversion method for different byproduct from microalgae such as bioethanol, biobutanol, biomethanol, biodiesel, syngas. Modified from (Ganesan et al. 2020)

difficult, while PBR has high capital expenditures (CAPEX) and high operating expense (OPEX cost). Lighting requirement is high (Egbo et al. 2018).

As shown in Table 5.4 compared PBR with hybrid system and raceway pond, economy of algae production is low in raceway pond due to high evaporating loss, contamination risk is high, and maintenance cost is low, while biomass quality is variable CO_2. In hybrid system, setup cost is moderate as compared to PBR, while except in maintenance cost, everything is comparable. PBR system is batch type while in hybrid system it's continuous.

Increasing lipid content is challenge for biodiesel production without losing the growth of algae. This is essential for economic viability and enhanced by biofuel production. Few work reported to enhance lipid production via two-stage cultivation strategy (Aziz et al. 2020).

One of the best part of microalgae is that they grow fast and produces high amount of lipid (Pugazhendhi et al. 2020). Yet biodiesel production via microalgae

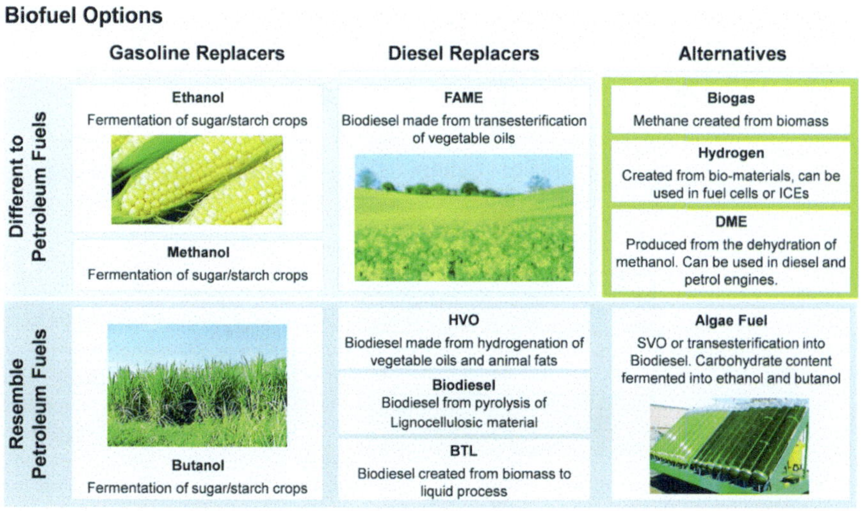

Fig. 5.4 Biofuel classification as fermentation based on (1) ethanol, methanol; (2) some diesel replacers are made by different route like FAME, HVO, and BTL, while some best alternatives are (3) biogas, hydrogen, DME, and algae biofuel-based biodiesel

Table 5.1 Microalgae and their lipid content

S. No.	Microalgae	Lipid %	Reference
1	*Thraustochytrium*	30–90%	Scott et al. (2011),
2	*Chlorella zofingiensis*	30–50%	Liu et al. (2012)
3	*Botryococcus braunii*	34%	Tibbetts et al. (2015)
4	*Scenedesmus*	33%	Lu et al. (2017)

Table 5.2 Feedstock source Net biodiesel produced per hectare per year. Adapted from (Medipally et al. 2015)

Feedstock source	Oil content (% oil by wt. in biomass)	Oil yield (oil in liter's/ha/year)	Biodiesel productivity (kg biodiesel/ha/year)
Oil palm	36	5366	4747
Maize	44	172	152
Physic nut	41–59	741	656
Caster	48	1307	1156
Microalgae with high oil content	70	136,900	121,104

at commercialization scale is not feasible due to various challenges such as (1) bioreactor, (2) water scarcity, and (3) suitable species.

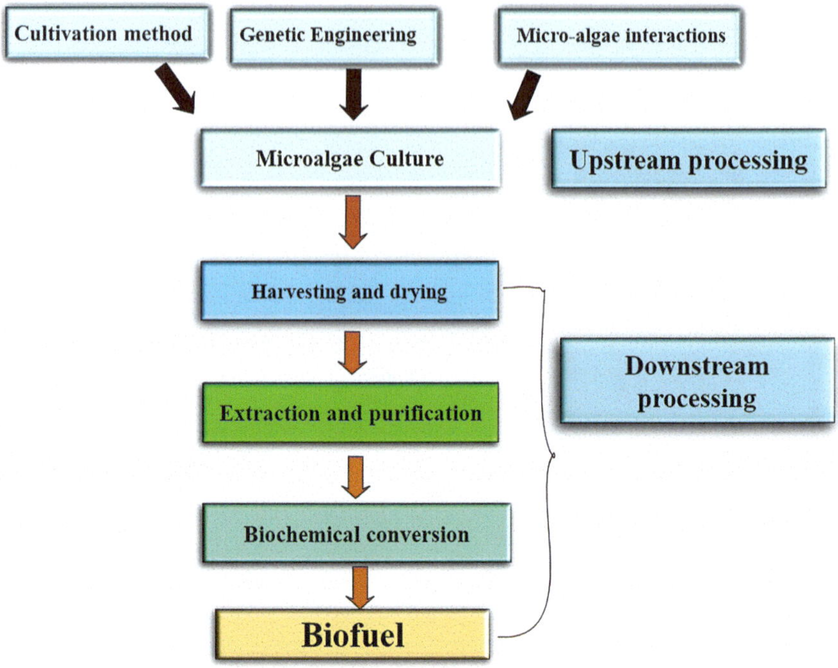

Fig. 5.5 Complete steps in biofuel production from algal based biofuel (Medipally et al. 2015)

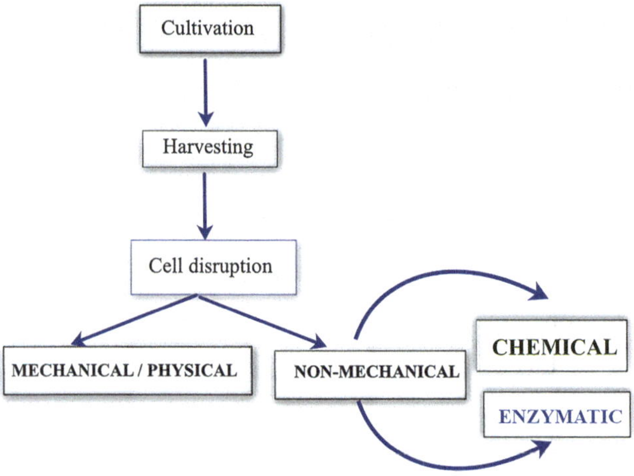

Fig. 5.6 Steps in cultivation harvesting of algal oil

Table 5.3 Difference between PBR and OP (Egbo et al. 2018)

Features	PBR's	OP
CAPEX cost	High	Low
OPEX cost	High	Low
Contamination chance	Low	High
H_2O required	Low	High
CO_2/nutrient loss	Low	High
Hydrodynamic stress	Low	Low
Lighting efficiency	High	Low
Physical footprint	Low	High
SA/V ratio	High	Low
Productivity	Very high	Low
Temperature control	Easy	Difficult
DO accumulation	High	Low
Scale up	Difficult	Difficult
Portability	Possible	Impossible
Energy requirement	High	Low
Commercial application	Low	High
Species grown	Numerous	Few
pH/light control	Easy	Difficult

Various other factor is also to be considered such as fresh water or polluted water where specific microalgae are to be screened out (Ganesan et al. 2020).

5.2.1.2 Growth Condition of Microalgae

Microalgae shown to grow in a variety of conditions where pH varies from 8 to 9 and temperature varies from 20 to 30 °C have the ability to grow in organic-rich polluted water. Light intensity required 380–500 nm (blue light) and 600–700 nm (red light) for optimal growth (Ganesan et al. 2020). Photoperiod varies from 16 to 18 h. Open raceway pond has cell densities more than 2.8×10^6 cells/mL after 32 h (Narala et al.

Table 5.4 Photobioreactor versus raceway pond and hybrid system (Mata et al. 2018)

Factors	Photobioreactors	Open ponds	Hybrid system
Space required	Moderate	High	High
Evaporation loss	Low	High	Moderate
CO_2 efficiency	High	Low	Moderate
Maintenance	Difficult	Easy	Moderate
Contamination risk	Low	High	low
Biomass quality	Reproducible	Variable	Reproducible
Energy input	High	Low	Moderate
Operation type	Batch	Batch	Continuous
Setup cost	High	Low	Moderate
Exponential phase maintaining	Difficult	Difficult	Easy

Fig. 5.7 Design structure of raceway pond

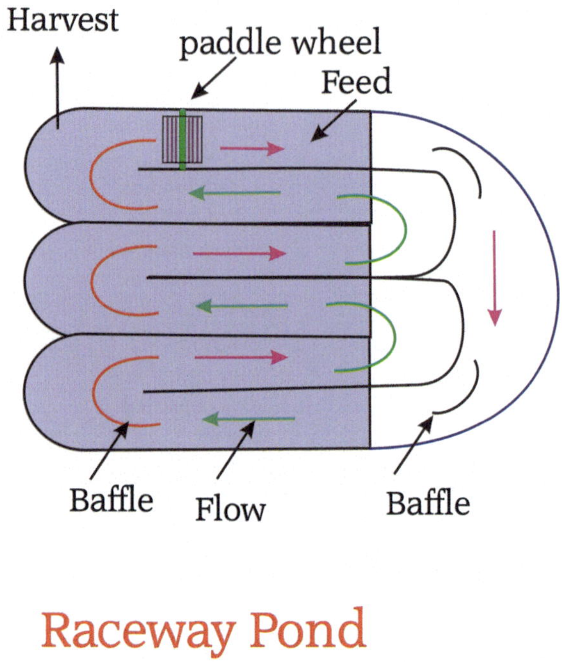

2016). *Spirulina* sp. has high concentration of sodium bicarbonate, while *Dunaliella salina* grow in extremely salty water.

In chloroplasts, lipids accumulate as TAG. Raceway pond design for maximum lipid accumulation is shown in Fig. 5.7. They resemble race track and with help of paddle biomass grow continuously, a kind of close re-circulation loop where paddle makes a uniform circulation pathway. But still getting high lipid from many species with high biomass is a challenge.

Excellent biomass is produced under closed PBR system.

5.2.1.3 Optimization of Lipid Content from Microalgae

Biodiesel production from bio-oil is only about 20% that can be increased under optimized condition which may be up to 80%. Some techniques that are fruitful worked out are response surface methodology which basically focused on increasing C/N ratio to increase lipid content (Ward and Rehmann 2019).

A two-stage cultivation and optimization of biomass have been carried out to increase the lipid content oleic acid accumulation under nitrate stress using Taguchi methodology in airlift bioreactor in *Chlorella minutissima* MCC (Chakraborty et al. 2016).

Lipid production optimization via response surface methodology (RSM) with central composite design (CCD), for example, in *Nannochloropsis gaditana lipid extraction, can be done by osmotic shock* by 0–20 KCI%, ultrasound (0–50 kHz), and lysozyme (0–20 U/mL) method.

Some species like *Ulothrix*, *Tribonema*, and *Euglena* have good potential for biodiesel production. Genetic engineering approaches have more opportunity for enhanced expression of gene.

For biodiesel production, various optimization strategies have been used like adaptive neuro-fuzzy inference system (ANFIS) and various photobioreactors like airlift, bubble column, and flat panel.

Algae have around 80% energy as compared to petroleum. Microalgae have around 30–40% more energy as compared to palm oil. *Botryococcus braunii* is supposed to have 40% more hydrocarbon. Production of algae is fast, convenient, and easy.

5.3 Culturing of Microalgae

Microalga is a promising and sustainable source of getting biodiesel using either open pond or any suitable photobioreactor system (Tan et al. 2020). Other advantages are that they can grow well in wastewater reducing organic pollutant and also helpful in reducing the green house CO_2 load in the environment (Yang et al. 2021).

The two most frequent techniques of microalgae cultivation are open cultivation systems, such as open ponds, tanks, and raceway ponds, and managed closed cultivation systems the use of one-of-a-kind sorts of bioreactors.

Raceway pond and open pond are the traditional way of growing microalgae where return on investment is very low as compared to photobioreactor (Chisti 2006; Nwoba et al. 2020; Sheets et al. 2014).

Open cultivation system has advantages that its installation requires minimum capital cost with very low maintenance, but on the other side, it requires large area and also difficult to control growth parameter.

In order to cultivate microalgae, several parameters must be controlled, such as gaseous transfer, mixing, light wavelength, constant temperature, pH, nutrients, and the culture period (Hossain and Mahlia 2019). Furthermore, diatoms and amoeba are very likely to cause contamination. Additionally, municipal wastewater should be suitable for growing various microalgae.

Closed system for growth of algae is also called as photobioreactor (Singh and Sharma 2012). Photobioreactor requires low space, has controlled condition, and can be designed as per need of strain.

However, various problems lie in PBRs such as (1) bio-fouling, (2) overheating, (3) benthic algae growth, (4) cleaning issues, and (5) A very high capital costs is required for designing and operating PBR. (Grima et al. 2003; Chisti 2006) (Table 5.5).

Table 5.5 Nutrition profile of microalgae biomass (Han et al. 2019)

Strain	Culture medium	Protein (%)	Lipid (%)	Carbohydrate (%)	Value-added compound
Thraustochytrium sp.	Medium with glycerol	NA	38.95	NA	EPA and DHA
Chlorella zofingiensis	Cane molasses	NA	30–50	NA	Polyunsaturated fatty acids
Scenedesmus sp.	Soybean oil	53.3	33.4	NA	EPA
Galdieria sulphuraria	Modified Allen medium	26.5	1.14	69.1	Dietary fiber
Galdieria sulphuraria	Modified Allen medium	32.5	1.77	62.9	Astaxanthin
Chlorella zofingiensis	Cane molasses	NA	NA	NA	Astaxanthin
Chlorella zofingiensis	Cane molasses	NA	30–50	NA	Astaxanthin
Haematococcus pluvialis	OHM medium	NA	NA	NA	Astaxanthin
Haematococcus pluvialis	Primary treated wastewater	NA	NA	NA	Astaxanthin
Botryococcus braunii	NA	39.9	34.4	18.5	Essential amino acids
Tetraselmis chuii	NA	46.5	12.3	25.0	Essential amino acids
Phaeodactylum tricornutum	NA	39.6	18.2	25.2	Essential amino acids
Porphyridium aerugineum	NA	31.6	13.7	45.8	Essential amino acids

5.4 Harvesting of Algae or Dewatering of Algae

Major challenges lie in harvesting of biomass (Ogbonna and Nwoba 2021).

Algal harvesting system has been explored extensively in various literatures (Bhatia et al. 2021; Branyikova et al. 2018; Show et al. 2019). Harvesting by flocculation (Branyikova et al. 2018), centrifugation, filtration, sedimentation, flocculation (Alam et al. 2016), and floatation or harvesting by attached system (Lee et al. 2014) where dewatering methodology is used increases the lipid content by >20%. Table 5.6 shows parameter for harvesting algal biomass by different methods.

Bioflocculation reported to be a low-cost separation technology (Alam et al. 2016; Ogbonna and Nwoba 2021; Vasistha et al. 2021) Eco-friendly harvesting technologies have been developed such as low-cost green coagulants and electrochemical harvesting (Ravindran et al. 2016). Chitosan-based flocculants also have been determined to be beneficial for algae harvesting (Yin et al. 2021).

Table 5.6 Parameter for harvesting algal biomass

SN		Harvesting cost	Safety level	Time Consumption
1	centrifugation	High	High	Short
2	Filtration	High	High	Short
3	Gravity driven sedimentation	Low	High	Long
4	Flocculation by chemicals	Low	Low	Short
5	Harvesting by edible fungi	Low	High	Short
6	Flotation	Low	High	Short

5.5 Centrifugation

Ideal method to release algal oil is centrifugation which under pressure lyses the cells without adding any chemical. A variety of centrifuges are available, including disk stack centrifuges, tubular centrifuges, nozzle centrifuges, decanters, and perforated and imperforated basket centrifuges (Ananthi et al. 2021). Harvesting of microalgae involve first bulk harvesting and second biomass thickening. Bulk harvesting means concentration of biomass using any one techniques such as flocculation, sedimentation, or floatation in which collection of biomass is the primary objective, whereas thickening of biomass can be done by filtration or centrifugation (Rakesh et al. 2020).

5.6 Sedimentation

Even though sedimentation is based on densities, separating algal biomass is difficult because of the small size of microalgae.

5.7 Filtration Technology

Various filtration techniques are very important to separate filter designs including (1) microfiltration (0.1–10 μm), (2) macrofiltration (10 μm), (3) ultrafiltration (0.01–0.1 μm), and (4) nanofiltration (0.001–0.01 μm). Osmotic pressure is great tool in this regard where biomass can be separated from liquid using reverse and forward osmosis.

Hydrophobic membrane used in filtration technology is based on hydrophobicity, surface charge, flow parameters, etc. The typical materials used for polymer preparation of membrane are polyethersulfone polyvinylpyrrolidone, polyvinyl chloride, cellulose acetate, polyvinylidene fluoride, polyacrylonitrile, polyethersulfone, and polyamide (Corrêa et al. 2021).

Centrifugation technique is cheap and is more efficient; biomass recovery >90% can be achieved.

Certain modern technique such as supercritical fluid extraction, or some other methods such as ultrasound-assisted extraction, or pulsed electric fields, can be used without solvents to extract the algal oil (Fig. 5.8).

Biodiesel production from bio-oil is a major challenge, but challenges can be solved by specific approaches; oil is basically present up to 20% and can be increased under specific condition, up to 80%. Some species like *Ulothrix*, *Tribonema*, and *Euglena* have good potential for biodiesel production. Genetic engineering approaches have more opportunity for enhanced expression of gene.

For biodiesel production, various optimization strategies have been used like adaptive neuro-fuzzy inference system (ANFIS) various photobioreactor like airlift, bubble column, and flat panel.

Algae have around 80% energy as compared to petroleum. Microalgae have around 30–40% more energy as compared to palm oil. *Botryococcus braunii* is

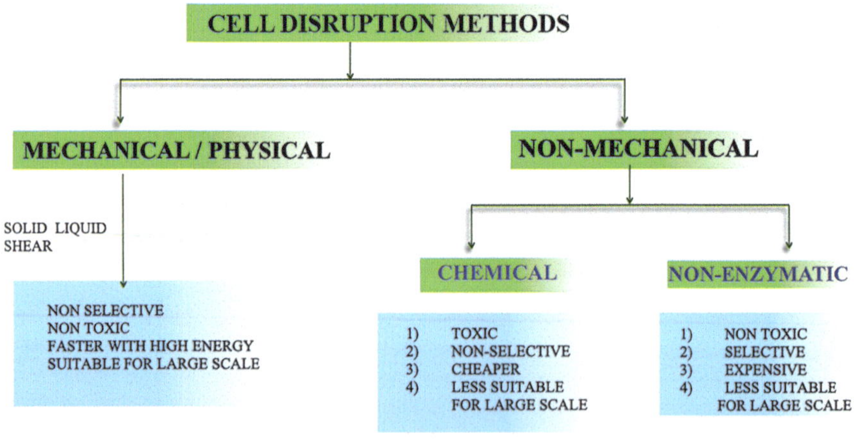

Fig. 5.8 Mechanical and nonmechanical method of cell disruption

supposed to have 40% more hydrocarbon. Production of algae is fast, convenient, and easy.

5.7.1 Photobioreactor

A photobioreactor can be portrayed as an encased, enlightened culture vessel intended for controlled biomass production. Photobioreactor closed photobioreactor systems that have close environment (with no direct exchange of gases and contaminants with the surroundings. Photobioreactors), in spite of their high expenses, and have a few significant benefits over open systems:

- Closed photobioreactors limit infection and allow monocultures cultivation.
- Closed photobioreactors offer better control of pH, CO_2, temperature, light, focus, and many others.
- Closed photobioreactors result in much less CO_2 loss.
- Closed photobioreactors prevent evaporation of water.
- Closed photobioreactors allow high cell concentration growth.
- Closed photobioreactors permit the creation of complex biopharmaceuticals (Tables 5.7 and 5.8).

5.7.1.1 Types of Bioreactors

1. **Vertical-column photobioreactors**

   A vertical photobioreactor is designed called a bubble bioreactor to grow algae where O_2 and CO_2 are bubbled by sparger and thus allow more photosynthesis.

   The bubble column reactor is more efficient than the airlift reactor. Mass transfer of CO_2 and removal of oxygen generated during photosynthesis allow more product formation.
2. **Bubble-column photobioreactor**

   As shown in Fig. 5.9, this type of reactor has more height than width and as a result, it has less cost and more mass transfer benefits. It is also easy to sterilize because of its good heat and mass exchange properties. Sparger allows the better exchange of CO_2 and O_2. Perforated plates help in better bubble formations (Fig. 5.10).
3. **Airlift photobioreactor**

   As the name suggests gas is sparged with a connecting pipe called a riser, while downcomers pipe doesn't allow any gas to pass. The facility of gas input-output can be changed by giving a split facility based on those two types of a reactor designed; one is internal loop, and the other is an external loop.

   A draft tube separates the region in the case of the internal loop reactor. Internal loop reactor is further classified into two types (1) internal loop split airlift reactor and (2) internal loop concentric tube reactor.

Table 5.7 Merits and demerits of different types of culture system

Type of culture system	Advantages	Disadvantages	Reference
Open ponds	Cheap to install, easy to clean up after cultivation, can be used for mass cultivation of algae	Culture conditions are difficult to control; productivity is poor. Only few strains of algae can be cultured, high risk of contamination	Liu et al. (2019)
Vertical-column photobioreactors	1. Can be scaled up 2. Sterilization is easy 3. Immobilization possible 4. Energy consumption is low 5. Photo inhibition and photo-oxidation are less seen	Small illumination surface area, their construction requires sophisticated materials, shear stress to algal cultures	Pawar (2016)
Flat-plate photobioreactors	Surface area for illumination is large; light pathway is good; can be used for outdoor cultures; immobilization of algae can be done; biomass productivity is efficient; Economical cleanup is easy	Difficulty in scale-up as it requires many additional accessories. Not easy to control culture temperature. Growth on wall can be seen. Some strains of algae seen to have hydrodynamic stress	Hamed et al. (2021), Hulatt et al. (2017)
Tubular photobioreactors	Economical, biomass productivity is efficient. Surface area for illumination is large	Difficult to manage change in pH, dissolved oxygen and CO_2 along the tubes. Problem of fouling and growth on walls. Large space is required	Ippoliti et al. (2016), Moraes et al. (2020)
Internally illuminated photobioreactors	Can use both solar and artificial light system, can be heat sterilized, so contamination can be prevented, can be used both day and night	Outdoor mass cultivation of algae requires more technical support	Deniz et al. (2019), Garbowski et al. (2020)

Table 5.8 Microalgae-based biofuel companies

Company name	Location	Biofuel type
Algenol Biofuels	Florida, USA	Ethanol
Sapphire Energy	Headquarters in San Diego, USA; green crude farm in New Mexico, USA	Similar to crude oil
Seambiotic	Israel	Biodiesel and bioethanol
Solazyme	South San Francisco, USA	Biodiesel
Solix BioSystems	Colorado, USA	Biodiesel

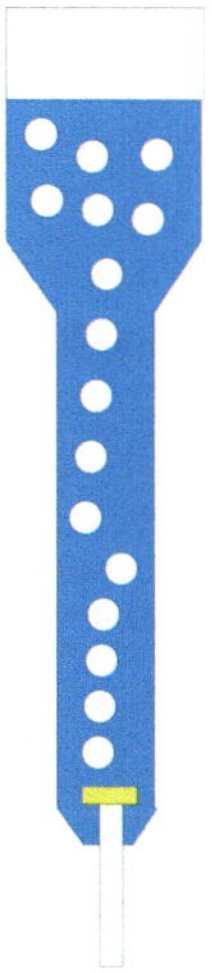

Fig. 5.9 Bubble-column photobioreactor

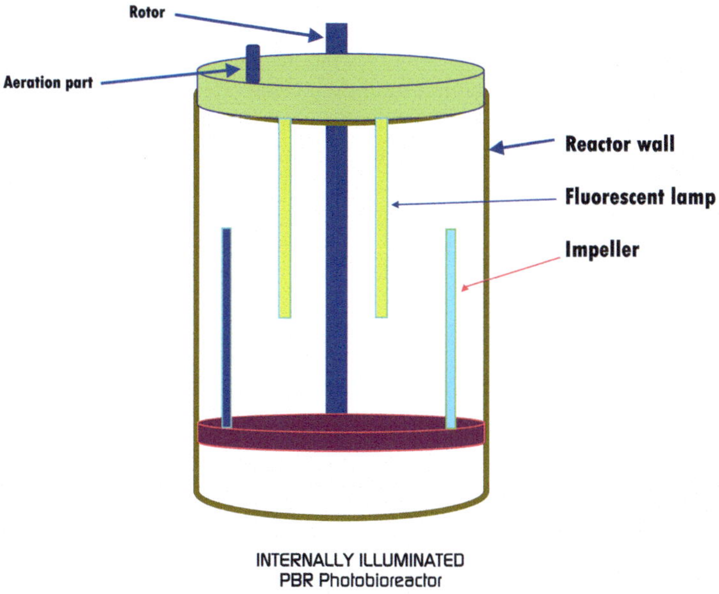

Fig. 5.10 Illuminated photobioreactor

The external loop reactor has the advantage of two different tubes physically separating the riser and downcomer. This has benefited both the bubble column and airlift reactor. Circulating liquid culture passes alternatively through the dark and light phases.

Bioethanol production from begasses has been used in airlift reactor recently found to show good results (Restiawaty et al. 2020).

4. **Flat-plate photobioreactors**

   Flat plate PBR is supposed to be best for mass production of algal biomass and is useful because of various reasons since its flat having a depth of 1–5 cm but length breadth doesn't exceed 1 m. Several worker has designed reactor for enhanced growth of microalgae (Santana et al. 2017; Sierra et al. 2008). Only disadvantages is that because of uncontrolled bubble release causes shear damage to cells.

   These types of photoreactors can be illuminated internally with the help of fluorescent lamps. These reactors are designed to work both in artificial and natural sunlight. When the natural sunlight intensity decreases, it can be switched on to an artificial light source. Thus this type of photobioreactor can be operated continuously (day and night). Some studies suggest the use of optic fibers which can collect and distribute solar light in these photobioreactors. This is the only type of photobioreactor that can be sterilized and thus avoiding huge contamination

5. **Internally illuminated photobioreactors**

   These types of photoreactors can be illuminated internally with the help of fluorescent lamps. These reactors are designed to work both in artificial and natural sunlight. When the natural sunlight intensity decreases, it can be switched on to an artificial light source. Thus this type of photobioreactor can be operated continuously day and night. Some studies reported the use of optic fibers which can collect and distribute solar light in these photobioreactors. This is the only type of photobioreactor that can be heat sterilized and thus avoiding contamination (Amaral et al. 2020; Deniz et al. 2019; Sutor et al. 2014).

5.8 Conclusion

In conclusion design of photobioreactor decides overall biomass production. Quality of lipid is decided by the type of bioreactor and environmental conditions. Overall commercial production is possible by some more modifications. In the coming future, technology can be adapted.

References

Alam A et al (2016) Bioflocculation as an innovative harvesting strategy for microalgae. Rev Environ Sci Biotechnol 15(4):573–583

Amaral M d S, Loures CCA, Silva MB, Prata AMR (2020) Adjustment of the operational parameters of an unconventional integrated and illuminated internally photobioreactor

(ILI-PBR) for the batch autotrophic cultivation of the Chlorella Minutissima, using the Taguchi method. Appl Biochem Biotechnol 191(1):245–257

Ananthi V et al (2021) A critical review on different harvesting techniques for algal based biodiesel production. Sci Total Environ 780:146467

Arora N, Philippidis GP (2021) Microalgae strain improvement strategies: random mutagenesis and adaptive laboratory evolution. Trends Plant Sci 26:1199

Aziz MMA et al (2020) Two-stage cultivation strategy for simultaneous increases in growth rate and lipid content of microalgae: a review. Renew Sust Energ Rev 119:109621

Banerjee S, Singh H, Das D, Atta A (2019) Process optimization for enhanced biodiesel production by Neochloris Oleoabundans UTEX 1185 with concomitant CO2 sequestration. Ind Eng Chem Res 58(35):15760–15771

Bhatia SK et al (2021) Wastewater based microalgal biorefinery for bioenergy production: progress and challenges. Sci Total Environ 751:141599

Branyikova I et al (2018) Harvesting of microalgae by flocculation. Fermentation 4(4):93

Chakraborty S, Mohanty D, Ghosh S, Das D (2016) Improvement of lipid content of Chlorella Minutissima MCC 5 for biodiesel production. J Biosci Bioeng 122(3):294–300

Cheng J, Zhu Y, Zhang Z, Yang W (2019) Modification and improvement of microalgae strains for strengthening CO2 fixation from coal-fired flue gas in power plants. Bioresour Technol 291: 121850

Chisti Y (2006) Microalgae as sustainable cell factories. Environ Eng Manag J 5(3):261–274

Corrêa PS et al (2021) Microalgae biomolecules: extraction, separation and purification methods. Processes 9(1):1–40

Deniz I, Demirel Z, Imamoglu E, Dalay MC (2019) Enhanced microalgal lipid production in internally illuminated airlift photobioreactor. Mar Technol Soc J 53(2):38–45

Egbo MK, Okoani AO, Okoh IE (2018) Photobioreactors for microalgae cultivation – an overview. Int J Sci Eng Res 9:65–74

Ganesan R et al (2020) A review on prospective production of biofuel from microalgae. Biotechnol Rep 27:e00509

Garbowski T, Charazińska S, Pulikowski K, Wiercik P (2020) Application of microalgae cultivated on pine bark for the treatment of municipal wastewater in cylindrical photobioreactors. Water Environ J 34(S1):949–959

González-Camejo J et al (2019) Optimising an outdoor membrane photobioreactor for tertiary sewage treatment. J Environ Manag 245:76–85

Grima EM et al (2003) Recovery of microalgal biomass and metabolites: process options and economics. Biotechnology Advances 20(7-8):491–515

Hamed I et al (2021) Influence of stress factors on growth and pigment production in three dunaliella species cultivated outdoors in flat-plate photobioreactors. Plant Biosyst 155(1): 179–187

Han P, Lu Q, Fan L, Zhou W (2019) A review on the use of microalgae for sustainable aquaculture. Applied Sciences 9(11):2377

Hossain N, Mahlia TMI (2019) Progress in physicochemical parameters of microalgae cultivation for biofuel production. Crit Rev Biotechnol 39(6):835–859

Hulatt CJ, Wijffels RH, Bolla S, Kiron V (2017) Production of fatty acids and protein by nannochloropsis in flat-plate photobioreactors. PLoS One 12(1):e0170440

Ippoliti D et al (2016) Outdoor production of tisochrysis lutea in pilot-scale tubular photobioreactors. J Appl Phycol 28(6):3159–3166

Khan S et al (2017) Biodiesel production from algae to overcome the energy crisis. HAYATI J Biosci 24(4):163–167

Khan MI, Shin JH, Kim JD (2018) The promising future of microalgae: current status, challenges, and optimization of a sustainable and renewable industry for biofuels, feed, and other products. Microb Cell Factories 17(1):36

Kumar S (2020) Estimation capabilities of biodiesel production from algae oil blend using adaptive neuro-fuzzy inference system (ANFIS). Energy Sour A Recov Utiliz Environ Effects 42(7): 909–917
Kumar G et al (2020) Bioengineering of microalgae: recent advances, perspectives, and regulatory challenges for industrial application. Front Bioeng Biotechnol 8:914
Lee SH et al (2014) Higher biomass productivity of microalgae in an attached growth system, using wastewater. J Microbiol Biotechnol 24(11):1566–1573
Liu J, Huang J, Jiang Y, Chen F (2012) Molasses-based growth and production of oil and astaxanthin by Chlorella zofingiensis. Bioresour Technol 107:393–398
Liu W, Yu C, Wang J, Liu T (2019) Biomass productivity of Scenedesmus dimorphus (Chlorophyceae) was improved by using an open pond–photobioreactor hybrid system. Eur J Phycol 54(2):127–134
Lu Q et al (2017) Exploration of a mechanism for the production of highly unsaturated fatty acids in Scenedesmus sp. at low temperature grown on oil crop residue based medium. Bioresour Technol 244:542–551
Mata TM et al (2018) Carbon footprint of microalgae production in photobioreactor. Energy Procedia 153:432–437
Medipally SR, Yusoff FM, Banerjee S, Shariff M (2015) Microalgae as sustainable renewable energy feedstock for biofuel production. Biomed Res Int 2015:519513
Moraes L et al (2020) Bioprocess strategies for enhancing the outdoor production of Nannochloropsis Gaditana: an evaluation of the effects of pH on culture performance in tubular photobioreactors. Bioprocess Biosyst Eng 43(10):1823–1832
Narala RR et al (2016) Comparison of microalgae cultivation in photobioreactor, open raceway pond, and a two-stage hybrid system. Front Energy Res 4:29
Nwoba EG et al (2020) Pilot-scale self-cooling microalgal closed photobioreactor for biomass production and electricity generation. Algal Res 45:101731
Ogbonna CN, Nwoba EG (2021) Bio-based flocculants for sustainable harvesting of microalgae for biofuel production. A review. Renew Sust Energ Rev 139:110690
Pawar SB (2016) Process engineering aspects of vertical column photobioreactors for mass production of microalgae. Chem Bio Eng Rev 3(3):101–115
Pugazhendhi A et al (2020) Various potential techniques to reduce the water footprint of microalgal biomass production for biofuel—a review. Sci Total Environ 749:142218
Rakesh S et al (2020) Sustainable cost-effective microalgae harvesting strategies for the production of biofuel and oleochemicals. Highlights BioSci 3
Ravindran B et al (2016) Microalgae potential and multiple roles-current progress and future prospects-an overview. Sustainability 8(12):1215
Restiawaty E et al (2020) Bioethanol production from sugarcane bagasse using neurospora intermedia in an airlift bioreactor. Int J Renew Energy Dev 9(2):247–253
Santana H et al (2017) Microalgae cultivation in sugarcane vinasse: selection, growth and biochemical characterization. Bioresour Technol 228:133–140
Scott SD, Armenta RE, Berryman KT, Norman AW (2011) Use of raw glycerol to produce oil rich in polyunsaturated fatty acids by a thraustochytrid. Enzym Microb Technol 48(3):267–272
Sheets JP, Ge X, Park SY, Li Y (2014) Effect of outdoor conditions on nannochloropsis salina cultivation in artificial seawater using nutrients from anaerobic digestion effluent. Bioresour Technol 152:154–161
Show K-Y, Yan Y-G, Lee D-J (2019) Algal biomass harvesting and drying. In: Pandey A et al (eds) Biofuels from Algae. Elsevier, Amsterdam, pp 135–166
Sierra E et al (2008) Characterization of a flat plate photobioreactor for the production of microalgae. Chem Eng J 138(1–3):136–147
Singh RN, Sharma S (2012) Development of suitable photobioreactor for algae production - a review. Renew Sust Energ Rev 16(4):2347–2353
Sutor A, Heining M, Lindenberger C, Buchholz R (2014) Method for optimizing the field coils of internally illuminated photobioreactors. IEEE Trans Magn 50(11):1
Tan JS et al (2020) A review on microalgae cultivation and harvesting, and their biomass extraction processing using ionic liquids. Bioengineered 11(1):116–129

Tibbetts SM, Milley JE, Lall SP (2015) Chemical composition and nutritional properties of freshwater and marine microalgal biomass cultured in photobioreactors. J Appl Phycol 27(3): 1109–1119

Vasistha S, Khanra A, Clifford M, Rai MP (2021) Current advances in microalgae harvesting and lipid extraction processes for improved biodiesel production: a review. Renew Sust Energ Rev 137:110498

Ward VCA, Rehmann L (2019) Fast media optimization for mixotrophic cultivation of chlorella vulgaris. Sci Rep 9(1):19262

Yang L, Li H, Lu Q, Zhou W (2021) Emerging trends of culturing microalgae for fish-rearing environment protection. J Chem Technol Biotechnol 96(1):31–37

Yin Z et al (2021) Application of chitosan-based flocculants to harvest microalgal biomass for biofuel production: a review. Renew Sust Energ Rev 145:111159

Chapter 6
Influence of Culture Conditions on the Microalgal Biomass and Lipid Accumulation

Manisha Verma and Vishal Mishra

Abstract In the current situation, almost every industrial and transportation activities are dependent on fossil fuels for their energy requirements. Fossil fuels are an overpriced and non-renewable source of energy. Besides these fossil fuels, combustion emits greenhouse gases that cause air pollution and global warming. Nowadays, microalgae-originated biofuels gained huge attention from researchers as they are expected to be a potential alternative to conventional fuels. Microalgal biofuel sector considers various strategies to enhance lipid content and biomass productivity in different habitats. To attain high lipid content from microalgae, lipid triggering circumstances are required to be optimized. This chapter summarizes various cultivation conditions, including pH, light intensity, and temperature for raised lipid accumulation inside microalgal cells. Different levels of physical factors affecting microalgal growth and lipid yield have been discussed in this chapter. The influence of the cultivation conditions such as CO_2 concentration, temperature, light colour, and light intensity on lipid accumulation is evaluated comprehensively. Also, very recent progress and research studies on microalgal biomass and biodiesel production are discussed and summarized.

Keywords Microalgae · Lipid accumulation · Algal cultivation · Algal biomass · Biodiesel

Abbreviations

ATP	Adenosine triphosphate
CO_2	Carbon dioxide
MUFA	Monounsaturated fatty acid
NADH	Nicotinamide adenine dinucleotide

M. Verma · V. Mishra (✉)
School of Biochemical Engineering, IIT (BHU), Varanasi, Uttar Pradesh, India
e-mail: vishal.bce@itbhu.ac.in

N. Srivastava, P. K. Mishra (eds.), *Technological Advancement in Algal Biofuels Production*, Clean Energy Production Technologies,
https://doi.org/10.1007/978-981-19-6806-8_6

PBR	Photobioreactor
PS I	Photosystem I
PS II	Photosystem II
PUFA	Polyunsaturated fatty acid
Q_{10}	Temperature coefficient
SFA	Saturated fatty acid
TAG	Triacylglycerol

6.1 Introduction

In the present, continuous depletion in fossil fuel reservoirs causes scarcity of energy resources. Conventional fuels will be going to become exhausted and unable to satisfy the growing demand for energy. Continuously inflated fuel prices, global warming, and greenhouse gas emission bring the requirement to upgrade sustainable energy resources. Algae-derived biofuels are alternative to replace conventional fuels. Third-generation biofuels include microalga biomass for the production of ethanol, butanol, and biodiesel. Algae have huge potential to replace first- and second-generation agricultural feedstocks as microalgae are rich in carbohydrate-lipid content and have a rapid growth rate and high biomass yield (Verma and Mishra 2020). In fourth-generation biofuels, upcoming advances bring the metabolic and gene engineering approaches to design custom microalgae to increase biofuel production (Abdullah et al. 2019; Moravvej et al. 2019). Brazil, Germany, France, Sweden, and the United States are leading countries in the consumption and development of biofuels (Adenle et al. 2013). It is observed that genetically modified algae shown significant enhancement in biomass, lipid accumulation, and carbon capturing capacity (Beacham et al. 2017; Shuba and Kifle 2018; Dutta et al. 2014; Levitan et al. 2014). Algal species are found in many shapes and sizes, extending from single-cell microalgae to multicellular filaments macroalgae in different aquatic habitats (Shafik et al. 2015). Most of the algal species are found in extreme environmental conditions in various aquatic habitats (Adeniyi et al. 2018). Based on their cellular organization and development, algae are classified into cyanobacteria and eukaryotes. Cyanobacteria are differentiated by the lack of chloroplasts and an adequate nucleus. Based on cell wall composition, storage products, and pigments, eukaryotic algae can be classified into Chrysophyceae, Chlorophyceae, Euglenophyceae, Phaeophyceae, Pyrrophyceae, Rhodophyceae, and Xanthophyceae. Algae achieve its economic significance in biofuel generation, bioremediation, wastewater treatment, nutraceuticals, animal feed, and biofertilizer (Sirakov et al. 2015; Suparmaniam et al. 2019; Pulz and Gross 2004). Additionally, algae are rich in several metabolites, including fatty acids, nutraceuticals, proteins, pigments, and vitamins (Sirakov et al. 2015; Pulz and Gross 2004). Under optimized conditions, algae can grow rapidly and produce two to ten-fold higher lipid content as compared to soybean, jatropha, and rapeseed (Rawat et al. 2013; Chisti 2007;

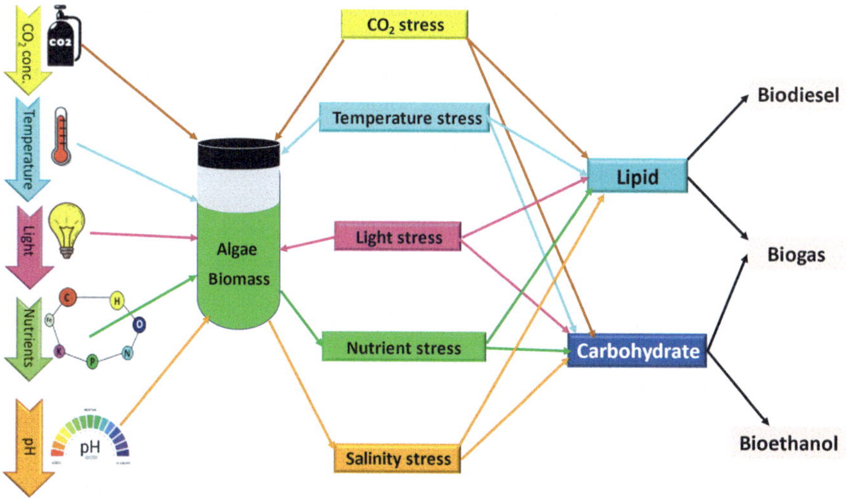

Fig. 6.1 Influence of culture condition on biofuel production

Lam and Lee 2011; Tsukahara and Sawayama 2005). This chapter summarizes various cultivation conditions, including pH, light intensity, and temperature for raised lipid accumulation inside microalgal cells. Different levels of physical factors affecting microalgal growth and lipid yield have been discussed in this chapter; the impact of the culture conditions like CO_2 concentration, temperature, light colour, and light intensity on lipid accumulation is evaluated comprehensively. Also, very recent progress and research studies on microalgal biomass and biodiesel production are discussed and summarized. Figure 6.1 represents crucial factors affecting microalgae culture and ultimately affect biofuel production.

6.2 Microalgal Cultivation

Algae have higher growth rates, photosynthetic levels, and CO_2 sequestering efficiency. Microalgae are significantly considered organisms for reducing the nutrient load (nitrogen and phosphorous) from agricultural, industrial, municipal, and domestic wastewater. They consist of considerable quantities of fatty acids that yield sustainable biodiesel after transesterification. Microalgal cultivation for lipid recovery and biofuel generation is a moderate process. Meanwhile, they need quite simple light and aeration arrangements to grow. However, some factors alter the optimum growth and total lipid yield in microalgae. These factors include macro- and micronutrients (concentration and availability), CO_2, pH, light (photoperiod/ intensity), and temperature. Algae have higher growth rates, photosynthetic levels, and CO_2 sequestering efficiency. Microalgae are significantly considered organisms for reducing the nutrient load (N and P) from industrial, agricultural, municipal, and domestic wastewater. They consist of considerable quantities of fatty acids that yield sustainable biodiesel after transesterification. Microalgal biomass cultivation for biodiesel production is a moderate process; meanwhile, they need quite simple light and aeration arrangements to grow. Some factors like nutrients availability, pH, temperature, CO_2, photoperiod, and light intensity might affect biomass yield and lipid accumulation inside microalgal cells. Among all these conditions, light and temperature are primarily significant for algal growth. A temperature that ranges from 20 to 30 °C is preferable for most microalgal species. Highly raised temperature results in physiological adaptations in microalgal cells that cause lesser unsaturated fatty acid production (Van Wagenen et al. 2012). Biodiesel properties like cetane number, heating point, oxidative stability, lubricity, and melting point are primarily based on the type of precursor fatty acid. Palmitoleic acid, stearic acid, linolenic acid, oleic acid, linoleic acid, and palmitic acid are the most suitable fatty acids for biodiesel generation (Zheng et al. 2013). The lipid synthesis pathway in microalgae might be altered in several limiting conditions and synthesize a higher amount of triacylglycerol (TAG) in lipid bodies (Hu et al. 2008). It is necessary to observe loads of microalgal species for their high lipid productivity; then selected algal species are optimized for different factors to attain maximum biodiesel production. The cultivation system is an essential factor as it affects the biofuel yield and photosynthesis efficiency of microalgae. Generally, algal cultivation is preferred in two ways: (1) open cultivation system and (2) closed cultivation system. Photobioreactors (PBR) are a closed system for microalgal cultivation. Inside PBR, microalgal cultural conditions as light intensity, aeration, and stirring are customized according to the microalgal species (Liao et al. 2018; Lee and Lee 2016). Glass or plastic material is used to construct PBR with different designs such as flat plate, airlift, and tubular bioreactor and bubble column (Brennan and Owende 2010). Airlift bioreactor is widely appreciated to eliminate the chance of contamination and to obtain high algal biomass (Huang et al. 2010). High cost, oxygen build-up, biofouling, overheating, cell damage, and difficulties in scale-up are major limitations of PBR (Adeniyi et al. 2018; Brennan and Owende 2010). In comparison, open cultivation systems like the shallow pond, unstirred pond, raceway pond, closed pond, and circular pond are less optimized but provide affordable, lower maintenance cost, less operational cost, and easy scale-up. Open systems are highly affected by water evaporation, temperature, and light variability. PBR exhibited highly customized growth conditions and ensured the availability of adequate light intensity; hence, PBR provides a higher biomass productivity rate than the open system (Leite et al. 2013). CO_2 consumption as a carbon source is the major pros of utilizing photoautotrophic systems. Furthermore, chances of contamination are significantly less while using photoautotrophs (Mata et al. 2010). A combined design of an open pond and closed PBR system is applied to achieve elevated biomass productivity and excellent nutrient removal. A combined setup of closed and open systems is termed as hybrid algal cultivation method, which is adequate for massive microalga cultivation (Razzak et al. 2017; Rawat et al. 2013). Hybrid systems are customized to eliminate drawbacks of open systems and to minimize the operational cost of photobioreactors. Such a hybrid cultivation system

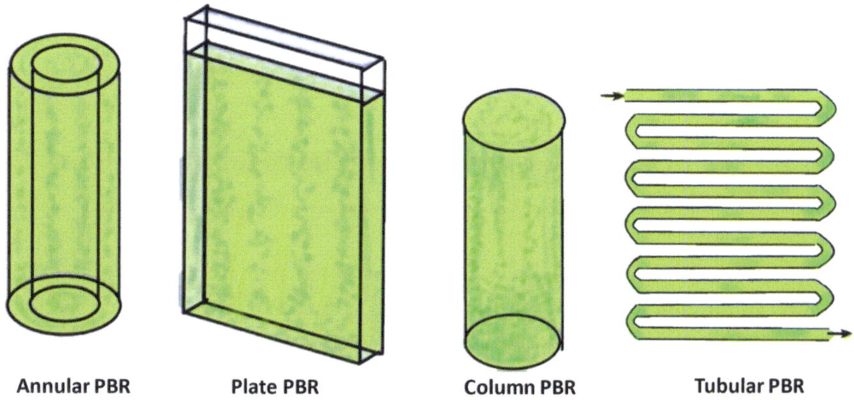

Fig. 6.2 Closed photobioreactor system for algal cultivation

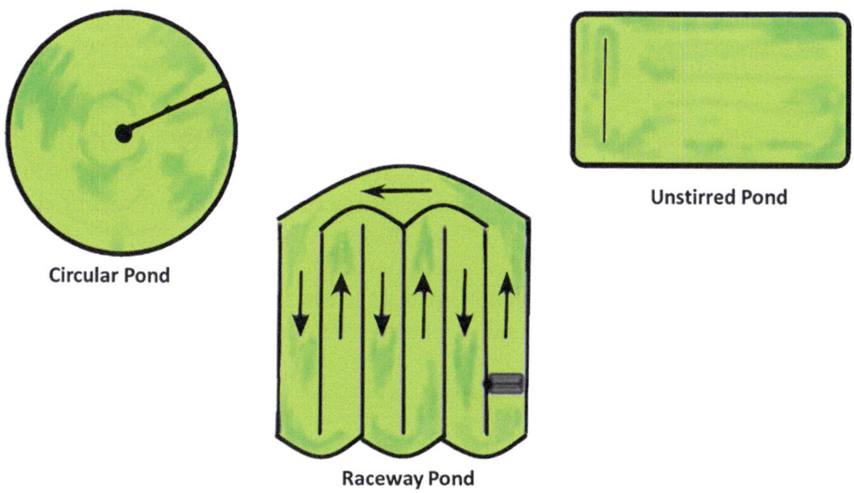

Fig. 6.3 Open pond system for algal cultivation

preferred the initial stage of microalgal culture inside photobioreactors, and then it will be transferred to ponds (Schenk et al. 2008) (Figs. 6.2 and 6.3).

Concerning metabolic activity, algal cells can be classified into main categories that are heterotrophic, photoautotrophic, autotrophic, and mixotrophic (Daliry et al. 2017). The heterotrophic mode of nutrition needs the presence of organic carbon as a substrate, whereas photoheterotrophs need both light and organic carbon to survive (Chen and Durbin 1994). Autotrophic microalgae convert inorganic source of carbon into chemical energy in the availability of sunlight, while mixotrophic algae can survive in both heterotrophic and autotrophic modes dependent on the presence/absence of carbon source and light (Burkholder et al. 2008). Scarsella et al. (2010) found that mixotrophy nutrition mode in microalgae is suitable for yielding maximum lipid and biomass. Heterotrophic metabolism has a better growth rate than autotrophic metabolism (Martinez et al. 1991).

6.3 Impact of Cultural Conditions on the Algal Lipid Content

Photoautotrophs like green algae can synthesize proteins, lipids, carbohydrates, and proteins. Lipid composition in algae is highly affected by its life cycle, whereas variation in lipid content depends on the cultivation medium and some other physical factors (Scarsella et al. 2010). Temperature, pH level, photoperiod, light intensity, and CO_2 highly affect photosynthesis in green algae (Deng et al. 2014). Figure 6.4 depicts impact of stress condition on algal cells.

6.3.1 Effect of Temperature

It is evident to evaluate the influence of temperature to achieve high microalgal growth rate at broad-scale outdoor cultivation systems. Microalgal culture in outdoor open ponds is expected to undergo extreme surrounding temperatures and produce high algal biomass for economic lipid production. Surrounding temperature is a crucial parameter that affects lipid composition and content in algal cells (Ras et al. 2013). Temperature modifies metabolic pathways and causes acceleration or retardation of biochemical reaction rate in algal cells (Afify et al. 2010). The dependency

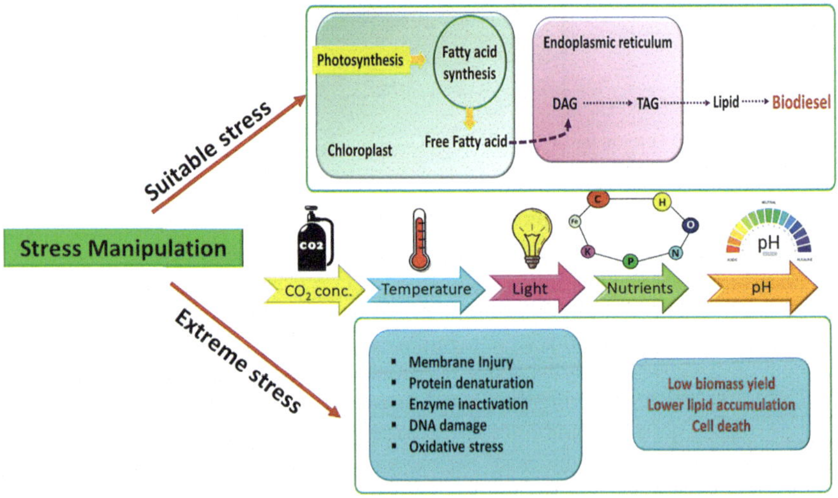

Fig. 6.4 Impact of stress manipulation on algal cells

of biochemical reactions rate on temperature can be described by Arrhenius function and Q_{10} (temperature coefficient). Q_{10} is a factor by which any biological processes or biochemical reaction rate increases if the temperature is raised by 10 units (Teoh et al. 2010). The relationship among temperature and algal growth rate follows the Arrhenius relationship. Q_{10} and Arrhenius function are equally expected to achieve accelerated growth rates with rising temperatures. However, microalgal growth rate accelerates up to an optimum temperature, but crossing that optimum value can cause retardate growth (Suzuki and Takahashi 1995; Montagnes and Franklin 2001). Table 6.1 present different studies observing the effect of temperature on different algal species. It has been seen that raised temperatures in *Nannochloropsis salina* and *Ochromonas Danica* cultivation cause correspondence increase in their lipid content, whereas *Chlorella sorokiniana* behave contradictory and show no variation in lipid content corresponding to raised temperature (Fakhry and El Maghraby 2015). Kalacheva et al. (2002) use *Botryococcus braunii* cultivation to observe the effect of different temperatures (18 °C, 25 °C, and 32 °C) on cellular lipid content. At 25 °C (optimum temperature), lipid content was estimated up to 22%, whereas crossing 32 °C cause 5% less cellular lipid content than optimum temperature (Kalacheva et al. 2002). Similarly, 25 °C is the optimum growth temperature for *Chlorella vulgaris* and *Nannochloropsis oculata*. A study found that increasing temperature from 20 to 25 °C *Nannochloropsis oculata* cultivation results in the twofold increment of lipid accumulation (7.90–14.92%) (Converti et al. 2009). However, in the case of *C. vulgaris*, raising temperature more than optimum (from 25 to 30 °C) causes approximately 7% reduction (14.71–5.90%) in lipid content (Converti et al. 2009). Researchers found that temperatures ranging from 15 to 35 °C might be effective in raising lipid accumulation inside microalgal cells (Converti et al. 2009; Kalacheva et al. 2002; Fakhry and El Maghraby 2015). Some studies were performed to observe the additive impact of nutrient availability and temperature on the cellular lipid contents of algae. To examine the synergistic effect of temperature and nutrient availability, microalgal species such as *Chromulina ochromonoides*, *Dunaliella tertiolecta*, *Isochrysis galbana*, *Nannochloropsis oculata*, *Odontella aurita*, and *Thalassiosira pseudonana* were cultivated at 10 and 20 °C with two different conditions (nutrient starved and sufficient) (Roleda et al. 2013; Shuping et al. 2010). Results revealed that adding multiple stress conditions would not be able to give additional impact to enhance lipid content, and only the nutrient-deficient condition is favourable to improve lipid accumulation (Shuping et al. 2010; Roleda et al. 2013). A study on microalgae was performed to analyse the impact of temperature on fatty acid composition. *Nannochloropsis oculata* and *Tetraselmis subcordiformis* were cultivated under several temperatures ranging from 15 to 35 °C (Wei et al. 2015). Results showed that high temperatures cause more MUFA accumulation than PUFA and neutral lipids (Wei et al. 2015). *Dunaliella salina*, a marine microalga, was studied under a temperature range shifting from 30 to 12 °C to analyse temperature shifting effect on lipid composition. Under lower temperatures, unsaturated fatty acid accumulation is raised by 20% (Sharma et al. 2012). Generally, low temperatures lead to high membrane fluidity, more unsaturated fatty acids, and less saturated fatty acids associated with algal cell

Table 6.1 Influence of temperature on different algal species

Algae	Temperature (°C)	Other culture conditions	Effects observed	References
N. salina	15	Photoperiod: 12 h:12 h (light: dark) Light intensity: 150 μmol m^{-2} s^{-1}, pH 7.5	Dry cell weight: 0.42 ± 0.0 gL^{-1} Lipid productivity: 0.31 ± 0 gL^{-1} day^{-1}	Fakhry and El Maghraby (2015)
N. salina	20	Photoperiod: 12 h:12 h (light: dark) Light intensity: 150 μmol m^{-2} s^{-1}, pH 7.5	Dry cell weight: 0.45 ± 0.0 gL^{-1} Lipid productivity: 0.35 ± 0 gL^{-1} day^{-1}	Fakhry and El Maghraby (2015)
N. salina	25	Photoperiod: 12 h:12 h (light: dark) Light intensity: 150 μmol m^{-2} s^{-1}, pH 7.5	Dry cell weight: 0.50 ± 0.04 gL^{-1} Lipid productivity: 0.37 ± 0.01 gL^{-1} day^{-1}	Fakhry and El Maghraby (2015)
N. salina	30	Photoperiod: 12 h:12 h (light: dark) Light intensity: 150 μmol m^{-2} s^{-1}, pH 7.5	Dry cell weight: 0.53 ± 0.02 gL^{-1} Lipid productivity: 0.43 ± 0.0 gL^{-1} day^{-1}	Fakhry and El Maghraby (2015)
N. salina	35	Photoperiod: 12 h:12 h (light: dark) Light intensity: 150 μmol m^{-2} s^{-1}, pH 7.5	Dry cell weight: 0.51 ± 0.02 gL^{-1} Lipid productivity: 0.41 ± 0.01 gL^{-1} day^{-1}	Fakhry and El Maghraby (2015)
B. braunii	18	Illumination: 20 W/ m^2 illumination period: 24 h (14 h day and 10 h night) 1 vol % CO_2	Dry biomass weight: 0.43 g/L Total intracellular lipids: 50% TGA content: 9.3%	Kalacheva et al. (2002)
B. braunii	25	Illumination: 20 W/ m^2 illumination period: 24 h (14 h day and 10 h night) 1 vol % CO_2	Dry biomass weight: 0.44 g/L Total intracellular lipids: 50% TGA content: 2%	Kalacheva et al. (2002)
B. braunii	32	Illumination: 20 W/ m^2 illumination period: 24 h (14 h day and 10 h night) 1 vol % CO_2	Dry biomass weight: 0.7 g/L Total intracellular lipids: 50% TGA content: 13.8%	Kalacheva et al. (2002)

(continued)

Table 6.1 (continued)

Algae	Temperature (°C)	Other culture conditions	Effects observed	References
C. vulgaris	25	14 days Photon flux density: 70.0 μE m^{-2} s^{-1}	Lipid productivity: 20.22 ± 0.60 mg lipids L^{-1} $days^{-1}$	Converti et al. (2009)
C. vulgaris	30	14 days Photon flux density: 70.0 μE m^{-2} s^{-1}	Lipid productivity: 8.16 ± 0.65 mg lipids L^{-1} $days^{-1}$	Converti et al. (2009)
C. vulgaris	35	14 days Photon flux density: 70.0 μE m^{-2} s^{-1}	Lipid productivity: 8.21 ± 0.17 mg lipids L^{-1} $days^{-1}$	Converti et al. (2009)
C. vulgaris	38	14 days Photon flux density: 70.0 μE m^{-2} s^{-1}	Lipid productivity: −2.72 ± 1.62 mg lipids L^{-1} $days^{-1}$	Converti et al. (2009)

adaptation in stress conditions (Mathimani and Nair 2016). Regardless of cellular organization, eukaryotic green microalgae *B. braunii*, *C. vulgaris*, and prokaryotic algae *S. platensis* showed variation in lipid yield over raised temperatures. *B. braunii*, *C. vulgaris*, and *S. platensis* accumulated a lower amount of unsaturated fatty acids as compared to saturated fatty acids at high temperatures (Sushchik et al. 2003). Researchers are not able to conclude a similar trend of temperature variation for its impact on lipid accumulation in algal cells. High temperature enhances lipid accumulation as a storage product of algal cells due to its maximum use (Sayegh and Montagnes 2011). Stress conditions are applied to the activity of photosystem II when microalgal cultivation is done at more than optimum temperature and results in a lower yield of biomass and lipid (Sheng et al. 2011), whereas at extreme temperatures, microalgal growth is obstructed due to enzyme denaturation and termination of metabolic reactions (Renaud et al. 2002).

6.3.2 Effect of Light Intensity

Alga is a widespread photosynthetic living being. Light is a vital aspect of the autotrophic activity, photosynthesis, and growth of microalgal species. Algal cells consist of numerous photon harvesting pigments, where chlorophyll 'a' and chlorophyll 'b' are most abundant pigments that have sensitivity towards red and blue wavelengths. Light is a prominent aspect for algal growth. Algae can grow at different photoperiods, wavelengths, and light intensities; however, different light conditions may cause the difference in their photosynthetic activity and lipid accumulation. In addition, light intensity might influence the lipid yield and alteration in the fatty acid composition (Liu et al. 2008). In a study, a *Cladophora* species (filamentous algae) was cultivated using extreme light intensities. Results revealed that the composition of polar phospholipids was significantly reduced, whereas triacylglyceride accumulation was enhanced (Napolitano 1994). Table 6.2 present different studies observing the influence of light intensity on several algal species. Similarly, *Desmarestia viridis* was cultivated under varying light intensities. When light intensity shift from 700 to 1500 μmol m^{-2} s^{-1}, 63% biomass reduction was observed (Gordillo et al. 1998a, b). Total lipid content is amplified when *Desmarestia viridis* was cultivated under dark conditions; however, triglyceride contents are reduced simultaneously (Smith et al. 1993). An increment in extracellular polysaccharide content was observed while raising proton flux density (Iqbal and Zafar 1993). Algae that are grown under enhanced intensity result in more lipid accumulation. *Scenedesmus* sp. was cultivated in 250–400 μmol m^{-2} s^{-1} range of light intensity and showed enhanced lipid accumulation (Liu et al. 2012). Despite that, two marine algal species *Nannochloropsis* and *Chlorella* sp. were grown under an extreme photon density of 10,000 lx, demonstrating lower lipid production (Cheirsilp and Torpee 2012). de Mooij et al. (2016) cultivate *C. reinhardtii* to study its productivity in different light colour. *C. reinhardtii* cultivation done inside a photobioreactor has light intensity of 1500 μmol m^{-2} s^{-1}, where yellow light results in highest productivity (54 g m^{-2} day^{-1}) as compared to red and blue light (de Mooij et al. 2016). Krzemińska et al. (2014) use five different algae (*B. braunii*, *N. conjuncta*, *N. texensis*, *N. terrestris*, and *S. obliquus*) to observe consequences of two different photoperiods (12:12 light/dark hour cycle and 24-h uninterrupted illumination) on biomass productivity and growth rate. 12:12 light/dark hour cycle was favourable to stimulate growth rate in *Neochloris* sp., while continuous stimulation is found effective for higher growth rates in case of *S. obliquus* and *B. braunii* (Krzemińska et al. 2014). Also, a study observes the effect of continuous illumination on the fatty acid composition of a green microalga *Chlorella minutissima*; however, there is no significant difference occurred in fatty acid composition at a light intensity of 200–400 μE m^{-2} s^{-1} (Tang et al. 2011). Although several findings relate the fatty acid composition to the light intensity, it was observed that PUFA content was dropped down with raised light intensity (Juneja et al. 2013). Continuous illumination, longer photoperiods, and high photon intensity reduce MUFA and PUFA content and elevate saturated fatty acid levels (Al-Qasmi et al. 2012). *Thalassiosira pseudonana* cultivated in 12:12 h light/dark photoperiod revealed lower PUFA and elevated SFA and MUFA levels (Brown et al. 1996). Inconsistently, intensified photon flux of 50–150 μmol m^{-2} s^{-1} light intensity increases fatty acid accumulation in algal cells (Schnurr et al. 2016). No considerable alteration occurs in the fatty acid profile when raising the photon flux beyond 300–-600 μmol m^{-2} s^{-1} (Schnurr et al. 2016). Cuellar-Bermudez et al. (2015) studied the combined effect of CO_2 and light intensity on the fatty acid composition of *Synechocystis* sp. *PCC6803*. Total lipid content enhanced when *Synechocystis* sp. cultivated under 1920 μmol m^{-2} s^{-1} light intensity with 3% CO_2 concentration; however increasing light illumination does not affect fatty acid composition (Cuellar-Bermudez et al. 2015). It is considered that growth and lipid yield is varying with different algal strains. Beyond the threshold limit of light, intensity triggers lamellae disruption within the chloroplast and destined photoinhibition or inactivation of CO_2 fixing enzymes (Juneja et al. 2013). Extreme illumination intensities

Table 6.2 Impact of light on various microalgal species

Algae	Light conditions	Other growth condition	Effects observed	References
C. vulgaris	Light intensity: 300, 150, and 50 μmol m^{-2} s^{-1}	16:8 h (light/dark) Temperature: 25 °C Shaking: 150 rpm	Higher light intensity results greater biomass	Nzayisenga et al. (2020)
Desmodesmus sp.	Light intensity: 300, 150, and 50 μmol m^{-2} s^{-1}	16:8 h (light/dark) Temperature: 25 °C Shaking: 150 rpm	Higher light intensity results in greater biomass and raised fatty acid content	Nzayisenga et al. (2020)
E. pseudoalveolaris	Light intensity: 300, 150, and 50 μmol m^{-2} s^{-1}	16:8 h (light/dark) Temperature: 25 °C Shaking: 150 rpm	Higher light intensity results in more biomass	Nzayisenga et al. (2020)
S. obliquus	Light intensity: 300, 150, and 50 μmol m^{-2} s^{-1}	16:8 h (light/dark) Temperature: 25 °C Shaking: 150 rpm	Higher light intensity results in high biomass and raised fatty acid content	Nzayisenga et al. (2020)
I. galbana LB987	Photoperiod: 12 h/12 h (day/night) Light intensity: 0–200 μmol m^{-2} s^{-1}, 10 days	pH 8.0, Temperature: 25 °C mixotrophic condition	Optimal light intensity: 80–150 μmol m^{-2} s^{-1} Stimulated lipid production and chlorophyll synthesis Total lipid content: 30%	Gim et al. (2016)
N. oculata CCAP849/1	Photoperiod: 12 h/12 h (day/night) Light intensity: 0–200 μmol m^{-2} s^{-1}, 10 days	pH 8.0, Temperature: 25 °C mixotrophic condition	At 150 μmol/m^2/s, fatty acid concentrations increased slightly Total lipid content: 37.3%	Gim et al. (2016)
D. salina	Photoperiod: 12 h/12 h (day/night) Light intensity: 0–200 μmol m^{-2} s^{-1}, 10 days	pH 8.0, Temperature: 25 °C, mixotrophic condition	At 150 μmol/m^2/s, fatty acid concentrations increased slightly Total lipid content: 31.3 %	Gim et al. (2016)
Scenedesmus sp.	Light intensity: 50 μmol/m^2/s	Temperature: 25 °C, CO_2: 2% v/v	Biomass yield: 2.55 g/L Lipid content: 26.2%	Liu et al. (2012)

(continued)

Table 6.2 (continued)

Algae	Light conditions	Other growth condition	Effects observed	References
Scenedesmus sp.	Light intensity: 250 μmol/m^2/s	Temperature: 25 °C, CO_2: 2% v/v	Biomass yield: 3.62 g/L Lipid content: 39.2%	Liu et al. (2012)
Scenedesmus sp.	Light intensity: 400 μmol/m^2/s	Temperature: 25 °C, CO_2: 2% v/v	Biomass yield: 3.88 g L^{-1} Lipid content: 41.1 %	Liu et al. (2012)

(beyond threshold level) disrupt the photosynthetic receptor system and trigger photoinhibition (Wahidin et al. 2013). Availability of essential fatty acids, different pigments, and dark respiration rate are the severe factors behind the physiological photo-acclimatization, whereas changes in volume, number, and density of photosynthetic apparatus can cause morphological photo-acclimation (Fábregas et al. 2004). It is found that desaturated chloroplast membranes of algae can exceed the limitation of low light intensity (Mock and Kroon 2002). Prevention of photoinhibition and dissipation of excess photon energy could assist by NADH and ATP generated during photosynthesis. Also, triacylglyceride synthesis needs high NADH and ATP levels; hence the carbon flux is generated during photosynthesis when microalgal cells are cultivated under extreme light conditions (Solovchenko et al. 2008; Liu et al. 2012). Despite this, when algal cells absorb the surplus level of excitation energy, it results in photodamage (destruction of photoreceptors of PS II); however, polypeptides of the reaction centre are more stable during extreme light exposure (He et al. 2001). When *Synechocystis* sp. *PCC6883* was acclimatized under extreme light intensities, plastoquinone oxidation takes place, and the PS I/PS II ratio reaches up to 5, which was advantageous to minimize photodamage (Cuellar-Bermudez et al. 2015). The light could provoke algal culture growth, chloroplast membrane development, and lipid synthesis; hence light is a crucial factor for lipid accumulation in microalgal cells, but light intensity/photoperiod requirements are dependent and varies with algal strains.

6.3.3 Effect of Salinity

Salinity (NaCl concentration) is the vital factor that modifies the composition of biochemical compounds inside microalgal cells. Too low or high salinity exposure of algal cells cause alteration in their natural growth rate and biochemical composition. Studies revealed that high salinity can cause increased lipid content in algae (Zhila et al. 2011; Renaud and Parry 1994). In a study, marine algae *Dunaliella* was cultivated under different saline concentrations ranging from 0.4 to 4 M; results reported an increase in SFA and MUFA with increased salinity up to 4 M (Xu and

Beardall 1997). Similarly, an increase in triglyceride (40–56%) and lipid content (60–67%) was reported when *Dunaliella tertiolecta* was cultivated under saline concentration raised from 0.5 to 1.0 M NaCl (Takagi and Yoshida 2006). *Botryococcus braunii* cultivation under high salinity level results in higher growth rate, with increase in lipid and carbohydrate content (Rao et al. 2007). Correspondingly, *Botryococcus braunii* was cultivated in two salinity conditions (without NaCl and with 0.50 NaCl). This study reported higher lipid content in a media with 0.50 NaCl; however, carbohydrates, proteins, and pigments were less (Ben-Amotz et al. 1985). A study reported reduced protein content whereas no variation in lipid and carbohydrate content in *Botryococcus braunii* under high saline levels. However, growth rate was reduced as algal cells was unable of adapt in high saline concentration (Vazquez-Duhalt and Arredondo-Vega 1991). Accordingly, another study supported reduced protein content per algal cells in *Tetraselmis suecica* with increased NaCl concentration (Fabregas et al. 1984).

6.3.4 Effect of Nutrients

Varying nutrient composition using different nutrient limitations could be helpful to identify which nutrient or substrate affect algal growth up to what extent. In general, Michaelis-Menten equation described nutrient uptake rate. At optimal culture conditions, algal growth is proportional to the utilization rate of highly limiting substrate. N and P are crucial macronutrients for metabolism and growth of microalgal cells. Nitrogen is basic element to form nucleic acid and proteins. Similarly, phosphate plays a major role as energy currency in the form of ATP in all living beings. Despite that, phosphorous is key element found in RNA, DNA, and phospholipids. Limitation of nitrogen and phosphorus will shift metabolic system of microalgal cells. Starvation of N and P shifts the lipid metabolism and tends towards lipid storage instead of membrane lipid synthesis. Hence the total lipid content increases in microalgae under N- and P-starved conditions (Hirano et al. 1997). Specific discussion on all limiting substrate is reviewed below.

6.3.4.1 Carbon

C, H, and O are crucial non-mineral nutrients. Carbon is most essential nutrient for algal growth, photosynthesis, and reproduction. Carbon uptake is generally fixed by three ways: (1) as energy resource; (2) as respiration; and (3) as a raw material for cell formation (Berman-Frank and Dubinsky 1999). Lower carbon fixing rate results in lesser growth rate in microalgal cells. Algal cells need inorganic carbon substrate for photosynthesis which is available in the form of carbon dioxide (CO_2) or carbonate in autotrophic mode of nutrition. CO_2 dissolved in water are convertible into carbonate or bicarbonate depending on nutrient composition, pH, and temperature. With rise in pH level, carbonate content rises as compared to carbon dioxide and bicarbonate (Chen and Durbin 1994). Riebesell et al. (2000) observe *Emiliania huxleyi* to analyse the influence of CO_2 concentration on lipid composition. CO_2 concentration significantly influences the composition of PUFA and alkenones. Study revealed low concentration of CO_2 tends to increase PUFA 22:6 (n-3), while high CO_2 concentration raises 14:0 fatty acids composition (Riebesell et al. 2000). A study reported that increased CO_2 concentration results in enhanced unsaturation and fatty acids content (Tsuzuki et al. 1990). Similarly, in *Dunaliella salina*, increased CO_2 levels leads to elevated level of fatty acid accumulation (Muradyan et al. 2004).

6.3.4.2 Nitrogen

Nitrogen is readily consumed by algal cell; algae consume the inorganic source of nitrogen existing in media and turn it into biochemical compounds to carry out their physiological processes appropriately (Ross et al. 2018; Yang et al. 2017). However, it was studied that nitrogen-starved condition leads to high lipid biosynthesis level and enhanced triglyceride accumulation with lesser protein component (Wei et al. 2020; Heraud et al. 2005; Wang et al. 2009). So, the higher lipid and lower protien content could be obtained at the disbursement of algal growth rate (Converti et al. 2009; Li et al. 2008), under nitrogen-limited conditions to redirect their photosynthetic carbon into carbohydrate synthesis (Hu 2004). Nitrogen-depleted media can also cause lower oxygen generation, lesser CO_2 fixation, and decreased tissue and chlorophyll content (Kolber et al. 1988). In a study, sugar phosphates and ammonium were additionally supplied into the growth media of *Chlorella pyrenoidosa* where an increase in amino acid was observed (Holm-Hansen et al. 1959). In algal photosystem II, phycobilisomes act as antennae for light harvesting. In red algae and cyanobacteria, nitrogen-limited culture media cause degradation of phycobilisomes (Collier and Grossman 1992). Photosynthesis remains persistent (with reduced rate) until nitrogen concentration depleted below a threshold level. *Spirulina platensis* cells were grown under high CO_2 availability with nitrogen-depleted medium and depict very less carbon fixation efficiency (Gordillo et al. 1998a, b). Nitrogen depletion may cause variation in enzyme levels inside algal cells and cause lipid synthesis however simultaneously reducing chlorophyll synthesis that generates higher carotenoid accumulation (Zhang et al. 2017).

6.3.4.3 Phosphorus

Phosphorus is a crucial element for adequate growth and metabolism in microalgal cells. Besides nitrogen, phosphorus is primary limiting macronutrient for microalgal cells (Ren et al. 2017). Phosphorus limitation generally cause lesser than needful rate of light assimilation for carbon fixation; also it will result in reduced level of substrate synthesis during Calvin-Benson cycle (Barsanti and Gualtieri 2005). Phosphorus limitation will be an advantageous condition to gain high lipid accumulation in algal cells. In a study, *Scenedesmus* sp. is grown at 2.0 mg L^{-1} of initial phosphorus concentration and then reduced it to 0.1 mg L^{-1} early phosphorus concentration. It was observed that reduction of phosphorus concentration attains increment in total lipid accumulation to 53% (at 0.1 mg L^{-1}) from 23% (at 2.0 mg L^{-1}) (Xin et al. 2010). Phosphatidylglycerol is a glycerolipid present in chloroplast membrane. It also has significant roles in cell growth, keeping satisfactory levels of chlorophyll protein complex and PS II function. In a study, *Chlamydomonas reinhardtii* was grown in a phosphorus-limited medium; results depict decreased phosphatidylglycerol content (Sato et al. 2000). In phosphorus-deficient medium, protein and *chlorophyll a* content tends to reduce and thus enhanced relative carbohydrates amount in algal cells (Healey and Hendzel 1979). Like nitrogen deficiency, phosphorus limitation tends to reduced phycobilisome content (Collier and Grossman 1992). In *Selenastrum minutum*, phosphorus-deprived condition causes reduction in respiration rate of algal cells (Theodorou et al. 1991).

6.3.4.4 Micronutrients

Trace metals are micronutrients required by living cells in very less quantity (less than 4 ppm), and these trace metals are very crucial for physiology of algal cells. Trace metals commonly include cobalt, copper, manganese, iron, zinc, and nickel (Facey et al. 2019). For algal cells, trace metal accessibility is dependent on free ion concentration, i.e. it is independent of dissolving trace metal concentration in the culture media (Parent et al. 1996). High level of trace metal in culture media can cause reduction in antioxidant levels, cell membrane damage, and photosynthesis impairment; similarly deficiency of these can also inhibit cell growth. Iron plays significant role as catalyst redox reactions of electron transport system, nitrogen assimilation, and photosynthesis in algal cells (Rueler and Ades 1987). Iron deficiency causes reduction in photosynthetic electron transport, which further reduces NADPH availability. Iron deficiency causes reduction in ferredoxin availability and substitutes ferredoxin with flavodoxin (McKay et al. 1999). Subsequently, ferredoxin substitution could be unfavourable because the catalytic activity of flavodoxin is very slow and not comparable and efficient as ferredoxin (McKay et al. 1999). Iron-limited media can bring chlorophyll-deficit condition inside algal cells (Greene et al. 1992). In *C. vulgaris*, effect of elevated iron concentration has been observed in a study and remarks enhanced lipid content (Liu et al. 2008), whereas iron-deficient condition can cause depletion in carotenoid composition (van Leeuwe and Stefels 1998). Algal cell membrane consists of many functional groups like phosphate, carboxylic, and sulfhydryl groups. Such functional groups act as binding site for metal ions. Most of the metabolic reactions will be inhibited by the presence of some non-essential elements like chromium, lead, and cadmium. A very small concentration of these metals can cause severe toxicity within algal cells. Similarly, presence of excess concentration of essential metals like zinc, nickel, and copper can also cause cellular toxicity (Campanella et al. 2001; Kennish 1992; Rai and Mallick 1993).

6.3.5 Effect of CO_2

Carbon dioxide (CO_2) is accessible in the atmosphere and emitted continuously from different industrial sources, vehicles, and power plants. Increasing levels of CO_2 in atmosphere will be harmful for living beings. Algal cells are naturally contributing to mitigate CO_2 by carbon capturing and carbon sequestration. At high CO_2 concentration, *Chlorella* sp. shows higher rate of photosynthesis (Singh and Singh 2014). Mainly CO_2 concentration affects lipid content and microalgal growth rate. Some researchers inspected the impact of CO_2 concentration on fatty acid, lipid composition, and biomass growth in different algae. In a study, *C. vulgaris*, *Scenedesmus* sp., and *B. braunii* are chosen to observe the effect of different CO_2 concentrations (Yoo et al. 2010). At 10% CO_2 level, after 2 weeks, *B. braunii* gain 26.55 mg L^{-1} biomass productivity and 5.51 mg L^{-1} day^{-1} of total lipid productivity, whereas *Chlorella vulgaris* obtain biomass productivity of 104.76 mg L^{-1} day^{-1} and lipid productivity of 6.91 mg L^{-1} day^{-1} (Yoo et al. 2010). On similar culture conditions and 10% CO_2 levels, *Scenedesmus* sp. obtains biomass productivity of 217.50 mg L^{-1} day^{-1} and lipid productivity of 20.65 mg L^{-1} day^{-1} (Yoo et al. 2010). In the same analysis, Yoo et al. (2010) investigate effect of flue gases comprising 5.5% CO_2 levels on *C. vulgaris*, *Scenedesmus* sp., and *B. braunii*. Results show enhanced biomass productivity in all three species, whereas *Scenedesmus* sp. shows 3.7-fold increment in the total lipid productivity (Yoo et al. 2010). In a study, *B. braunii* 765 was cultured with 20% CO_2 availability; findings reported increasing CO_2 levels from 2% to 20% cause enhanced biomass productivity and total lipid productivity varies from 10.41% to 12.71% by weight (Ge et al. 2011). With an environmental perspective, *Phaeodactylum tricornutum* and *Nannochloropsis salina* were cultured using desulphated and untreated flue gas with inorganic source of CO_2. Result observed no significant changes in biomass obtained from *P. tricornutum* and *N. salina*, although the presence of some toxic non-essential metals like mercury, vanadium, and nickel in flue gas can affect the biomass productivity (Moheimani 2013). A study was performed on *D. viridis* which was cultured in nitrogen deficit and high CO_2 availability simultaneously. Result has shown that lipid composition in *D. viridis* is very sensitive towards nitrogen-deficient condition (in presence of high CO_2 concentration) (Gordillo et al. 1998a, b).

6.3.6 Effect of pH

pH of any culture media is the essential factor in any metabolic process, so in the microalgal growth, pH plays major role in metal speciation, nutrient availability, and enzymatic activities within the microalgal cells. Any changes in pH levels may directly influence enzymatic actions and further affected different metabolic pathways like photosynthesis and lipid synthesis (Jin et al. 2016). However, response of algal cells and its metabolic process towards optimum pH or at other pH range is different and strain specific (Moheimani 2013). Varying pH affects lipid and biomass yield in microalgal cells. In a study, during inoculation initial pH of algal growth media was kept neutral because availability of carbon species is depending on the pH. After some time, algal cells consume inorganic carbon present in media, and hence pH of the growth media gradually increases (Rai et al. 2015). Effect of numerous inorganic carbon sources and pH on *Chlorella* sp. and *T. suecica CS-187* was observed for lipid and biomass productivity. At pH 7.5, *T. suecica* gained a biomass productivity of 320 mg L^{-1} day^{-1} and lipid productivity of 92 mg L^{-1} day^{-1}, whereas at pH 7.0, *Chlorella* sp. obtained biomass productivity of 407 mg L^{-1} day^{-1} and lipid productivity of 99 mg L^{-1} day^{-1} (Moheimani 2013). In a study, growth media was set with different pH ranging from 6.0 to 10.0 to observe effect of pH on the lipid accumulation and biomass productivity in *N. salina* (Bartley et al. 2014). Outcomes revealed that pH 8 to pH 9 depicts maximum growth rates in *N. salina* (Bartley et al. 2014). In a study, *C. vulgaris* cultivated in heterotrophic mode at optimum pH (7.5) in sulphur-deficient condition obtain lipid content of 53.43% and maximum specific growth rate of 0.541 $days^{-1}$. Cell death occurs in *C. vulgaris* at pH 3, 4, and 11, whereas at pH 9.5 algal cells start to aggregate (Sakarika and Kornaros 2016). In a different study *Chlorella* sp. was grown at varying pH levels ranging from 5 to 11 to observe impact of pH on biomass growth and lipid content (Zhang et al. 2014). After a month, highest lipid content was 32.8% (lipid yield of 168 mg L^{-1}) and was noticed at pH 7.0, whereas at pH 5.0, maximum triacylglyceride content was 63% (Zhang et al. 2014). For *Dunaliella salina* pH 11.5 and *Dunaliella acidophila* pH 3 is found as optimum pH (Varshney et al. 2015).

6.4 Future Aspects

In this chapter, we discussed influence of different culture conditions like nutrients, and environmental factors like light, pH, CO_2 availability, temperature, etc., on algal biomass and lipid content. Tremendous efforts have been done by the researchers to understand behaviour of algal biomass growth for lipid production and nutrient recovery from wastewater. These efforts include enhanced optimization of the environmental factors and its effects on algal cell growth and composition. With the progress made by biotechnology research and bioinformatics software, it is feasible to acquire an enormous information from genetic data. It is possible now to trace molecular interaction and its effect on metabolism. Genome sequences of *Chlorella vulgaris*, *Chlamydomonas reinhardtii*, and *Dunaliella salina* are known (Smith et al. 2010; Merchant et al. 2007; Blanc et al. 2010). Predictive models and automated control systems for optimum culture condition would be effective solution for huge scale algal biomass and fatty acid production (Coşgun et al. 2021).

Understanding of critical factors affecting algal production will be beneficial for successful and sustainable biomass production for renewable fuel generation.

6.5 Conclusions

In present scenario, increasing demands of conventional fuels bring necessity of alternative sustainable energy resources. Biodiesel, biobutanol, biohydrogen, and bioethanol production from microalgae is promising alternative for fossil fuels. However, lipid yield is still not sufficient to meet biodiesel production as per demand. Hence, most research work is inclined towards enhanced lipid production by altering different culture conditions such different light colours, light intensities, photoperiods, temperature, pH, and inorganic carbon sources. So, this chapter give a comprehensive glance by discussing influence of critical factors for increasing lipid content, fatty acid composition, and biomass growth of microalgae.

Acknowledgement The authors of the manuscript are thankful to the Indian Institute of Technology (BHU) Varanasi, Varanasi, for extending their technical and financial support.

Conflict of Interest The authors have declared no conflict of interest.

References

Abdullah B, Muhammad SAFAS, Shokravi Z, Ismail S, Kassim KA, Mahmood AN, Aziz MMA (2019) Fourth generation biofuel: a review on risks and mitigation strategies. Renew Sust Energ Rev 107:37–50

Adeniyi OM, Azimov U, Burluka A (2018) Algae biofuel: current status and future applications. Renew Sust Energ Rev 90:316–335

Adenle AA, Haslam GE, Lee L (2013) Global assessment of research and development for algae biofuel production and its potential role for sustainable development in developing countries. Energy Policy 61:182–195

Afify AEMM, Shalaby EA, Shanab SM (2010) Enhancement of biodiesel production from different species of algae. Grasas Aceites 61(4):416–422

Al-Qasmi M, Raut N, Talebi S, Al-Rajhi S, Al-Barwani T (2012) A review of effect of light on microalgae growth. In: Proceedings of the world congress on engineering, vol 1(4)

Barsanti L, Gualtieri P (2005) Algae: anatomy, biochemistry, and biotechnology. CRC Press

Bartley ML, Boeing WJ, Dungan BN, Holguin FO, Schaub T (2014) pH effects on growth and lipid accumulation of the biofuel microalgae Nannochloropsis salina and invading organisms. J Appl Phycol 26(3):1431–1437

Beacham TA, Sweet JB, Allen MJ (2017) Large scale cultivation of genetically modified microalgae: a new era for environmental risk assessment. Algal Res 25:90–100

Ben-Amotz A, Tornabene TG, Thomas WH (1985) Chemical profile of selected species of microalgae with emphasis on lipids 1. J Phycol 21(1):72–81

Berman-Frank I, Dubinsky Z (1999) Balanced growth in aquatic plants: Myth or reality? Phytoplankton use the imbalance between carbon assimilation and biomass production to their strategic advantage. Bioscience 49(1):29–37

Blanc G, Duncan G, Agarkova I, Borodovsky M, Gurnon J, Kuo A et al (2010) The Chlorella variabilis NC64A genome reveals adaptation to photosymbiosis, coevolution with viruses, and cryptic sex. Plant Cell 22(9):2943–2955

Brennan L, Owende P (2010) Biofuels from microalgae—a review of technologies for production, processing, and extractions of biofuels and co-products. Renew Sust Energ Rev 14(2):557–577

Brown MR, Dunstan GA, Norwood SJ, Miller KA (1996) Effects of harvest stage and light on the biochemical composition of the diatom Thalassiosira pseudonana 1. J Phycol 32(1):64–73

Burkholder JM, Glibert PM, Skelton HM (2008) Mixotrophy, a major mode of nutrition for harmful algal species in eutrophic waters. Harmful Algae 8(1):77–93

Campanella L, Cubadda F, Sammartino MP, Saoncella AJWR (2001) An algal biosensor for the monitoring of water toxicity in estuarine environments. Water Res 35(1):69–76

Cheirsilp B, Torpee S (2012) Enhanced growth and lipid production of microalgae under mixotrophic culture condition: effect of light intensity, glucose concentration and fed-batch cultivation. Bioresour Technol 110:510–516

Chen CY, Durbin EG (1994) Effects of pH on the growth and carbon uptake of marine phytoplankton. Mar Ecol Prog Ser 109:83–83

Chisti Y (2007) Biodiesel from microalgae. Biotechnol Adv 25(3):294–306

Collier JL, Grossman AR (1992) Chlorosis induced by nutrient deprivation in Synechococcus sp. strain PCC 7942: not all bleaching is the same. J Bacteriol 174(14):4718–4726

Converti A, Casazza AA, Ortiz EY, Perego P, Del Borghi M (2009) Effect of temperature and nitrogen concentration on the growth and lipid content of Nannochloropsis oculata and Chlorella vulgaris for biodiesel production. Chem Eng Process Process Intensif 48(6):1146–1151

Coşgun A, Günay ME, Yıldırım R (2021) Exploring the critical factors of algal biomass and lipid production for renewable fuel production by machine learning. Renew Energy 163:1299–1317

Cuellar-Bermudez SP, Romero-Ogawa MA, Vannela R, Lai YS, Rittmann BE, Parra-Saldivar R (2015) Effects of light intensity and carbon dioxide on lipids and fatty acids produced by Synechocystis sp. PCC6803 during continuous flow. Algal Res 12:10–16

Daliry S, Hallajisani A, Mohammadi RJ, Nouri H, Golzary A (2017) Investigation of optimal condition for Chlorella vulgaris microalgae growth. Global J Environ Sci Manage 3(2):217–223

de Mooij T, de Vries G, Latsos C, Wijffels RH, Janssen M (2016) Impact of light color on photobioreactor productivity. Algal Res 15:32–42

Deng YL, Kuo MY, Juang YJ (2014) Development of flow through dielectrophoresis microfluidic chips for biofuel production: Sorting and detection of microalgae with different lipid contents. Biomicrofluidics 8(6):064120

Dutta K, Daverey A, Lin JG (2014) Evolution retrospective for alternative fuels: First to fourth generation. Renew Energy 69:114–122

Fabregas J, Abalde J, Herrero C, Cabezas B, Veiga M (1984) Growth of the marine microalga Tetraselmis suecica in batch cultures with different salinities and nutrient concentrations. Aquaculture 42(3-4):207–215

Fábregas J, Maseda A, Domínguez A, Otero A (2004) The cell composition of Nannochloropsis sp. changes under different irradiances in semicontinuous culture. World J Microbiol Biotechnol 20(1):31–35

Facey JA, Apte SC, Mitrovic SM (2019) A review of the effect of trace metals on freshwater cyanobacterial growth and toxin production. Toxins 11(11):643

Fakhry EM, El Maghraby DM (2015) Lipid accumulation in response to nitrogen limitation and variation of temperature in Nannochloropsis salina. Bot Stud 56(1):1–8

Ge Y, Liu J, Tian G (2011) Growth characteristics of Botryococcus braunii 765 under high CO2 concentration in photobioreactor. Bioresour Technol 102(1):130–134

Gim GH, Ryu J, Kim MJ, Kim PI, Kim SW (2016) Effects of carbon source and light intensity on the growth and total lipid production of three microalgae under different culture conditions. J Ind Microbiol Biotechnol 43(5):605–616

Gordillo FJ, Goutx M, Figueroa FL, Niell FX (1998a) Effects of light intensity, CO_2 and nitrogen supply on lipid class composition of Dunaliella viridis. J Appl Phycol 10(2):135–144

Gordillo FJ, Jiménez C, Figueroa FL, Niell FX (1998b) Effects of increased atmospheric CO_2 and N supply on photosynthesis, growth and cell composition of the cyanobacterium Spirulina platensis (Arthrospira). J Appl Phycol 10(5):461–469

Greene RM, Geider RJ, Kolber Z, Falkowski PG (1992) Iron-induced changes in light harvesting and photochemical energy conversion processes in eukaryotic marine algae. Plant Physiol 100(2):565–575

He Q, Dolganov N, Björkman O, Grossman AR (2001) The high light-inducible polypeptides in Synechocystis PCC6803: expression and function in high light. J Biol Chem 276(1):306–314

Healey FP, Hendzel LL (1979) Indicators of phosphorus and nitrogen deficiency in five algae in culture. J Fish Board Canada 36(11):1364–1369

Heraud P, Wood BR, Tobin MJ, Beardall J, McNaughton D (2005) Mapping of nutrient-induced biochemical changes in living algal cells using synchrotron infrared microspectroscopy. FEMS Microbiol Lett 249(2):219–225

Hirano A, Ueda R, Hirayama S, Ogushi Y (1997) CO_2 fixation and ethanol production with microalgal photosynthesis and intracellular anaerobic fermentation. Energy 22(2-3):137–142

Holm-Hansen O, Nishida K, Moses V, Calvin M (1959) Effects of mineral salts on short-term incorporation of carbon dioxide in Chlorella. J Exp Bot 10(1):109–124

Hu Q (2004) Environmental effects on cell composition, vol 1. Blackwell Science Ltd., Oxford, pp 83–93

Hu Q, Sommerfeld M, Jarvis E, Ghirardi M, Posewitz M, Seibert M, Darzins A (2008) Microalgal triacylglycerols as feedstocks for biofuel production: perspectives and advances. Plant J 54(4): 621–639

Huang G, Chen F, Wei D, Zhang X, Chen G (2010) Biodiesel production by microalgal biotechnology. Appl Energy 87(1):38–46

Iqbal M, Zafar SI (1993) Effects of photon flux density, CO_2, aeration rate, and inoculum density on growth and extracellular polysaccharide production by Porphyridium cruentum. Folia Microbiol 38(6):509–514

Jin J, Dupré C, Legrand J, Grizeau D (2016) Extracellular hydrocarbon and intracellular lipid accumulation are related to nutrient-sufficient conditions in pH-controlled chemostat cultures of the microalga Botryococcus braunii SAG 30.81. Algal Res 17:244–252

Juneja A, Ceballos RM, Murthy GS (2013) Effects of environmental factors and nutrient availability on the biochemical composition of algae for biofuels production: a review. Energies 6:4607–4638

Kalacheva GS, Zhila NO, Volova TG, Gladyshev MI (2002) The effect of temperature on the lipid composition of the green alga Botryococcus. Microbiology 71(3):286–293

Kennish MJ (1992) Ecology of Estuaries. Anthropogenic effects. CRC Press Inc., Boca Raton, FL

Kolber Z, Zehr J, Falkowski P (1988) Effects of growth irradiance and nitrogen limitation on photosynthetic energy conversion in photosystem II. Plant Physiol 88(3):923–929

Krzemińska I, Pawlik-Skowrońska B, Trzcińska M, Tys J (2014) Influence of photoperiods on the growth rate and biomass productivity of green microalgae. Bioprocess Biosyst Eng 37(4): 735–741

Lam MK, Lee KT (2011) Renewable and sustainable bioenergies production from palm oil mill effluent (POME): win–win strategies toward better environmental protection. Biotechnol Adv 29(1):124–141

Lee OK, Lee EY (2016) Sustainable production of bioethanol from renewable brown algae biomass. Biomass Bioenergy 92:70–75

Leite GB, Abdelaziz AE, Hallenbeck PC (2013) Algal biofuels: challenges and opportunities. Bioresour Technol 145:134–141

Levitan O, Dinamarca J, Hochman G, Falkowski PG (2014) Diatoms: a fossil fuel of the future. Trends Biotechnol 32(3):117–124

Li Y, Horsman M, Wang B, Wu N, Lan CQ (2008) Effects of nitrogen sources on cell growth and lipid accumulation of green alga Neochloris oleoabundans. Appl Microbiol Biotechnol 81(4): 629–636

Liao Q, Chang JS, Herrmann C, Xia A (eds) (2018) Bioreactors for microbial biomass and energy conversion. Springer

Liu ZY, Wang GC, Zhou BC (2008) Effect of iron on growth and lipid accumulation in Chlorella vulgaris. Bioresour Technol 99(11):4717–4722

Liu J, Yuan C, Hu G, Li F (2012) Effects of light intensity on the growth and lipid accumulation of microalga Scenedesmus sp. 11-1 under nitrogen limitation. Appl Biochem Biotechnol 166(8): 2127–2137

Martinez F, Ascaso C, Orus MI (1991) Morphometric and stereologic analysis of Chlorella vulgaris under heterotrophic growth conditions. Ann Bot 67(3):239–245

Mata TM, Martins AA, Caetano NS (2010) Microalgae for biodiesel production and other applications: a review. Renew Sust Energ Rev 14(1):217–232

Mathimani T, Nair BB (2016) Evaluation of microalga for biodiesel using lipid and fatty acid as a marker—a central composite design approach. J Energy Inst 89(3):436–446

McKay RML, La Roche J, Yakunin AF, Durnford DG, Geider RJ (1999) Accumulation of ferredoxin and flavodoxin in a marine diatom in response to Fe. J Phycol 35(3):510–519

Merchant SS, Prochnik SE, Vallon O, Harris EH, Karpowicz SJ, Witman GB, Grossman AR (2007) The Chlamydomonas genome reveals the evolution of key animal and plant functions. Science 318(5848):245–250

Mock T, Kroon BM (2002) Photosynthetic energy conversion under extreme conditions—II: the significance of lipids under light limited growth in Antarctic Sea ice diatoms. Phytochemistry 61(1):53–60

Moheimani NR (2013) Inorganic carbon and pH effect on growth and lipid productivity of Tetraselmis suecica and Chlorella sp (Chlorophyta) grown outdoors in bag photobioreactors. J Appl Phycol 25:387–398

Montagnes DJS, Franklin DJ (2001) Effect of temperature on diatom volume, growth rate, and carbon and nitrogen content: reconsidering some paradigms. Limnol Oceanogr 46:2008–2018

Moravvej Z, Makarem MA, Rahimpour MR (2019) The fourth generation of biofuel. In: Second and third generation of feedstocks. Elsevier, pp 557–597

Muradyan EA, Klyachko-Gurvich GL, Tsoglin LN, Sergeyenko TV, Pronina NA (2004) Changes in lipid metabolism during adaptation of the Dunaliella salina photosynthetic apparatus to high CO_2 concentration. Russ J Plant Physiol 51(1):53–62

Napolitano GE (1994) The relationship of lipids with light and chlorophyll measurements in freshwater algae and periphyton 1. J Phycol 30(6):943–950

Nzayisenga JC, Farge X, Groll SL, Sellstedt A (2020) Effects of light intensity on growth and lipid production in microalgae grown in wastewater. Biotechnol Biofuels 13(1):1–8

Parent L, Twiss MR, Campbell PG (1996) Influences of natural dissolved organic matter on the interaction of aluminum with the microalga Chlorella: a test of the free-ion model of trace metal toxicity. Environ Sci Technol 30(5):1713–1720

Pulz O, Gross W (2004) Valuable products from biotechnology of microalgae. Appl Microbiol Biotechnol 65(6):635–648

Rai LC, Mallick N (1993) Heavy metal toxicity to algae under synthetic microcosm. Ecotoxicology 2(4):231–242

Rai MP, Gautom T, Sharma N (2015) Effect of salinity, pH, light intensity on growth and lipid production of microalgae for bioenergy application. OnLine J Biol Sci 15(4):260

Rao AR, Dayananda C, Sarada R, Shamala TR, Ravishankar GA (2007) Effect of salinity on growth of green alga Botryococcus braunii and its constituents. Bioresour Technol 98(3):560–564

Ras M, Steyer JP, Bernard O (2013) Temperature effect on microalgae: a crucial factor for outdoor production. Rev Environ Sci Biotechnol 12(2):153–164

Rawat I, Kumar RR, Mutanda T, Bux F (2013) Biodiesel from microalgae: a critical evaluation from laboratory to large scale production. Appl Energy 103:444–467

Razzak SA, Ali SAM, Hossain MM, deLasa H (2017) Biological CO_2 fixation with production of microalgae in wastewater—a review. Renew Sust Energ Rev 76:379–390

Ren L, Wang P, Wang C, Chen J, Hou J, Qian J (2017) Algal growth and utilization of phosphorus studied by combined mono-culture and co-culture experiments. Environ Pollut 220:274–285

Renaud SM, Parry DL (1994) Microalgae for use in tropical aquaculture II: effect of salinity on growth, gross chemical composition and fatty acid composition of three species of marine microalgae. J Appl Phycol 6(3):347–356

Renaud SM, Thinh LV, Lambrinidis G, Parry DL (2002) Effect of temperature on growth, chemical composition and fatty acid composition of tropical Australian microalgae grown in batch cultures. Aquaculture 211(1–4):195–214

Riebesell U, Revill AT, Holdsworth DG, Volkman JK (2000) The effects of varying CO_2 concentration on lipid composition and carbon isotope fractionation in Emiliania huxleyi. Geochim Cosmochim Acta 64(24):4179–4192

Roleda MY, Slocombe SP, Leakey RJ, Day JG, Bell EM, Stanley MS (2013) Effects of temperature and nutrient regimes on biomass and lipid production by six oleaginous microalgae in batch culture employing a two-phase cultivation strategy. Bioresour Technol 129:439–449

Ross ME, Davis K, McColl R, Stanley MS, Day JG, Semião AJ (2018) Nitrogen uptake by the macro-algae Cladophora coelothrix and Cladophora parriaudii: Influence on growth, nitrogen preference and biochemical composition. Algal Res 30:1–10

Rueler JG, Ades DR (1987) The role of iron nutrition in photosynthesis and nitrogen assimilation in SCENEDESMUS QUADRICAUDA (Chlorophyceae) 1. J Phycol 23(3):452–457

Sakarika M, Kornaros M (2016) Effect of pH on growth and lipid accumulation kinetics of the microalga Chlorella vulgaris grown heterotrophically under sulfur limitation. Bioresour Technol 219:694–701

Sato N, Hagio M, Wada H, Tsuzuki M (2000) Environmental effects on acidic lipids of thylakoid membranes. Biochem Soc Trans 28(6):912–914

Sayegh FA, Montagnes DJ (2011) Temperature shifts induce intraspecific variation in microalgal production and biochemical composition. Bioresour Technol 102(3):3007–3013

Scarsella M, Belotti G, De Filippis P, Bravi M (2010) Study on the optimal growing conditions of Chlorella vulgaris in bubble column photobioreactors. Chem Eng 20:85–90

Schenk PM, Thomas-Hall SR, Stephens E, Marx UC, Mussgnug JH, Posten C, Hankamer B (2008) Second generation biofuels: high-efficiency microalgae for biodiesel production. Bioenergy Res 1(1):20–43

Schnurr PJ, Espie GS, Allen GD (2016) The effect of photon flux density on algal biofilm growth and internal fatty acid concentrations. Algal Res 16:349–356

Shafik HM, Saad MG, El-Serehy HA (2015) Impact of nitrogen regime on fatty acid profiles of Desmodesmus quadricaudatus and Chlorella sp. and ability to produce biofuel. Acta Bot Hungar 57(1-2):205–218

Sharma KK, Schuhmann H, Schenk PM (2012) High lipid induction in microalgae for biodiesel production. Energies 5(5):1532–1553

Sheng J, Kim HW, Badalamenti JP, Zhou C, Sridharakrishnan S, Krajmalnik-Brown R, Vannela R (2011) Effects of temperature shifts on growth rate and lipid characteristics of Synechocystis sp. PCC6803 in a bench-top photobioreactor. Bioresour Technol 102(24):11218–11225

Shuba ES, Kifle D (2018) Microalgae to biofuels: 'promising' alternative and renewable energy, review. Renew Sust Energ Rev 81:743–755

Shuping Z, Yulong W, Mingde Y, Kaleem I, Chun L, Tong J (2010) Production and characterization of bio-oil from hydrothermal liquefaction of microalgae Dunaliella tertiolecta cake. Energy 35(12):5406–5411

Singh SP, Singh P (2014) Effect of CO_2 concentration on algal growth: a review. Renew Sust Energ Rev 38:172–179

Sirakov I, Velichkova K, Stoyanova S, Staykov Y (2015) The importance of microalgae for aquaculture industry. Review. Int J Fish Aquat Stud 2(4):81–84

Smith RE, Cavaletto JF, Eadie BJ, Gardner WS (1993) Growth and lipid composition of high Arctic ice algae during the spring bloom at Resolute, Northwest Territories, Canada. Mar Ecol Prog Ser:19–29

Smith DR, Lee RW, Cushman JC, Magnuson JK, Tran D, Polle JE (2010) The Dunaliella salina organelle genomes: large sequences, inflated with intronic and intergenic DNA. BMC Plant Biol 10(1):1–14

Solovchenko AE, Khozin-Goldberg I, Didi-Cohen S, Cohen Z, Merzlyak MN (2008) Effects of light intensity and nitrogen starvation on growth, total fatty acids and arachidonic acid in the green microalga Parietochloris incisa. J Appl Phycol 20(3):245–251

Suparmaniam U, Lam MK, Uemura Y, Lim JW, Lee KT, Shuit SH (2019) Insights into the microalgae cultivation technology and harvesting process for biofuel production: a review. Renew Sust Energ Rev 115:109361

Sushchik NN, Kalacheva GS, Zhila NO, Gladyshev MI, Volova TG (2003) A temperature dependence of the intra-and extracellular fatty-acid composition of green algae and cyanobacterium. Russ J Plant Physiol 50(3):374–380

Suzuki Y, Takahashi M (1995) Growth response of several diatom species isolated from various environments to temperature. J Phycol 31:880–888

Takagi M, Yoshida T (2006) Effect of salt concentration on intracellular accumulation of lipids and triacylglyceride in marine microalgae Dunaliella cells. J Biosci Bioeng 101(3):223–226

Tang H, Chen M, Garcia MED, Abunasser N, Ng KS, Salley SO (2011) Culture of microalgae Chlorella minutissima for biodiesel feedstock production. Biotechnol Bioeng 108(10): 2280–2287

Teoh ML, Chu WL, Phang SM (2010) Effect of temperature change on physiology and biochemistry of algae: a review. MJS 29(2):82–97

Theodorou ME, Elrifi IR, Turpin DH, Plaxton WC (1991) Effects of phosphorus limitation on respiratory metabolism in the green alga Selenastrum minutum. Plant Physiol 95(4):1089–1095

Tsukahara K, Sawayama S (2005) Liquid fuel production using microalgae. J Jpn Pet Inst 48(5): 251–259

Tsuzuki M, Ohnuma E, Sato N, Takaku T, Kawaguchi A (1990) Effects of CO_2 concentration during growth on fatty acid composition in microalgae. Plant Physiol 93(3):851–856

van Leeuwe MA, Stefels J (1998) Effects of iron and light stress on the biochemical composition of Antarctic Phaeocystis sp. (Prymnesiophyceae). II. Pigment composition. J Phycol 34(3): 496–503

Van Wagenen J, Miller TW, Hobbs S, Hook P, Crowe B, Huesemann M (2012) Effects of light and temperature on fatty acid production in Nannochloropsis salina. Energies 5(3):731–740

Varshney P, Mikulic P, Vonshak A, Beardall J, Wangikar PP (2015) Extremophilic micro-algae and their potential contribution in biotechnology. Bioresour Technol 184:363–372

Vazquez-Duhalt R, Arredondo-Vega BO (1991) Haloadaptation of the green alga Botryococcus braunii (race A). Phytochemistry 30(9):2919–2925

Verma M, Mishra V (2020) An introduction to algal biofuels. In: Microbial strategies for techno-economic biofuel production. Springer, Singapore, pp 1–34

Wahidin S, Idris A, Shaleh SRM (2013) The influence of light intensity and photoperiod on the growth and lipid content of microalgae Nannochloropsis sp. Bioresour Technol 129:7–11

Wang ZT, Ullrich N, Joo S, Waffenschmidt S, Goodenough U (2009) Algal lipid bodies: stress induction, purification, and biochemical characterization in wild-type and starchless Chlamydomonas reinhardtii. Eukaryot Cell 8(12):1856–1868

Wei L, Huang X, Huang Z (2015) Temperature effects on lipid properties of microalgae Tetraselmis subcordiformis and Nannochloropsis oculata as biofuel resources. Chin J Oceanol Limnol 33(1):99–106

Wei Z, Wang H, Li X, Zhao Q, Yin Y, Xi L et al (2020) Enhanced biomass and lipid production by co-cultivation of Chlorella vulgaris with Mesorhizobium sangaii under nitrogen limitation. J Appl Phycol 32(1):233–242

Xin L, Hong-Ying H, Ke G, Ying-Xue S (2010) Effects of different nitrogen and phosphorus concentrations on the growth, nutrient uptake, and lipid accumulation of a freshwater microalga Scenedesmus sp. Bioresour Technol 101(14):5494–5500

Xu XQ, Beardall J (1997) Effect of salinity on fatty acid composition of a green microalga from an Antarctic hypersaline lake. Phytochemistry 45(4):655–658

Yang J, Gao H, Glibert PM, Wang Y, Tong M (2017) Rates of nitrogen uptake by cyanobacterially-dominated assemblages in Lake Taihu, China, during late summer. Harmful Algae 65:71–84

Yoo C, Jun SY, Lee JY, Ahn CY, Oh HM (2010) Selection of microalgae for lipid production under high levels carbon dioxide. Bioresour Technol 101(1):S71–S74

Zhang Q, Wang T, Hong Y (2014) Investigation of initial pH effects on growth of an oleaginous microalgae Chlorella sp. HQ for lipid production and nutrient uptake. Water Sci Technol 70(4): 712–719

Zhang P, Li Z, Lu L, Xiao Y, Liu J, Guo J, Fang F (2017) Effects of stepwise nitrogen depletion on carotenoid content, fluorescence parameters and the cellular stoichiometry of Chlorella vulgaris. Spectrochim Acta A Mol Biomol Spectrosc 181:30–38

Zheng Y, Li T, Yu X, Bates PD, Dong T, Chen S (2013) High-density fed-batch culture of a thermotolerant microalga Chlorella sorokiniana for biofuel production. Appl Energy 108:281–287

Zhila NO, Kalacheva GS, Volova TG (2011) Effect of salinity on the biochemical composition of the alga Botryococcus braunii Kütz IPPAS H-252. J Appl Phycol 23(1):47–52

Chapter 7
Advanced Genetic Approaches Toward Custom Design Microalgae for Fourth-Generation Biofuels

Manisha Verma and Vishal Mishra

Abstract Fourth-generation (4G) algae biofuels have been gone well with machine engines and transport vehicles. With biotechnology advances in genetic and metabolic engineering tools, significant advancement has been created to 4G biofuels production like long-chain hydrocarbons, bioethanol, and fatty acid. Microalgae are gaining more interest for being a suitable candidate for biodiesel and biohydrogen production; apart from that, their biotechnological applications significantly increased because most of them are autotrophic photosynthetic microorganisms, producers of wider varieties of the value-added compounds like vitamins, proteins, nutraceuticals, secondary metabolites, carotenoids, and some other pigments. Hence, microalgae are found to be a promising source for biofuels and other valuable metabolite productions. This chapter includes different genetic engineering efforts made to obtain compatible fourth-generation microalgae biofuel.

Keywords 4G biofuels · Genomics · Metabolic engineering · Algae biofuels · Microalgae

Abbreviations

3G-B	Third-generation biofuels
4G-B	Fourth-generation biofuels
CRISPR/Cas9	Clustered regularly interspaced short palindromic repeats associated protein 9
FAMEs	Fatty acid methyl esters
GHG	Greenhouse gas
PEG	Polyethylene glycol
TAG	Triacylglycerol

M. Verma · V. Mishra (✉)
School of Biochemical Engineering, IIT (BHU), Varanasi, Uttar Pradesh, India
e-mail: vishal.bce@itbhu.ac.in

N. Srivastava, P. K. Mishra (eds.), *Technological Advancement in Algal Biofuels Production*, Clean Energy Production Technologies,
https://doi.org/10.1007/978-981-19-6806-8_7

TALENs	Transcription activator like effector nucleases
ZFN	Zinc finger nucleases

7.1 Introduction

Algae can be found naturally in ponds or other places where moisture and light are available, even though a massive amount of algae biomass is necessary for biofuel production. So, for large algae, biomass production is associated with highly efficient and optimized cultivation methods owing to tolerate the macro-panel band of light, pH, temperature, nutrient availability, salinity, etc. In third-generation biofuels, dry algal biomass and lipid are used as feedstock for biodiesel, bioethanol, and biohydrogen production (Verma and Mishra 2020). This algae biofuel generation has become more advanced and optimized in fourth-generation algae biofuels. Genetic engineering tools help create customized microalgae suitable for renewable energy generation, producing biodiesel at lesser production cost and being a potential and valid source of energy (ElFar et al. 2021). Microalgal biofuels can serve well as a viable alternative to fossil fuels. Numerous researches described the use of algae biofuels as a replacement for gasoline or petrol, and it might be helpful to reduce particulate matter and greenhouse gas emissions. Similarly, algae cells utilized atmospheric carbon dioxide via carbon capturing and simultaneously can be used for nutrient recovery (C, N, P, K) from wastewater. Despite having huge potential for biofuel production, there are several limitations such as lesser energy content as compared to petroleum, less efficient to run engine properly, less biomass production, shortage of feed, lower lipid accumulation, and high production costs (ElFar et al. 2021). Fourth-generation biofuels (4G-B) are generated from genetically modified microalgae biomass. Genetic modification in algae grabs significant consideration from scientific and industrial circles because switching fossil fuel resources with third-generation biofuels (3G-B) is still imitated by a lot of challenges (Shokravi et al. 2021). Algae biomass cultivation is a simultaneous cure for various environmental problems like fuel scarcity, carbon dioxide emission, GHG emission, and water pollution. So, algae-derivative biofuels are being the truly sustainable option to petroleum fuels. Biodiesel has very similar chemical characteristics to petroleum fuels, and so this has emerged as one of the promising green alternatives. Microalgae with high biomass yield and lipid content will turn into a resource for higher scale productivity of biodiesel (Jagadevan et al. 2018). 4G-B combines the 3G-B properties with an advantage of genetic optimization of microalgae cells. Cyanobacteria is a photosynthetic bacterium that resembles microalgae and owes ease of genetic manipulations. In cyanobacteria, genetic modification allows the fixation of atmospheric carbon and its conversion into crude biomass and desired end products. Most of the metabolic end products (free fatty acids, alkenes, and fatty alcohols) secreted into culture media by genetically engineered cyanobacteria are suitable for biofuel production. Product secretion outside the algae cell within

Table 7.1 Evaluation and advancement of genetic engineering in microalgae

In 1988	Delivery of foreign gene into a chloroplast's genome through microparticle bombardment
In 1990	Nuclear transformation by the help of glass bead agitation of algal cells and DNA mixture
In 1991	Sonication in the presence of PEG (polyethylene glycol)
In 1993	Suitable for a nuclear transformation with silicon whiskers
In 1999–2001	Got success in applying antisense and inverted repeats containing RNA
In 2004	Cocultivation with pathogens of plants like *Agrobacterium tumefaciens*
In 2013–2014	Targeting specific DNA sequences by TALENs and ZFNs
In 2014	Provides the success of transient expression of sgRNA gene and Cas9 gene
In 2016	Genome editing by delivering the Cas9-gRNA RNPs

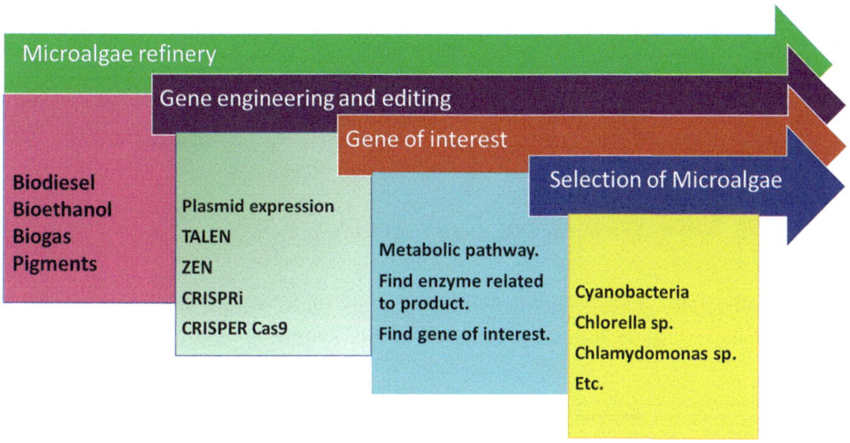

Fig. 7.1 Schematic insights of fourth-generation microalgal biorefinery

culture media skips costly stages of algae biodiesel processing such as cell harvesting, cell disruption, and lipid extraction. Therefore, 4G-B has the most effective processing technologies (Sarsekeyeva et al. 2015). Table 7.1 depicts a timeline regarding the evolution of genetic engineering.

4G-B generation is based on the genetic alteration in microalgae genome. It is committed to creating such microalgae strains that should capture a higher amount of carbon dioxide, increased TAG production, and adaptability to produce huge biomass by recovering nutrients (N, P, and COD) from wastewater effluents for treatment purposes. Figure 7.1 represents schematic insight of 4G-B microalgal refinery. The genetic manipulation approach in microalgae ranges from conventional transformation methods to genome editing systems like RNA interference, ZFN (zinc finger nucleases), CRISPR/Cas9, and TALENs (transcription activator like effector nucleases). However, genetic modification is limited to several species of microalgae as it depends on genomic sequences of the species, and it is still unknown for maximum algal species. *Chlamydomonas reinhardtii*, *Chlorella sorokiniana*, and *Chlorella variabilis* are explored in different studies for genetic modification as the genomic information is accessible for these algal species (Mat Aron et al. 2020).

7.2 Fatty Acids and Triacylglycerol Accumulation in Microalgae

Using the transesterification process, the triacylglycerol (TAG) forms fatty acid methyl esters (FAMEs), which are an eminent and adaptable variety of biodiesel and build the foundation for biofuel production (Hossain et al. 2008). Algae cells are capable of producing fatty acids and TAG that make them a potential resource for biodiesel production. Lipid metabolism within algae cells specifically includes fatty acid and TAG biosynthesis pathways. Most of the genes involved in lipid metabolism of higher plants and algae cells are the same and mark sequence homology or similar biochemical properties. Therefore, it is generally considered that the basic pathway of algal lipid synthesis is analogous to plants.

7.2.1 *De Novo Synthesis of Fatty Acids*

In algal cells, fatty acid de novo synthesis takes place within the chloroplast. A comprehensive outline for fatty acid de novo synthesis is given in Fig. 7.2.

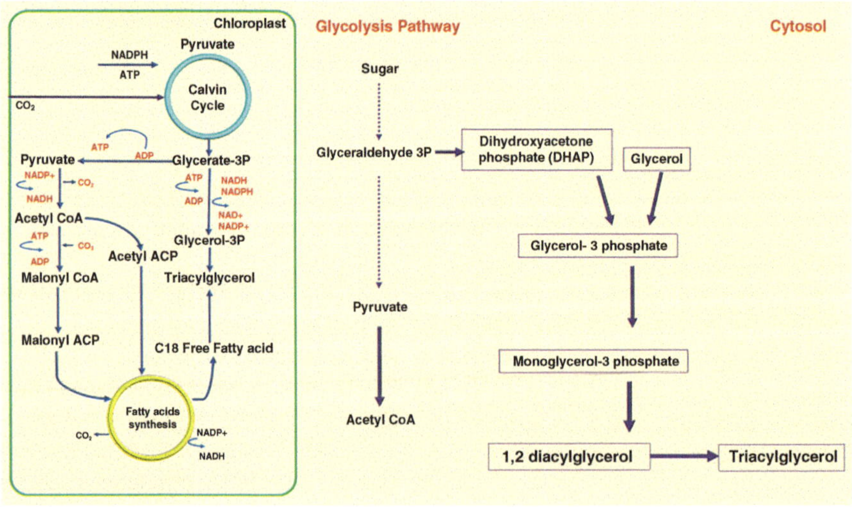

Fig. 7.2 Triacylglycerol biosynthesis in algae

Generally, fatty acid synthesis results in 16C or 18C fatty acids. These fatty acids are further used as a precursor for the synthesis of cellular and chloroplast membranes; despite this, under adverse culture conditions, neutral lipids are synthesized and stored within the algal cell. Conversion of acetyl CoA into malonyl CoA is a crucial step in de novo synthesis of fatty acid where acetyl CoA carboxylase (ACCase) acts as a catalyst (Ohlrogge and Browse 1995). Within chloroplast, the photosynthetic pathway maintains the supply of acetyl CoA. In algae, pyruvate generated during glycolysis is the main photosynthate that is converted into acetyl CoA using pyruvate kinase (PK) (Andre et al. 2007). Formation of malonyl CoA (catalyzed by ACCase) takes place in two steps: (1) the first step requires ATP—carbon dioxide is supplied by biotin carboxylase group (prosthetic) of ACCase to an N of the amino group of a lysine residue within biotin (prosthetic group); (2) biotin transfers CO_2 (carboxylation reaction) to acetyl CoA and forms malonyl CoA—enzyme carboxyltransferase involves in catalysis of this second step. Malonyl CoA acts as the chief carbon donor during fatty acid synthesis (Ohlrogge and Browse 1995). Further, the malonyl group transfers to an acyl carrier protein (ACP). After that, ACP involves in all successive reactions of the pathway till the final product is formed. Malonyl ACP contributes to a chain of condensation reactions involving acyl ACP acceptor. 3-Ketoacyl ACP synthase III is a condensing enzyme and forms a 4C compound (Jaworski et al. 1989). 3-Ketoacyl ACP synthase I is another enzyme for condensation reactions and forms a varying length of (6C to 16C) carbon chain. After condensation, the product undergoes three other reactions to form saturated fatty acid. These reactions are as follows (Ohlrogge and Browse 1995): (1) 3-ketoacyl ACP synthase-catalyzed product undergoes to reduction reaction using 3-ketoacyl ACP reductase enzyme; (2) dehydration is catalyzed by hydroxyacyl ACP dehydratase; and (3) again reduction reaction is catalyzed with enoyl ACP reductase. The final step of fatty acid biosynthesis construct saturated 16:0-ACP and 18:0-ACP. For unsaturated fatty acids, double bonds are established by the involvement of stearoyl ACP desaturase (Ohlrogge and Browse 1995).

7.2.2 Triacylglycerol Biosynthesis

The direct glycerol pathway is involved in TAG biosynthesis in algae. Fatty acids generated at the chloroplast are transmitted from CoA to the first and second position of glycerol-3-phosphate (G3P) to form phosphatidic acid (PA). A phosphatase enzyme catalyzes the dephosphorylation reaction with PA and make diacylglycerol (DAG) (Ratledge 1988; Ohlrogge and Browse 1995). For final TAG synthesis, diacylglycerol acyltransferase catalyzes the shift of third fatty acid at the third place of DAG. In TAG synthesis, acyltransferase enzyme might have specificity for different acyl CoA molecules (Roessler et al. 1994); so, the ultimate acyl composition varies in TAG.

7.3 Biotechnological Strategies to Engineer Microalgae Genetically

Recently, genetic engineering grabs too much attention because advanced and more accurate genetic tools are progressively available for genome editing (Song et al. 2015). Similarly, the ability to integrate and express a desired metabolic system in an organism where it was previously absent brings new opportunities to design customized algae cells for biofuel production (Andrianantoandro et al. 2006). In gene engineering, selection of appropriate genome editing tools and transformation approaches are significant steps to achieve a successful genetic modification (Gimpel et al. 2015). In general, the genetic manipulation approach includes: (1) selection of host organism, (2) desired gene isolation, (3) construction of suitable plasmid, and (4) selection of transformation and DNA editing tools. On the other hand, metabolic engineering and gene interfering systems are significant for effective gene targeting. Recently, TALEN, CRISPR-Cas9, silence RNA (siRNA), ZFN, and micro RNA (miRNA) are common gene interfering tools (Sizova et al. 2013; Shin et al. 2016; Kasai et al. 2015). These gene interfering tools are able to achieve desired phenotypes in algae by activating or suppressing a gene expression (Yao et al. 2016; Vieler et al. 2012; Huang et al. 2016). Genetic engineering in algae contains two vital points: (1) gene delivery tools and (2) screenable and selectable marker genes. In the next section, conventional approaches to gene delivery have been discussed in brief.

7.3.1 Conventional Approaches of Gene Delivery

Algae are grouped into many phyla and species that have varying chemical, genomics, physiological, structural, and physical characteristics. In the present scenario, such gene manipulation technique that has been applicable to all algal species as well as having an industrial interest is unavailable. Therefore, many efforts have been made for transgene stability and its successful delivery into microalgae cells by applying advancements in traditional genetic tools. Additionally, recent developments in genome sequencing projects and broad information about chloroplast, mitochondrial, and nuclear genomes significantly enhanced the advancement of novel genome editing methods (Jeon et al. 2017; ElFar et al. 2021). The most extensively used transformation methods for the genetic modification of algae are concisely explained in the upcoming section (Leon and Fernandez 2007). Basically, five transformation approaches have been used for DNA delivery in microalgae: (1) glass bead-mediated agitation, (2) electroporation, (3) amino clay nanoparticle-mediated transformation, (4) particle bombardment, and (5) *Agrobacterium*-mediated transformation. Each technique has its pros and cons depending on the transgene's stability, integration, and efficiency. On the other hand, the selection system may depend on reporter or antibiotic resistance gene selection and might affect the selection efficiency (Leon and Fernandez 2007).

7.3.1.1 Biolistic

Biolistic method of transformation in algae chloroplast is reported in *Haematococcus*, *Tetraselmis*, and *Dunaliella* species (Doron et al. 2016; Purton et al. 2013). This method involves application of a device termed as gene gun for shooting DNA fragments directly into the algae cells. DNA or RNA adheres on inert particle-like tungsten or gold to prepare a DNA/particle complex. This DNA/particle complex is projected and bombarded in vacuum at higher velocities into the microalgal cell with a microprojectile gene gun tool. For transformation in algae cells, biolistic is the most suitable method for chloroplast and host microalgae cells. The intense penetration capacity of the particles and their ability to crossing all physical barriers like cell membranes or cell walls enhance the efficiency rate of transformation (Mosey et al. 2021). Figure 7.3 depicts the working mechanism of gene guns for biolistic transformation.

7.3.1.2 Glass Beads Agitation

This method is considered to be the straightforward method for transgene delivery into the cells without any advanced equipment. This method has frequently been applicable in laboratories because of its reproducibility and simplicity. *C. reinhardtii*, *C. vulgaris*, and *D. salina* are the most frequently transformed microalgae using glass bead agitation. In this method, algal cells are exposed in front of the desired DNA and agitated with glass beads (0.5 mm). The study suggests that the addition of a surfactant like PEG (polyethylene glycol) during agitation

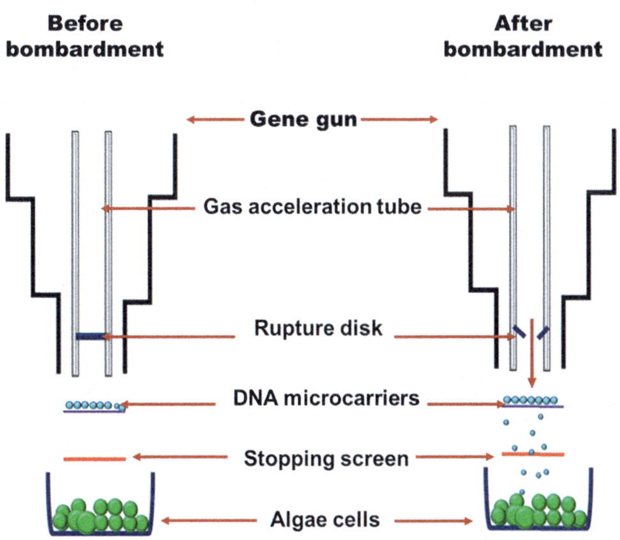

Fig. 7.3 The working mechanism of gene gun for biolistic transformation

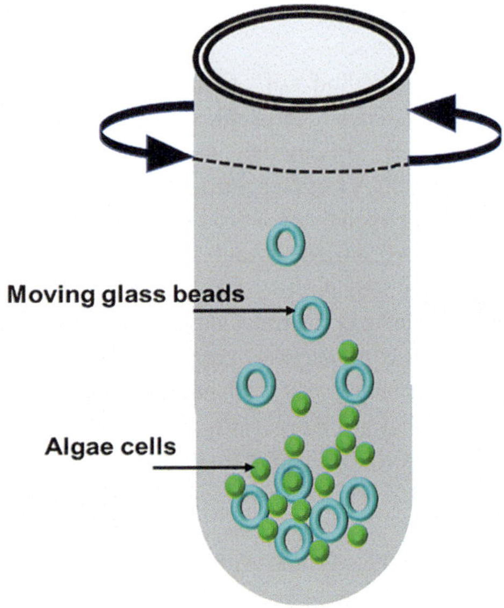

Fig. 7.4 Schematic diagram of glass bead agitation

enhances the number of transformants by 5- to 12-fold (Coll 2006). The transformation efficiency of glass bead agitation is based on various factors such as cell size, the velocity of agitation, duration of agitation, and surfactant concentration (Gutiérrez and Lauersen 2021). A schematic diagram of glass bead agitation is shown in Fig. 7.4.

7.3.1.3 Electroporation

In electroporation, it is compulsory to provide an electric field for the development of interim pores in the cell surface so that the transgene can enter within the algae cell. This method is considered the most convenient method for effective nuclear transformation (Shimogawara et al. 1998). The transformation efficiency of electroporation depends on ionic strength, electric field strength, pulse length, the presence of the physical barrier, and other host cell characteristics (Gutiérrez and Lauersen 2021). This method applied electrodes to establish a voltage gradient across the cell membrane (Fig. 7.5). The cell membrane temporarily upsets the phospholipid bilayer and allows target DNA molecules to cross into the cell. Electroporation techniques were widely used to increase the yields of biofuel and biorefinery processing as they are useful for feedstock pretreatment as well as microalgal cell destruction (Chen and Lee 2019).

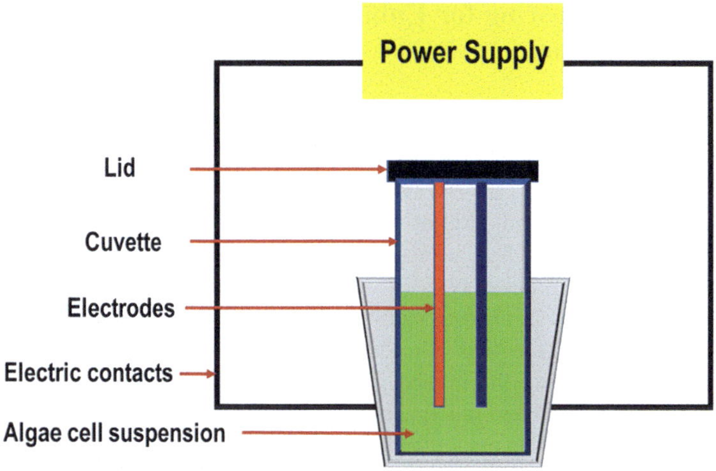

Fig. 7.5 Schematic diagram of electroporation

7.3.1.4 Amino Clay Nanoparticle-Facilitated Transformation

In this current nonconventional DNA delivery approach, 3-aminopropyl molecules attached to magnesium phyllosilicate as a functional group to form Mg-amino clay nanoparticles. These nanoparticles are mixed along with an imported DNA for nuclear transformation in the host algae cells. The dimension of Mg-amino clay nanoparticles (45 nm) is more effective for transformation compared to gold or tungsten particles (1–2 μm) used in the gene gun method. Kim et al. (2014) utilize the Mg-amino clay nanoparticles-mediated transformation method in *C. reinhardtii* successfully.

7.3.1.5 *Agrobacterium tumefaciens*-Facilitated Transformation

It is a biological transfection approach in which the fragment of a Ti plasmid (T-DNA) from *A. tumefaciens* is transferred to the plant cells and randomly combined into chromosomal genes. *A. tumefaciens*-mediated DNA delivery is very common in plant cells; however, studies have been described transformation in several microalgae such as *Chlorella* sp., *C. reinhardtii*, *D. bardawil*, *H. lacustris*, *Nannochloropsis* sp. *P. kessleri*, and *Symbiodinium* sp., but it is not widely accepted for algae transformation. Still, the exact mechanism of *A. tumefaciens*-mediated DNA delivery to microalgae cells is not clear. It is still unknown whether the *A. tumefaciens* infection requires specific recognition machinery in algae or is target independent (Pratheesh et al. 2014; Mini et al. 2018; Prasad et al. 2019; Kumar et al. 2004).

7.4 Gene Engineering for Enhanced Lipid Biosynthesis in Microalgae

Genetic engineering approaches involve overexpression of enzymes engaged in the triacylglycerides, lipid, and fatty acid synthesis pathway (Yunus and Jones 2018; Xin et al. 2017; Fukuda et al. 2018). Apart from this, focusing on transcriptional factors engaged in the regulation of the lipid synthesis and knocking out parallel competitive pathways like lipid catabolism or starch synthesis are proven strategies for enhanced lipid accumulation (Kao and Ng 2017; Nomaguchi et al. 2018; Wei et al. 2017; Ajjawi et al. 2017). In a study, *P. tricornutum* (a diatom) was engineered to enhance the NADH (nicotinamide adenine dinucleotide (NAD) + hydrogen (H)) supply by overexpressing the related genes (Xue et al. 2017). However, genetic approaches are suspectable in most cases, and results are not as per the expectations. The results obtained after genetic manipulation are reported as per fold changes that occur in lipid, fatty acid, fatty alcohol, and starch productivity of modified strain with respect to wild strains. Starch biosynthesis suppression is assumed to enhance lipid production. In a study, *Coccomyxa* sp., the TALEN-based mutation, was used to obtain a starchless mutant (*Coccomyxa* sp. AG125). The mutant strain has been shown to reduce biomass yield but enhance lipid productivity. However, reduced biomass indicates inhibition of starch synthesis might be a risky way to enhance lipid production (Takahashi et al. 2018). In a study, five metabolic pathways were introduced into two green algae, *Synechocystis* sp. PCC 6803 and *C. reinhardtii* strain UVM4 (Yunus et al. 2018). These synthetic pathways are the following: (1) carboxylic acid reductase (CAR), (2) acyl ACP synthase (Daas) and thioesterase (TesA), (3) fatty acid photo-decarboxylase (FAP), (4) olefin biosynthesis (OleT), and (5) undecene biosynthesis (UndA, UndB). Results revealed 19- and 8-fold increased buildup of fatty alcohols, fatty acids, and other hydrocarbons in PCC 6803 and UVM4, respectively (Yunus et al. 2018).

7.4.1 Genome Editing

Apart from the traditional transformation methods, genome editing is an emerging approach in past years for genetic customization in algal species. Genome editing requires different recombinant nucleases enzymes to identify specific cleavage sites in a genome sequence. Cleavage in a genome sequence results in double-strand breaks. Error-prone DNA repair and homology-independent DNA repair mechanisms are involved in repairing such double-strand breaks. These two DNA repair mechanisms are collectively resulting nonhomologous end joining and cause mutation at the site of cleavage (Jeon et al. 2017). ZFN, RNA interference, CRISPR/Cas9, and TALENs are some genome editing approaches. Such nuclease-based genome editing techniques could be a substitute for conventional DNA transformation methods in microalgae. However, these genetic manipulation techniques are in

Table 7.2 Different algal modification approach and their consequences for enhanced biofuel production

Algal strains	Modification	Targeted gene	Consequences	References
C. reinhardtii strain 704	Transformation	ACS2 (acetyl CoA synthetase)	Enhance biosynthesis of neutral lipids	Rengel et al. (2018)
C. vulgaris and *C. sorokiniana*	Electroporation	CA (carbonic anhydrase)	Enhanced biomass production, protein content, and lipid accumulation	Lin et al. (2018)
C. reinhardtii	CRISPRi silencing	PEPC1 (phosphoenolpyruvate carboxylase)	Lipid productivity increased (94.2% more than the wild strain)	Kao and Ng (2017)
C. reinhardtii	Transformation	*DtTE* (*D. tertiolecta* fatty acyl-ACP thioesterase)	Significant increase in neutral lipid and total lipid content	Tan and Lee (2017)
C. merolae	Knockout and overexpression	*CmFAX1* (chloroplast inner membrane protein	Enhance cell yield and TAG content	Takemura et al. (2019)
P. tricornutum	Electroporation	*G6PD* (glucose-6-phosphate dehydrogenase)	Lipid content raised by 2.7-fold	Xue et al. (2017)
C. reinhardtii strain CC-124	Polyploidization	Diploid genome (duplication)	Increased FAME yields and biomass yield, improved stress tolerance	Kwak et al. (2017)
C. vulgaris UV715	Random mutagenesis	Mutagenesis by ethyl methanesulfonate	Mutant displays high lipid productivity and higher amounts of monounsaturated and saturated fatty acid	Sarayloo et al. (2018)
Coccomyxa sp. Obi	Mutagenesis by TALEN	AGPL (ATP: glucose-1-phosphate adenyltransferase)	Growth and lipid productivity in certain conditions was lesser than wild type	Takahashi et al. (2018)
C. reinhardtii CC-4349	CRISPR-Cas9-based knockout	PLA2 (phospholipase)	Increases TAG accumulation and total lipid productivity	Shin et al. (2019)

progress and have several limitations. Table 7.2 presents different algal modification approach and their consequences for enhanced biofuel production.

Recently, too many studies have been reported applying genome editing techniques over microalgae cells. CRISPRs technology is most frequently used for genome editing. *C. reinhardtii* was the first algae to successfully implement the CRISPR/Cas9 approach to enhance lipid synthesis (Kao and Ng 2017). Some more studies are available, where CRISPR/Cas9 is implemented to boost lipid production in *Synechocystis* sp., *S. elongatus* UTEX 2973, *N. gaditana*, *P. tricornutum*, and *N. oceanica* (Ng et al. 2017). Daboussi et al. (2014) utilize the TALENs technique to enhance the triacylglycerol (TGA) accumulation.

7.4.1.1 RNAi Silencing Approaches

RNAi silencing/interference is a process in which dsRNA induces degradation of homologous mRNA and causes gene silencing (Cerutti 2003). RNAi silencing can be a source to get a better understanding and insights about the gene expression, which could be established as eventual gene target to rise TAG accumulation in algal cells. Phosphoenolpyruvate carboxykinase (PEPC) catalyzes the conversion of phosphoenolpyruvate (PEP) into oxaloacetate. Therefore, PEPC brings an oxaloacetate pool to the TCA cycle and regulates proteins synthesis. Accordingly, silencing PEPC might cause increased carbon flux for lipid biosynthesis in microalgae. In *P. tricornutum*, silencing PEPC form two cell lines, PEPCK19 and PEPCK21, enhances lipid accumulation (25%) than control cells (21%) (Kao and Ng 2017; Yang et al. 2017). Additionally, the PEPC enzyme complex has two classes: PEPC1 and PEPC2 (Deng et al. 2012). PEPC2 silencing affects lipid accumulation negatively, whereas silencing of PEPC1 raised the lipid accumulation significantly (Deng et al. 2012). Diacylglycerol acyltransferase (DGAT) is an enzyme in TAG biosynthesis that catalyzes the final reaction of TAG formation. DGAT silencing will affect TAG biosynthesis negatively. In *C. reinhardtii*, five homologous DGAT genes were silenced (DGAT1, DGAT2, DGAT3, DGAT4, and DGAT5). Results revealed significant TAG accumulation reduction up to 16–24% for DGAT1 silencing and 28–37% for DGAT2 silencing (Deng et al. 2012). Also, under the nitrogen-limited condition in *N. oceanica*, silencing DGAT1 led to a 25% reduction in TAG accumulation, even though overexpression of DGAT1 boosted TAG accumulation by 39% compared to wild cell lines (Wei et al. 2017).

7.4.1.2 Editing Transcriptional Factors

Transcriptional factors bind to particular DNA motifs and control the expression of various enzymes and components in a metabolic pathway (Bajhaiya et al. 2017). Editing transcriptional factors might affect the expression of many enzymes, and this is how it enhances the possibility of developing new strains. In *C. reinhardtii*, nearly 87 putative transcriptional regulators and 147 putative transcriptional factors have been identified (Courchesne et al. 2009). Some of the most common transcriptional factors found in different algae are given in Table 7.3.

PSR1 is essential for phosphorus regulation by Pi transporters and phosphatases upregulation (Bajhaiya et al. 2016, 2017). In phosphorus starved conditions, *C. reinhardtii* mutants (PSR1 lacking) results in lipid inhibition and starch accumulation, whereas overexpression of PSR1 promotes starch accumulation but decreases neutral lipid content. GmDOF4 bind to the cis-DNA elements at promoter region

Table 7.3 Common transcriptional factors found in algae

Transcriptional factors	Relevance	References
PSR1	Pi starvation response	Bajhaiya et al. (2016)
GmDOF4	Glycine maxDNA binding with one finger 4	Zhang et al. (2014)
CHT7	Compromised hydrolysis of triacylglycerol 7	Tsai et al. (2014)
WRI 1	Wrinkled 1	Shang et al. (2016)
bHLH	Basic helix–loop–helix	Kang et al. (2015)
ROC40	The rhythm of chloroplast 40	Goncalves et al. (2016)
NRR-1	Nitrogen response regulator 1	Boyle et al. (2012)
TAR1	TAG accumulation regulator	Kajikawa et al. (2015)
Zn(II)2Cys6	Zn(II)2Cys6-encoding genes	Ajjawi et al. (2017)

and activate ACCase. Overexpression of GmDOF4 in *C. ellipsoidea* enhances lipid accumulation by 46–52% (Zhang et al. 2014). Kang et al. (2017) perform overexpression of WRI in *N. salina* and observe that total lipid content enhanced by nearly 44% as compared to a wild one. Again, in *N. salina*, overexpression of bHLH1 and bHLH2 increased growth rate (60%) and lipid content by 24.5%, 46%, and 32.5% in bHLH1, bHLH2, and wild strain, respectively (Kang et al. 2015). In *Chlorella* sp., proteomic studies revealed the role of ROC40 for inducing a protein involved in circadian rhythm regulation, which is important for the existence of microalgae in varying conditions (Goncalves et al. 2016).

7.4.1.3 CRISPR Technology

In the last two decades, most genetic manipulation in microalgae was focused on DNA delivery transformation techniques and transformation efficiency in both nuclear and chloroplast genomes. Nowadays, CRISPR/Cas9 approach seems to be the future of most gene editing as CRISPRi technology is much manageable, so it has huge potential application for genome editing. For microalgae, genetic engineering, a CRISPR/Cas9-based approach, was first reported by Jiang et al. (2014) in *C. reinhardtii* and began new doors for genome editing. The major challenge of the CRISPR/Cas9 approach is to reduce the toxic action of Cas9 nuclease so that the mutation rate will increase (Jiang et al. 2014). Utilizing Cas9 gRNA ribonucleoproteins is an effective approach to eliminate the toxic result of Cas9. Cas9 gRNA ribonucleoproteins bring mutations at three loci and enhance mutation up to 100-fold (Shin et al. 2016; Baek et al. 2016). However, the number of effective clones is very less and requires more optimization for effective applicability. The first trial to validate the CRISPR-Cas9 gene editing in *C. reinhardtii* provides an insight about Cas9/sgRNA system's ability to be well expressed in microalgae, but it is less efficient and has a very less surviving ratio (Jiang et al. 2014). Still, several other CRISPR/Cas9-based mutations in microalgae were demonstrated. The genome of *P. tricornutum* was effectively edited by utilizing a CRISPR/Cas9-optimized vector (Nymark et al. 2016). In *Nannochloropsis* sp., CRISPR technology has been applied to develop a mutant microalgae variety aimed for CO_2 capturing and lipid synthesis (Wang et al. 2016). CRISPRi approach was reported first time in the year 2017 for the repression of the CrPEPC1 gene (regulate PEP carboxylase) in *C. reinhardtii* CC400 and effectively increase lipid accumulation (Kao and Ng 2017). In *Synechocystis* sp. PCC 6803, CRISPRi system is used to repress multiple genes (Yao et al. 2016). Additionally, some research includes both CRISPRi and CRISPR systems simultaneously to enhance succinate production and target the gene regulation in *S. elongatus* PCC 7942 (Li et al. 2016; Huang et al. 2016). Similarly, the CRISPRi system is applied in *Synechococcus* sp. PCC 7002 for the repression of the *glnA* gene (regulate glutamine synthetase) to enhance lactate production (Gordon et al. 2016).

7.4.1.4 Metabolic Pathways Manipulation

Microalgae cells are biological factories to produce different biochemical, nutrients, and biofuels. Being a potential source of biofuels, many studies concerning microalgae genomics and metabolic engineering were conducted using different genetic tools. Metabolic engineering concerns understanding and alteration in metabolic pathways of the microorganisms using genetic engineering approaches like overexpressing or silencing a gene. Metabolic engineering approaches involve selecting the desired gene from another microorganism, then addition of the desired gene into the host genome, optimization of enzymatic reactions, and improvements in production efficiency. The metabolic engineering approach includes optimization of photosynthesis reaction rate and lipid synthesis pathway in microalgae for biodiesel production. Phototrophic microalgae participate in capturing and mitigating atmospheric CO_2. Microalgae use CO_2 as a carbon source and use to assimilate it during photosynthesis and produce organic compounds. Therefore, microalgae biomass is a truly sustainable and eco-friendly source for green fuel technology. For carbon fixation, microalgae use the Calvin cycle. During the light-dependent period, Calvin cycle takes place in chloroplast and utilizes inorganic carbon source, ATP, and NADH to form carbohydrates. Calvin cycle includes many steps and a series of reactions like carboxylation and reduction (Beardall and Raven 2016). Ribulose-1,5-biphosphate carboxylase oxygenase (RuBisCo) is the primary catalyst (enzyme) in the Calvin cycle. RuBisCo catalyzes carboxylation (by incorporating CO_2) of ribulose-1,5-biphosphate and results in 3-phosphoglyceraldehyde. Several strategies, such as velocity, selectivity, overexpression, have been suggested to expand the efficiency of the RuBisCo enzyme (Ng et al. 2017). Considering metabolic engineering in microalgae, aldolases enzymes are targeted to upgrade carbon fixation rate. Aldolases are enzymes participated in the middle of the Calvin cycle and catalyze two reversible reactions as follows: (1) glyceraldehyde 3-phosphate (G3P) and dihydroxyacetone phosphate (DHAP) conversion into fructose-1,6-biphosphate (FBP) and (2) erythrose 4-phosphate and DHAP conversion into sedoheptulose-1,7-biphosphate. In a study, the efficiency of photosynthesis is increased by 1.2-fold, expressing cyanobacterial aldolase into the chloroplast of

C. vulgaris (Yang et al. 2017). Carbonic anhydrase is an enzyme involved in the fixation of CO_2 and water into bicarbonate ions, protons, and carbonic acid. In a research, *Synechococcus elongatus* PCC7942 (genetically modified cyanobacteria for carbonic anhydrase) revealed up to 41% increased rate of carbon fixation (Chen et al. 2012). Another study has been done to enhance the carbon fixation rate by photorespiratory metabolism redirecting phosphoglycolate (Kebeish et al. 2007). In microalgae cells, carbon flux regulates photosynthetic pathways and their efficiency, which further is a cause of lipid accumulation and biofuel production. Triacylglycerides (TAG) production is the major concern for biodiesel generation. After the transesterification process, TAG is converted into fatty acid methyl esters (FAME). Some researches were conducted to obtain FAMEs directly via the lipid biosynthesis pathway in algae cells (Jeon et al. 2017). Diacylglycerol acyltransferase (DGAT) catalyzes the esterification of diacylglycerol (DAG) into TAG in lipid biosynthesis. In *B. braunii* and *Dunaliella salina*, up to 60% of TAG accumulation occur (Gangl et al. 2015). In *S. obliquus*, DGAT overexpression results 128% increment in the lipid content (Chen et al. 2012). Acetyl-CoA carboxylase is an enzyme that involves in the carboxylation of acetyl-CoA and converts it into malonyl-CoA during TAG synthesis (Ng et al. 2017). Lipid biosynthesis is competed by phospholipid, starch, and oxaloacetate synthesis pathways. Suppression of such competing pathways will cause enhanced lipid accumulation within microalgae cells. In *C. reinhardtii*, oxaloacetate synthesis is suppressed by gene editing in phosphoenolpyruvate carboxylase genes using CRISPRi and knockdown oxaloacetate production (Radakovits et al. 2010). Similarly, suppression of starch synthesis in *C. pyrenoidosa* results in increased lipid content in starch-deficit mutants (Radakovits et al. 2010).

7.5 Considerations for the Future of Algal Modification

Robust microalgae strains will design to increase lipid accumulation and easier recovery using a combined approach that includes adoptive evolution, metabolic engineering, and membrane engineering. The selection of suitable transformation techniques will enhance the efficiency of DNA delivery to the target host cell line. These transformation techniques might bring new paths of efficient DNA delivery and transformed algal strains. The following factors are important to evaluate whichever technique is selected for genetic modification in target algae: (1) molecular tools, (2) target gene expression efficiency, (3) transgene copy number, (4) nuclear or chromosomal gene target, (5) integration and expression of chromosomal plasmid, and (6) reproducibility of the system. It also appreciates that biotechnology techniques improve microalgae potential every year. New research studies and technology bring advanced future possibilities to develop potential microalgae strains. With the progress of computational bioinformatics, the application of transcriptomics, proteomics, and metabolomic disciplines helps in the development of metabolic pathway simulations, gene annotation, and genome sequencing of an organism. These strategies combinedly help to obtain very efficient microalgal strains for industrial biofuel production. Future studies should make efforts toward integrating omics technology to understand and develop common genome editing targets.

7.6 Conclusion

Industrial-scale microalgae biodiesel production depends on numerous factors comprising the algae cell tolerance to outdoor culture conditions, high lipid and productivity per unit area, simple and effective harvesting, and sustainable extraction techniques for lipid recovery. Certainly, optimization of these factors needs well interpretation of microalgal biology accompanied by the development of genetic engineering tools that will create ease to obtain valuable algal strains. Genetically modified algae strains are expected to have the auto-flocculating ability, able to tolerate outdoor stress conditions, and have easy lipid recovery. With the progress of computational bioinformatics, the application of transcriptomics, proteomics, and metabolomics disciplines helps to improve our understanding of metabolic pathway simulations, gene annotation, and genome sequencing of microalgae. Transcriptomics, proteomics, and metabolomics help evaluate different algal species under different stress conditions and provide insight by recognizing the crucial regulatory genes for enhanced TAG accumulation. In microalgae genome editing, future studies should not limit till TAG-regulating genes, but it should also be focused on genes involved in stress tolerance, photosynthesis, and CO_2 fixation. Eventually, integrating microalgal biology, omics systems, and genome editing will help accomplish sustainable microalgal biofuels very soon.

Acknowledgment The authors are thankful to the Indian Institute of Technology (BHU) Varanasi, Varanasi, for extending their technical and financial support.

Conflict of Interest The authors have declared no conflict of interest.

References

Ajjawi I, Verruto J, Aqui M, Soriaga LB, Coppersmith J, Kwok K, Moellering ER (2017) Lipid production in Nannochloropsis gaditana is doubled by decreasing expression of a single transcriptional regulator. Nat Biotechnol 35(7):647–652

Andre C, Froehlich JE, Moll MR, Benning C (2007) Aheteromeric plastidic pyruvate kinase complex involved in seed oil biosynthesis in Arabidopsis. Plant Cell 19:2006–2022

Andrianantoandro E, Basu S, Karig DK, Weiss R (2006) Synthetic biology: new engineering rules for an emerging discipline. Mol Syst Biol 2(1):2006-0028

Baek K, Kim DH, Jeong J, Sim SJ, Melis A, Kim JS et al (2016) DNA-free two-gene knockout in Chlamydomonas reinhardtii via CRISPR-Cas9 ribonucleoproteins. Sci Rep 6(1):1–7

Bajhaiya AK, Dean AP, Zeef LA, Webster RE, Pittman JK (2016) PSR1 is a global transcriptional regulator of phosphorus deficiency responses and carbon storage metabolism in Chlamydomonas reinhardtii. Plant Physiol 170(3):1216–1234

Bajhaiya AK, Moreira JZ, Pittman JK (2017) Transcriptional engineering of microalgae: prospects for high-value chemicals. Trends Biotechnol 35(2):95–99

Beardall J, Raven JA (2016) Carbon acquisition by microalgae. In: The physiology of microalgae. Springer, Cham, pp 89–99

Boyle NR, Page MD, Liu B, Blaby IK, Casero D, Kropat J et al (2012) Three acyltransferases and nitrogen-responsive regulator are implicated in nitrogen starvation-induced triacylglycerol accumulation in Chlamydomonas. J Biol Chem 287(19):15811–15825

Cerutti H (2003) RNA interference: traveling in the cell and gaining functions? Trends Genet 19(1): 39–46

Chen Z, Lee WG (2019) Electroporation for microalgal biofuels: a review. Sustain Energy Fuels 3(11):2954–2967

Chen PH, Liu HL, Chen YJ, Cheng YH, Lin WL, Yeh CH, Chang CH (2012) Enhancing CO_2 bio-mitigation by genetic engineering of cyanobacteria. Energy Environ Sci 5(8):8318–8327

Coll JM (2006) Methodologies for transferring DNA into eukaryotic microalgae. Span J Agric Res 4:316–330

Courchesne NMD, Parisien A, Wang B, Lan CQ (2009) Enhancement of lipid production using biochemical, genetic and transcription factor engineering approaches. J Biotechnol 141(1–2): 31–41

Daboussi F, Leduc S, Marechal A, Dubois G, Guyot V, Perez-Michaut C et al (2014) Genome engineering empowers the diatom Phaeodactylum tricornutum for biotechnology. Nat Commun 5(1):1–7

Deng XD, Gu B, Li YJ, Hu XW, Guo JC, Fei XW (2012) The roles of acyl-CoA: diacylglycerol acyltransferase 2 genes in the biosynthesis of triacylglycerols by the green algae Chlamydomonas reinhardtii. Mol Plant 5(4):945–947

Doron L, Segal NA, Shapira M (2016) Transgene expression in microalgae—from tools to applications. Front Plant Sci 7:505

ElFar OA, Chang CK, Leong HY, Peter AP, Chew KW, Show PL (2021) Prospects of Industry 5.0 in algae: customization of production and new advance technology for clean bioenergy generation. Energy Convers Manag X 10:100048

Fukuda S, Hirasawa E, Takemura T, Takahashi S, Chokshi K, Pancha I, Imamura S (2018) Accelerated triacylglycerol production without growth inhibition by overexpression of a glycerol-3-phosphate acyltransferase in the unicellular red alga Cyanidioschyzon merolae. Sci Rep 8(1):1–12

Gangl D, Zedler JA, Rajakumar PD, Martinez EMR, Riseley A, Włodarczyk A et al (2015) Biotechnological exploitation of microalgae. J Exp Bot 66(22):6975–6990

Gimpel JA, Henríquez V, Mayfield SP (2015) In metabolic engineering of eukaryotic microalgae: potential and challenges come with great diversity. Front Microbiol 6:1376

Goncalves EC, Wilkie AC, Kirst M, Rathinasabapathi B (2016) Metabolic regulation of triacylglycerol accumulation in the green algae: identification of potential targets for engineering to improve oil yield. Plant Biotechnol J 14(8):1649–1660

Gordon GC, Korosh TC, Cameron JC, Markley AL, Begemann MB, Pfleger BF (2016) CRISPR interference as a titratable, trans-acting regulatory tool for metabolic engineering in the cyanobacterium Synechococcus sp. strain PCC 7002. Metab Eng 38:170–179

Gutiérrez S, Lauersen KJ (2021) Gene delivery technologies with applications in microalgal genetic engineering. Biology 10(4):265

Hossain AS, Salleh A, Boyce AN, Chowdhury P, Naqiuddin M (2008) Biodiesel fuel production from algae as renewable energy. Am J Biochem Biotechnol 4(3):250–254

Huang CH, Shen CR, Li H, Sung LY, Wu MY, Hu YC (2016) CRISPR interference (CRISPRi) for gene regulation and succinate production in cyanobacterium S. elongatus PCC 7942. Microb Cell Factories 15(1):1–11

Jagadevan S, Banerjee A, Banerjee C, Guria C, Tiwari R, Baweja M, Shukla P (2018) Recent developments in synthetic biology and metabolic engineering in microalgae towards biofuel production. Biotechnol Biofuels 11(1):1–21

Jaworski JG, Clough RC, Barnum SR (1989) A cerulenin insensitive short chain 3-ketoacyl acyl carrier protein synthase in Spinacia oleracea leaves. Plant Physiol 90:41–44

Jeon S, Lim JM, Lee HG, Shin SE, Kang NK, Park YI et al (2017) Current status and perspectives of genome editing technology for microalgae. Biotechnol Biofuels 10(1):1–18

Jiang W, Brueggeman AJ, Horken KM, Plucinak TM, Weeks DP (2014) Successful transient expression of Cas9 and single guide RNA genes in Chlamydomonas reinhardtii. Eukaryot Cell 13(11):1465–1469.3

Kajikawa M, Sawaragi Y, Shinkawa H, Yamano T, Ando A, Kato M et al (2015) Algal dual-specificity tyrosine phosphorylation-regulated kinase, triacylglycerol accumulation regulator1, regulates accumulation of triacylglycerol in nitrogen or sulfur deficiency. Plant Physiol 168(2): 752–764

Kang NK, Jeon S, Kwon S, Koh HG, Shin SE, Lee B et al (2015) Effects of overexpression of a bHLH transcription factor on biomass and lipid production in Nannochloropsis salina. Biotechnol Biofuels 8(1):1–13

Kang NK, Kim EK, Kim YU, Lee B, Jeong WJ, Jeong BR, Chang YK (2017) Increased lipid production by heterologous expression of AtWRI1 transcription factor in Nannochloropsis salina. Biotechnol Biofuels 10(1):1–14

Kao PH, Ng IS (2017) CRISPRi mediated phosphoenolpyruvate carboxylase regulation to enhance the production of lipid in Chlamydomonas reinhardtii. Bioresour Technol 245:1527–1537

Kasai Y, Oshima K, Ikeda F, Abe J, Yoshimitsu Y, Harayama S (2015) Construction of a self-cloning system in the unicellular green alga Pseudochoricystis ellipsoidea. Biotechnol Biofuels 8(1):1–12

Kebeish R, Niessen M, Thiruveedhi K, Bari R, Hirsch HJ, Rosenkranz R et al (2007) Chloroplastic photorespiratory bypass increases photosynthesis and biomass production in Arabidopsis thaliana. Nat Biotechnol 25(5):593–599

Kim S, Lee YC, Cho DH, Lee HU, Huh YS, Kim GJ, Kim HS (2014) A simple and non-invasive method for nuclear transformation of intact-walled Chlamydomonas reinhardtii. PLoS One 9(7): e101018

Kumar SV, Misquitta RW, Reddy VS, Rao BJ, Rajam MV (2004) Genetic transformation of the green alga—Chlamydomonas reinhardtii by Agrobacterium tumefaciens. Plant Sci 166(3): 731–738

Kwak M, Park WK, Shin SE, Koh HG, Lee B, Jeong BR, Chang YK (2017) Improvement of biomass and lipid yield under stress conditions by using diploid strains of Chlamydomonas reinhardtii. Algal Res 26:180–189

Leon R, Fernandez E (2007) Nuclear transformation of eukaryotic microalgae. In: Transgenic microalgae as green cell factories, pp 1–11

Li H, Shen CR, Huang CH, Sung LY, Wu MY, Hu YC (2016) CRISPR-Cas9 for the genome engineering of cyanobacteria and succinate production. Metab Eng 38:293–302

Lin WR, Lai YC, Sung PK, Tan SI, Chang CH, Chen CY, Ng IS (2018) Enhancing carbon capture and lipid accumulation by genetic carbonic anhydrase in microalgae. J Taiwan Inst Chem Eng 93:131–141

Mat Aron NS, Khoo KS, Chew KW, Show PL, Chen WH, Nguyen THP (2020) Sustainability of the four generations of biofuels—a review. Int J Energy Res 44(12):9266–9282

Mini P, Demurtas OC, Valentini S, Pallara P, Aprea G, Ferrante P, Giuliano G (2018) Agrobacterium-mediated and electroporation-mediated transformation of Chlamydomonas reinhardtii: a comparative study. BMC Biotechnol 18(1):1–12

Mosey M, Douchi D, Knoshaug EP, Laurens LM (2021) Methodological review of genetic engineering approaches for non-model algae. Algal Res 54:102221

Ng IS, Tan SI, Kao PH, Chang YK, Chang JS (2017) Recent developments on genetic engineering of microalgae for biofuels and bio-based chemicals. Biotechnol J 12(10):1600644

Nomaguchi T, Maeda Y, Liang Y, Yoshino T, Asahi T, Tanaka T (2018) Comprehensive analysis of triacylglycerol lipases in the oleaginous diatom Fistulifera solaris JPCC DA0580 with transcriptomics under lipid degradation. J Biosci Bioeng 126(2):258–265

Nymark M, Sharma AK, Sparstad T, Bones AM, Winge P (2016) A CRISPR/Cas9 system adapted for gene editing in marine algae. Sci Rep 6:24951

Ohlrogge J, Browse J (1995) Lipid biosynthesis. Plant Cell 7(7):957

Prasad B, Lein W, Lindenberger CP, Buchholz R, Vadakedath N (2019) Stable nuclear transformation of rhodophyte species Porphyridium purpureum: advanced molecular tools and an optimized method. Photosynth Res 140(2):173–188

Pratheesh PT, Vineetha M, Kurup GM (2014) An efficient protocol for the Agrobacterium-mediated genetic transformation of microalga Chlamydomonas reinhardtii. Mol Biotechnol 56(6):507–515

Purton S, Szaub JB, Wannathong T, Young R, Economou CK (2013) Genetic engineering of algal chloroplasts: progress and prospects. Russ J Plant Physiol 60(4):491–499

Radakovits R, Jinkerson RE, Darzins A, Posewitz MC (2010) Genetic engineering of algae for enhanced biofuel production. Eukaryot Cell 9(4):486–501

Ratledge C (1988) Products of hydrocarbon-microorganism interaction. In: Biodeterioration 7. Springer, Dordrecht, pp 219–236

Rengel R, Smith RT, Haslam RP, Sayanova O, Vila M, Leon R (2018) Overexpression of acetyl-CoA synthetase (ACS) enhances the biosynthesis of neutral lipids and starch in the green microalga Chlamydomonas reinhardtii. Algal Res 31:183–193

Roessler PG, Brown LM, Dunahay TG, Heacox DA, Jarvis EE, Schneider JC, Talbot SG, Zeiler KG (1994) Genetic engineering approaches for enhanced production of biodiesel fuel from microalgae. In: Himmel ME, Baker J, Overend RP (eds) Enzymatic conversion of biomass for fuels production. American Chemical Society, p 25

Sarayloo E, Simsek S, Unlu YS, Cevahir G, Erkey C, Kavakli IH (2018) Enhancement of the lipid productivity and fatty acid methyl ester profile of Chlorella vulgaris by two rounds of mutagenesis. Bioresour Technol 250:764–769

Sarsekeyeva F, Zayadan BK, Usserbaeva A, Bedbenov VS, Sinetova MA, Los DA (2015) Cyanofuels: biofuels from cyanobacteria. Reality and perspectives. Photosynth Res 125(1): 329–340

Shang C, Bi G, Yuan Z, Wang Z, Alam MA, Xie J (2016) Discovery of genes for production of biofuels through transcriptome sequencing of Dunaliella parva. Algal Res 13:318–326

Shimogawara K, Fujiwara S, Grossman A, Usuda H (1998) High-efficiency transformation of Chlamydomonas reinhardtii by electroporation. Genetics 148(4):1821–1828

Shin SE, Lim JM, Koh HG, Kim EK, Kang NK, Jeon S et al (2016) CRISPR/Cas9-induced knockout and knock-in mutations in Chlamydomonas reinhardtii. Sci Rep 6(1):1–15

Shin YS, Jeong J, Nguyen THT, Kim JYH, Jin E, Sim SJ (2019) Targeted knockout of phospholipase A2 to increase lipid productivity in Chlamydomonas reinhardtii for biodiesel production. Bioresour Technol 271:368–374

Shokravi H, Shokravi Z, Heidarrezaei M, Ong HC, Koloor SSR, Petrů M et al (2021) Fourth generation biofuel from genetically modified algal biomass: challenges and future directions. Chemosphere 285:131535

Sizova I, Greiner A, Awasthi M, Kateriya S, Hegemann P (2013) Nuclear gene targeting in C. hlamydomonas using engineered zinc-finger nucleases. Plant J 73(5):873–882

Song CW, Lee J, Lee SY (2015) Genome engineering and gene expression control for bacterial strain development. Biotechnol J 10(1):56–68

Takahashi K, Ide Y, Hayakawa J, Yoshimitsu Y, Fukuhara I, Abe J et al (2018) Lipid productivity in TALEN-induced starchless mutants of the unicellular green alga Coccomyxa sp. strain Obi. Algal Res 32:300–307

Takemura T, Imamura S, Tanaka K (2019) Identification of a chloroplast fatty acid exporter protein, CmFAX1, and triacylglycerol accumulation by its overexpression in the unicellular red alga Cyanidioschyzon merolae. Algal Res 38:101396

Tan KWM, Lee YK (2017) Expression of the heterologous Dunaliella tertiolecta fatty acyl-ACP thioesterase leads to increased lipid production in Chlamydomonas reinhardtii. J Biotechnol 247:60–67

Tsai CH, Warakanont J, Takeuchi T, Sears BB, Moellering ER, Benning C (2014) The protein Compromised Hydrolysis of Triacylglycerols 7 (CHT7) acts as a repressor of cellular quiescence in Chlamydomonas. Proc Natl Acad Sci 111(44):15833–15838

Verma M, Mishra V (2020) An introduction to algal biofuels. In: Microbial strategies for techno-economic biofuel production. Springer, Singapore, pp 1–34

Vieler A, Wu G, Tsai CH, Bullard B, Cornish AJ, Harvey C et al (2012) Genome, functional gene annotation, and nuclear transformation of the heterokont oleaginous alga Nannochloropsis oceanica CCMP1779. PLoS Genet 8(11):e1003064

Wang Q, Lu Y, Xin Y, Wei L, Huang S, Xu J (2016) Genome editing of model oleaginous microalgae Nannochloropsis spp. by CRISPR/Cas9. Plant J 88(6):1071–1081

Wei L, Xin Y, Wang Q, Yang J, Hu H, Xu J (2017) RNA i-based targeted gene knockdown in the model oleaginous microalgae Nannochloropsis oceanica. Plant J 89(6):1236–1250

Xin Y, Lu Y, Lee YY, Wei L, Jia J, Wang Q, Xu J (2017) Producing designer oils in industrial microalgae by rational modulation of co-evolving type-2 diacylglycerol acyltransferases. Mol Plant 10(12):1523–1539

Xue J, Balamurugan S, Li DW, Liu YH, Zeng H, Wang L, Li HY (2017) Glucose-6-phosphate dehydrogenase as a target for highly efficient fatty acid biosynthesis in microalgae by enhancing NADPH supply. Metab Eng 41:212–221

Yang B, Liu J, Ma X, Guo B, Liu B, Wu T et al (2017) Genetic engineering of the Calvin cycle toward enhanced photosynthetic CO_2 fixation in microalgae. Biotechnol Biofuels 10(1):1–13

Yao L, Cengic I, Anfelt J, Hudson EP (2016) Multiple gene repression in cyanobacteria using CRISPRi. ACS Synth Biol 5(3):207–212

Yunus IS, Jones PR (2018) Photosynthesis-dependent biosynthesis of medium chain-length fatty acids and alcohols. Metab Eng 49:59–68

Yunus IS, Wichmann J, Wördenweber R, Lauersen KJ, Kruse O, Jones PR (2018) Synthetic metabolic pathways for photobiological conversion of CO_2 into hydrocarbon fuel. Metab Eng 49:201–211

Zhang J, Hao Q, Bai L, Xu J, Yin W, Song L et al (2014) Overexpression of the soybean transcription factor GmDof4 significantly enhances the lipid content of Chlorella ellipsoidea. Biotechnol Biofuels 7(1):1–16

Chapter 8
Algal Biofuel Production from Municipal Waste Waters

Navodita Maurice

Abstract Supply of energy and quality of water are the two major crunches humans are struggling with in the last decade. The global population at present demands healthy food items followed by efficient energy supply, potable drinking water, and an environment with reduced CO_2 levels in order to ensure social and environmental safety. Treatment of the polluted water is one of the biggest challenges nowadays as the contaminated water bodies render water unsuitable for human consumption. In the developing countries, pollution of water by organic contaminants, heavy metals, sewage, and eutrophication is decreasing the human lifespan by increasing the risk of deadly diseases. Researchers all over the world are trying their best to convert the biowaste into bioenergy and one such approach is treatment of wastewater by microalgae in order to produce biofuels. Microalgae can thrive in diverse environments ranging from freshwater to polluted wastewaters. They are ecologically important as their structure and metabolic reactions are comparable to the higher plant species. They can utilize inorganic and organic matter from different types of wastewaters for biomass conversion that can be used to generate biofuel. Selection of microalgal species, wastewaters, and pretreatment methods are of prime importance in order to generate biofuel. This chapter focuses on the algal biofuel production from municipal wastewaters.

Keywords Microalgae · Biofuel · Municipal wastewater · Biomass

Abbreviations

ASP	Activated sludge process
CFPP	Cold filter plugging point
CN	Cetane number
COD	Chemical oxygen demand

N. Maurice (✉)
Prophyl Ltd., Mohács, Hungary

N. Srivastava, P. K. Mishra (eds.), *Technological Advancement in Algal Biofuels Production*, Clean Energy Production Technologies,
https://doi.org/10.1007/978-981-19-6806-8_8

CWW	Campus wastewater
DU	Degree of unsaturation
EOCs	Emerging organic contaminants
EPA	Eicosapentaenoic acid
FAME	Fatty acids methyl esters
GHG	Greenhouse gas
HRAPs	High-rate algal ponds
HRT	Hydraulic residence time
IV	Iodine value
LBs	Lipid bodies
MUFA	Monounsaturated fatty acid
MW	Municipal wastewater
N	Nitrogen
P	Phosphorus
PBRs	Photobioreactors
PTS	Primary-treated sewage
PUFA	Polyunsaturated fatty acid
PWW	Paper mill wastewater
SFA	Saturated fatty acid
STS	Secondary-treated sewage
SV	Saponification value
TAG	Triacylglycerol
TOC	Total organic carbon
WW	Wastewater

8.1 Introduction

Tremendous increase in the industrialization and urbanization has elevated the risk of water contamination due to the accumulation of harmful chemicals that are not good for human utilization, and the risk of dangerous diseases has also increased in the past decade (Sonune and Ghate 2004). In the developing countries, 80% of diseases are waterborne; therefore, appropriate methods for treatment of the contaminated water are needed in order to make water suitable not only for aquatic animals but for humans as well. Treatment of wastewater (WW) is a crucial process of environmental conservation where the quality of the wastewater is checked in order to avert pollution of water. The application of microalgae for cleaning the wastewaters has become much popular in the past few years where microalgal strains are cultivated in the wastewater as they are able to uptake the nutrients present in wastewaters and utilize them for their metabolic activities. Several researchers have investigated the cultivation of different microalgal strains in different wastewaters coming from different resources such as agriculture fields or factories, industries as well as from the municipal sectors (Mehrabadi et al. 2015). Sufficient research has

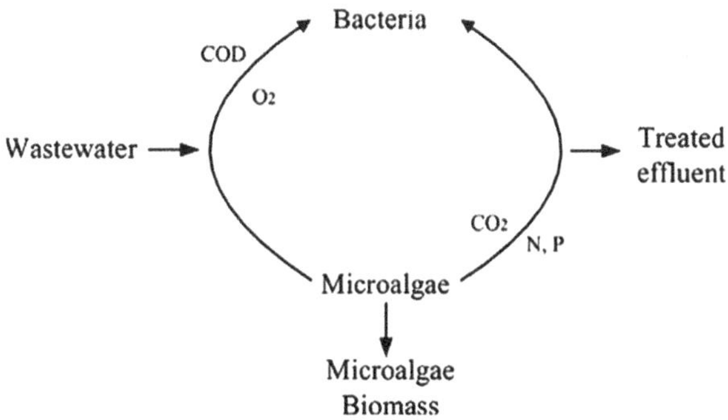

Fig. 8.1 Symbiotic association of microalgae with bacteria in the HRAP-WW treatment plant. (Adapted from Kadir et al. 2018)

already been done and many research projects are still in progress so as to test the appositeness of high-rate algal ponds (HRAPs) systems for the WW treatment (Park et al. 2011). Microalgal strains display symbiotic relationship with the heterotrophic bacteria present in wastewater where bacteria break the organic compounds by taking up oxygen produced by the microalgal strains through photosynthesis process (Fig. 8.1) (Passos and Ferrer 2014). Fossil fuel resources are limited and to cover up the energy demand of the everyday increasing world population, their amounts are also decreasing and these fossil fuels cause a negative effective on the environmental balance. So, there is a huge demand for eco-friendly and sustainable sources of energy and therefore biofuels (bioethanol, biomethane, and biodiesel) have gained much consideration. Several European countries have started to use biodiesel for transportation purposes due to the similar features of biodiesel with conventional fossil fuels. Biodiesel is regarded as a clean fuel as it emits lower amounts of sulphur dioxide (SO_2), environmentally friendly biodegradable, and nontoxic nature (Rincón et al. 2014).

Human population is expanding day by day, resulting in an enhanced discharge of WW; for example, daily human activities (cooking of food, bathing and washing of dishes, clothes, etc.) as well as schools and medical facilities generate a huge amount of municipal wastewater (MW) every day that needs to be treated properly. Treatment of MW by biological method, namely activated sludge process (ASP) is well known where organic material undergoes oxidation by the microbes through accelerated aeration in a reactor. Although ASP efficiently removes chemical oxygen demand (COD) from the municipal wastewaters, it drains a large proportion of energy (Gikas 2017). Aeration alone in the reactors consumes up to 70% of the overall energy and therefore ameliorations in the aeration schemes can lessen the energy consumption to an extent. Waste-activated sludge (WAS) generated by ASP also demands proper discharge and management (Liu et al. 2018). WAS is rich in ecologically degradable organic content that can be transformed into bioenergy. Bioenergy can be generated from the sludge of the WW treatment plants by anaerobic digestion but sludge management itself demands a large proportion of energy (Zhang et al. 2017). ASP wastewater treatment plants also contribute to a significant amount of GHG (greenhouse gas) emissions. They release GHG directly by biological operation, while indirect discharge comes from the power utilized for the treatment plant functioning (Singh et al. 2016). Consequently, there is a compelling demand for methods that are cheap and easy to handle in order to treat municipal wastewaters and their conversion to bioenergy. Researchers all over the globe are trying their best to solve this issue. Microbial fuel cells (MFCs), anaerobic treatment of MW, and growing microalgae are among the popular techniques adopted for WW treatment, resulting in bioenergy production (Daverey et al. 2019).

8.2 Algal Technology

Microalgae have been regarded as an auspicious substrate for the production of biofuels as they have the ability to fix GHGs, can grow faster, and can accumulate higher amounts of lipids (Alam et al. 2012). Microalgae produce about 15–300 times more biofuel than the traditional crops as it can be propagated throughout the year without hindrance and can be cultivated throughout the year due to its shorter production time (approximately 12 days depending on species and culture techniques). Although biofuel production by microalgae is eco-friendly, technical difficulties, higher production costs, and energy demands are limiting factors (Christenson and Sims 2011). To solve these issues, researchers are working to establish methods that are cheap, easy to handle as well microalgal species that are capable of agglomerating lipid concentrations (Ranjith Kumar et al. 2015). Harvesting of microalgae from the WWs is a derogatory step of the WW treatment as it covers a fraction of 30% of microalgae based on species, density, and the required end product (Barros et al. 2015). A compelling microalgal harvesting method is capable of producing a higher amount of biofuel by the removal of nutrients. Pond system WWs are inefficient for microalgal harvesting and as a result it is decomposed at the floor of the pond, resulting in methane emission (Abdel-Raouf et al. 2012). Recovery of microalgae from WWs combined with the extraction of lipids increases the energy production to 90% by increased biodiesel production (Lardon et al. 2009). Microalgae have a compact and inflexible cell wall made up of glycoproteins and composite carbohydrates having high mechanical stability and chemical withstanding; therefore, a supplementary pretreating step is required for the disruption of the algal cell wall for increased recovery of lipid. Disruption of the cell wall by pretreatment improves the contact between chemical solvents and fatty acids of microalgae, resulting in higher lipid extraction (Dong et al. 2016). Microalgal pretreatment falls under four major categories, namely (1) chemical, (2) thermal, (3) biological, and (4) mechanical (Passos et al. 2014). Pretreatment efficiency for biofuel recovery from microalgae is based upon the stability, strength, and structure of the cell wall and at present a wide variety of pretreatment methods have been tested (Passos and Ferrer 2014) (Fig. 8.2). Treatment of WWs by microalgae is a profitable, sustainable, and lesser energy demanding method. Microalgae are fast-growing photoautotrophs capable of producing oxygen by utilizing sunlight and

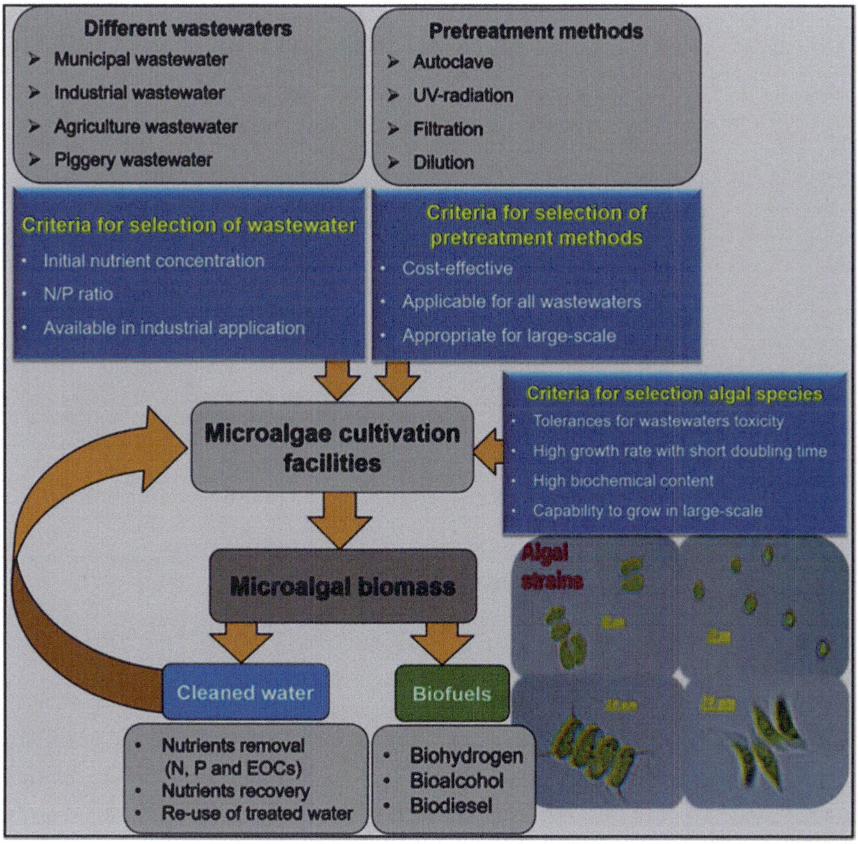

Fig. 8.2 Biofuel generation from microalgal biomass. (Adapted from Salama et al. 2017)

CO_2. Their oxygen-producing potential is rather superior to the photosynthetic species (Langley et al. 2012). In the last decades, lots of experiments have been conducted globally for the wastewater treatment by microalgae in order to remove nutrients and eliminate COD in the closed photobioreactors (PBRs) and high-rated algal ponds (HRAPs) (Olguín 2012).

Seventy percent COD has been found to be eliminated by *Chlorella* species cultivated in PBR while treating WWs. Similarly, *C. vulgaris* along with *Scenedesmus obliquus* can remove 64% COD in the PBR (Gouveia et al. 2016). Municipal wastewaters serve as ideal substrates for the cultivation of microalgae as they are rich in phosphorus (P) and nitrogen (N) levels (Craggs et al. 2013). Cultivation of microalgae for the treatment of WWs focuses on the bacterial and microalgal coculturing where the former converts organic material into inorganic contents by utilizing O_2, while the latter uses CO_2 and inorganic material for biomass production and generation of O_2 (Gonçalves et al. 2017). The cocultivation of microalgae and bacteria has resulted in >90% of nutrient removal from the WWs

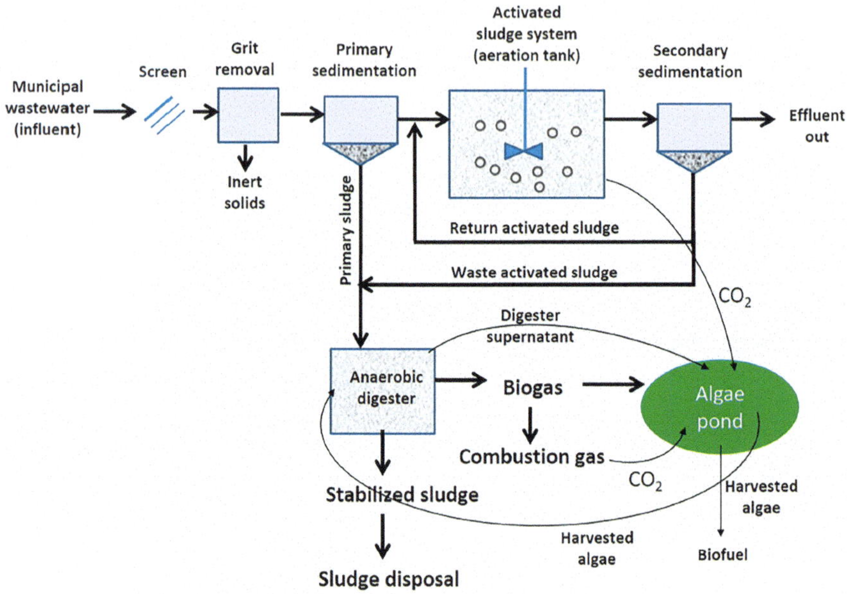

Fig. 8.3 Municipal wastewater treatment plant by algal technology. (Adapted from Daverey et al. 2019)

and decreased aeration costs (Khaldi et al. 2017). The biomass produced by microalgal species contains 50–70% lipids and is therefore ideal for the biofuel production (Chew et al. 2017). Coculturing-activated sludge and microalgae together in the PBR is beneficial for both bacterial and algal growth (Zhu et al. 2019). Microalgae–bacterial biomass valorization by anaerobic digestion or enzymatic treatment produces higher amounts of biogas but cultivation of microalgae in the wastewaters is a challenging practice (Arashiro et al. 2019). Climatic factors and growth of other microbes can affect the biofuel production from the microalgae in wastewaters (Razzak et al. 2017). Algal cell size and their growth in smaller colonies require intensive energy-utilizing techniques; for example, centrifugation and flocculation methods have been used to overcome energy consumption, which increases the application cost of the treatment units. Therefore, WW treatment by cultivation of microalgae is a competent technology for removing nutrients from WWs as well as for the bioenergy generation and is superior to traditional WW treatment methods (Fig. 8.3) (Branyikova et al. 2018).

Cultivation of microalgae and biomass production by anaerobic fermentation, esterification, pyrolysis, hydrothermal liquefaction, and gasification techniques were tested by Allnutt and Kessler (2015). Microalgal strains, presence of nutrients as well as PBRs and open-pond reactors influence the production of biofuel by microalgae and recycling of the biomass results in reduced GHG emission (Davis et al. 2016). Energy return on investment (EROI) suggests that the biofuel quality determines the consumption of biofuel produced by the microalgal cultivation (Hall et al. 2014).

The unicellular green microalgal species were considered as renewable energy sources decades ago as they can produce a higher proportion of lipid and biomass for biofuel production. Compared to plant substrates, microalgae can be either used directly or transformed into biofuels by thermo- or biochemical methods (Brennan and Owende 2010). Production of energy by direct combustion of dried biomass is produced by algae but this method is generally not used. Hydrogenation, gasification, liquefaction, and pyrolysis are some of the most commonly used thermochemical methods for biofuel production from algal biomass, while anaerobic digestion and fermentation are the common biochemical methods. Microalgal strains release hydrogen by bio-photolysis reaction, while extraction of lipid (triacylglycerol) as well as its biodiesel transformation is accomplished by transesterification (Hu et al. 2008). Many researchers have centralized their research in search of the microalgal strains and cultivation settings that can result in higher lipid production, leading to higher biofuel production (Griffiths and Harrison 2009). Many researchers have also targeted their studies in search of conditions that lead to higher lipid accumulation within the microalgal cells, for example, under P or N stress; however, higher lipid yield does not affect the total biomass yield (Dean et al. 2010). Higher lipid production can be achieved by utilizing the methods that result in increased microalgal biomass as well as increased biofuel yield (Griffiths and Harrison 2009). Microalgae can serve as substrates for the generation of biofuel as they can be cultivated in freshwaters with minimal amount of nutrients in comparison to plant feedstocks; for example, salty or brackish waters serve as excellent media for their cultivation as a lot of research has already been conducted in relation to microalgal growth in salty waters (Rodolfi et al. 2009). WWs are tremendously rich in essential nutrients required for the growth of microalgal strains and thus can be used as culture media for microalgal cultivation and harvesting, as microalgae have been regarded as eco-friendly substrates for the WW treatment by different methods in recent decades (Green et al. 1995). However, the majority of wastewaters have elevated levels of total P and N content along with toxic metals that demand expensive chemical treatment in order to get rid of them and municipal wastewaters are quite rich in these elements. Microalgae can effectively grow in these wastewaters that have a higher accumulation of nutrients and therefore serve as sustainable weapons against the treatment of WWs in a cost-effective manner (de-Bashan and Bashan 2010). It has already been proposed that the microalgae cultivated in the wastewaters must be utilized for the production of energy but still there is a contradiction regarding the economic workability of biofuels produced by microalgal strains. Although enormous efforts of researchers all over the world indicate that algal biofuel is eco-friendly and economically beneficial, still there are questions regarding their long-term sustainability (Stephens et al. 2010). Researchers still contradict that setting up of wastewater plants for the cultivation of algae in order to remove nutrients by algal biomass production and finally bioenergy production is expensive and demands more energy input on the one hand, while biofuel production by algae cultivated in wastewaters provides an option for decreasing water pollution on the other (van Beilen 2010).

8.3 Microalgae and Wastewater

For more than 75 years, several strains of microalgae, such as *Dunaliella* and *Chlorella*, have been used in WW treatment units working for the generation of microalgal biomass (Bux 2013). Around 60 genera and 80 species of microalgae have been listed by Palmer (1969) and other researchers that can tolerate pollutants present in wastewaters. Chlorophyta (taxonomic algal group) comprises the most abundant number of microalgal species with outspread dissemination. Several researchers have focused their research on different species of *Scenedesmus*, *Chlorella*, and *Chlamydomonas* as these species can remove P and N from different types of wastewaters at different concentrations (Gao et al. 2016). These species not only remove heavy metals and other toxic pollutants from wastewaters but are also capable of solving secondary pollution issues and can therefore produce higher proportions of lipids and biomass (Matamoros et al. 2015). Microalgal biomass is rich in the lipid triacylglycerol (TAG) and transesterification process can transform the TAG to biodiesel (especially fatty acids methyl esters, or FAME) which in turn produces minimal amounts of GHGs in the atmosphere (Chisti 2007). Microalgae grow faster in comparison to plants and therefore their potential to produce biomass is much higher with a little demand for their cultivation (Cho et al. 2011). Researchers have emphasized that microalgal cultivation under the conditions of stress can produce about 80% of dry biomass for the effective biodiesel conversion (Dasgupta et al. 2015). Microalgae can accumulate around 75% of lipids within their cells but it slows down their growth rate in comparison to the strains that have lower accumulation potential (Mata et al. 2010). *C. vulgaris*, *C. protothecoides*, and *Nannochloropsis* strains have been found superior for biodiesel production as they can accumulate higher proportions of lipids but still algal biofuel production is low because it is very expensive (Rincón et al. 2014). Researchers all over the globe have concentrated their research for the reduction of biofuel production costs by the utilization of microalgal strains as well as phytoremediation by the organic nutrient removal from the WWs by microalgae by photosynthesis (Green et al. 1995). Majority of the microalgal species can be cultivated in the wastewaters with higher concentrations of organic nutrients that can be easily absorbed by microalgae for their metabolic activities (de-Bashan and Bashan 2010), but the ratio of N to P must be favorable as this ratio marks the success of microalgal cultivation in the WWs (Salama et al. 2017). *Chlorella* species have been tested by several researchers for the treatment of wastewaters as they are capable of producing higher proportions of biomass and can accumulate higher lipid levels (Rawat et al. 2011). Wastewaters without treatment are a threat to the environment as they accumulate higher amounts of toxic pollutants and heavy metals that are unfit for human consumption as well as for the aquatic flora and therefore demand treatment; however, chemical treatment of wastewaters is expensive and can result in eutrophication (Mulbry et al. 2008). Researchers all over the globe have agreed to the fact that WW treatment (all types of) by microalgal strains is safe to environment, sustainable, and a cheaper method as the microalgal biomass can be directly or indirectly converted into

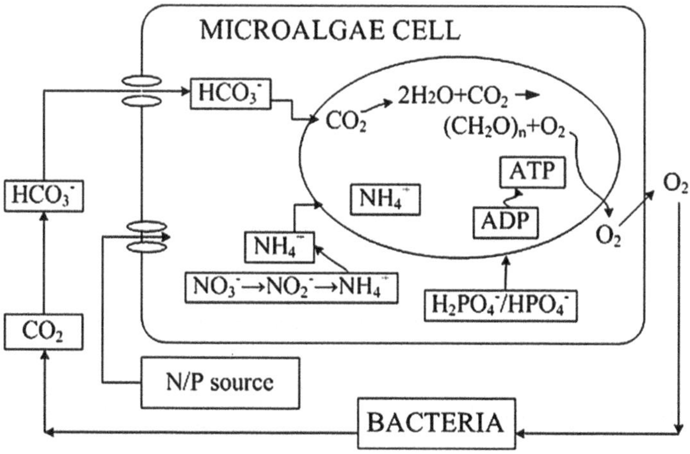

Fig. 8.4 Bacterial and microalgal nutrient exchange in WWs. (Adapted from Kadir et al. 2018)

biofuels (Rawat et al. 2011) (Fig. 8.4). Microalgal culture in WW serves a dual role; for example, the oxygen produced by microalgal strains is useful for the metabolic activity of the bacterial species present in WWs, as they digest these nutrients and generate CO_2 that can be taken up by the microalgae for their metabolic activities.

Considerably good microalgal growth has been reported in the wastewater from piggeries as it has a higher proportion of nitrogen and phosphorus elements in comparison to municipal, domestic, and wastewater from anaerobic digestion. Utilization of HRAPs for the WW treatment has already been in practice for many years. HRAPs are special types of ponds about 0.3–0.6 m deep where nutrients from wastewater and microalgae are mixed continuously with the help of a paddle wheel with a fixed velocity in order to avoid stratification and sedimentation. The shallow depth of HRAPs allows paramount penetration of light, thereby favoring photosynthesis resulting in the production of microalgal biomass (Vargas e Silva and Monteggia 2015). Studies have indicated that HRAPs not only result in increased algal biodiesel production but also improve the quality of the wastewaters. Piggery wastewater treatment and microalgae cultivation in HRAPs have been proved to be cost-effective, as the obtained biomass can be transformed into valuable products based upon the microalgal strains (Olguí 2003). Researchers also suggest that microalgae cultivation in wastewaters must focus on higher amounts of biomass production rather than lipid production and therefore abiotic factors such as temperature, pH, and light intensity must be optimized (Rawat et al. 2011). Despite having many advantages, treatment of WWs by HRAPs as well as harvesting of microalgal biomass for biofuel generation is restricted as wastewaters are loaded with higher levels of toxic pollutants that can obstruct microalgal growth; therefore, further investigation is needed for the optimization of the process (Sharma et al. 2012). Wastewaters are cheap, abundant resources rich in both micro and macronutrients that can replenish microalgal growth. Major elements present in wastewaters are N,

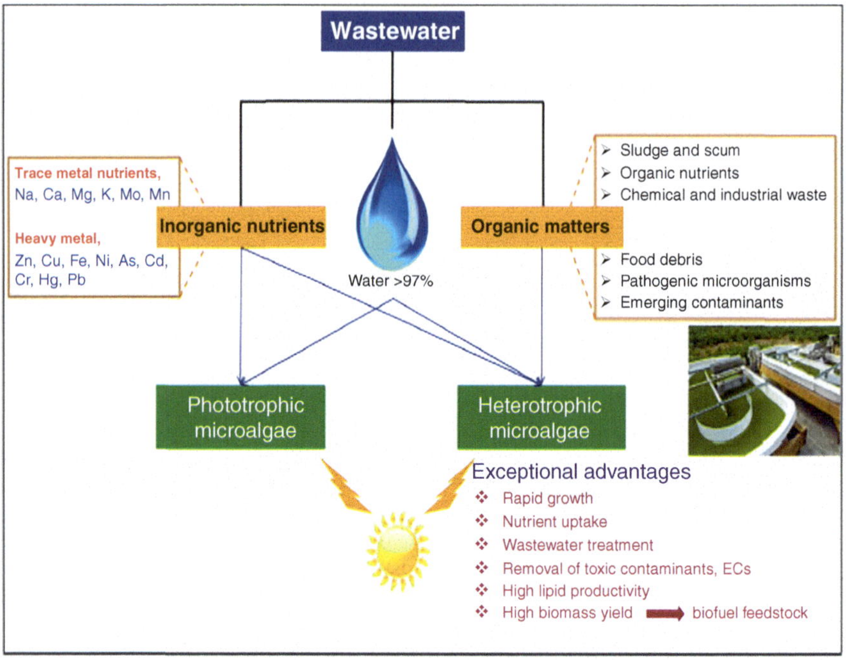

Fig. 8.5 Microalgal cultivation from wastewater resources. (Adapted from Salama et al. 2017)

P, ammonia (NH_3), urea, carbon (C), and trace elements that support metabolic activities of microalgae, with N, P, and C being the most essential ones for microalgal growth and cultivation (Ding et al. 2015). The benefits of microalgal biomass harvesting in the wastewaters are three-fold: first, microalgal biomass serves as an excellent substrate for producing biofuel; second, recovery of valuable nutrients from the WWs can also be performed, followed by the treatment of WWs by microalgae (Ajayan et al. 2015) (Fig. 8.5). Municipal wastewater (MW) appears to be an excellent source for the growth of microalgal species as they have higher proportion of nitrogen and phosphorus. Nutrient recovery from the MWs not only renders them clean but also prevents the formation of dangerous blooms formed by microbes (Pienkos and Darzins 2009).

Municipal wastewater can be supplied to the algal farms for algal cultivation, and combining algal culture with treatment of WW not only reduces biofuel expenses but also increases the efficiency of WW treatment units. Treatment of MW has the following phases: (a) solid matter sedimentation by primary treatment, (b) removal of dissolved and suspended organic nutrients by secondary treatment, and (c) tertiary water treatment before being discharged out. MW is rich in inorganic and organic nutrients followed by harmful elements as well as coliform bacteria. However, a major fraction of MW nutrients is occupied by NO_3^- (nitrate), NH_4^+ (ammonium), PO_4^{3-} (orthophosphate), and NO_2^- (nitrite). Treatment of MW basically comprises removal of sedimented solids, which occupy around 40% of the total biochemical oxygen demand (BOD), followed by removal of organic and inorganic elements, followed by toxic metals and other contaminants. *Scenedesmus* and *Chlorella* species are responsible for the natural treatment of MWs without human interference (Bhatnagar et al. 2010). Microalgae belonging to the genus *Chlorophyta* have been well known for their tolerance against the harmful pollutants of wastewaters (Aslan and Kapdan 2006). Wastewater treatment by tertiary methods is rather risky due to changing nutrient removal rates, cleaning costs, and ineffective utilization of available resources. Microalgae can remove N and P from the wastewaters both in suspension and mutilated forms. Many strains can efficiently remove N, P, and ammonia from wastewaters after secondary treatment; for example, *C. vulgaris* has been reported to eliminate a fraction of about 80–90% from the sedimentary WWs (Woertz et al. 2009). Redfield ratio can determine the total amount of N and P required by microalgae for their normal growth (Kesaano and Sims 2014). N, P, and silicon stress, followed by lower temperature ranges and pH, enhance the lipid accumulation in microalgae; however, this phenomenon is strain and culture technique specific (Hena et al. 2015). Majority of MWs differ in their C:N:P ratios, and nitrogen stress causes higher accumulation of lipids in the microalgae. Similarly, studies have shown that mixing microalgae with CO_2 removes a higher fraction of ammonia and orthophosphate while lipid accumulation remains constant (Woertz et al. 2009). The results of Hena et al. (2015) are in agreement with the findings of Woetz and colleagues (2009). Hena et al. (2015) treated the dairy wastewater with native microalgal species resulting in 98% nutrient removal while the lipid content remained only 17%. The requirement of CO_2 similar to that in the atmosphere limits growth of the microalgae; however, supplying CO_2 from an external source can minimize this problem (Demirel 2016). An interdependent model for algal biofuel generation coupled with the treatment of MW and agricultural wastewater offers many advantages, such as the recovery of other essential materials in addition to biofuel generation, and majority of the large wastewater plants are generally located close to the urban areas, whereas small treatment plants are located far from the populated landscapes (Muylaert et al. 2015). Microalgal cultivation in the wastewaters also reduces the need of freshwater and its production costs (Harun et al. 2013). Besides biofuels, microalgal cultivation also produces materials for food and other industrial purposes, such as carbohydrates, proteins, and fatty acids, which can be utilized for the manufacturing of pharmaceutical products, food supplements, and cosmetic items. So, it can be concluded that the products derived from the algal biorefinery fall under two major categories, namely energy and non-energy products (Trivedi et al. 2015). Biofuel and value-added material production from the microalgae can be achieved by the bio- and thermochemical processes. Algal carbohydrates (cellulose, starch, glucose, and other polysaccharides) can be transformed into other products by fermentation or other chemical methods. Different types of photosynthetic pigments (carotenoids, chlorophylls, and phycobiliproteins) are present in microalgae that can be used in the production of nutraceutical and pharmaceutical products. Astaxanthin, a pigment obtained from *Haematococcus pluvialis*, is used in the treatment of cancers, eye problems, and cardiovascular diseases. Several microalgal pigments are used in food and cosmetic industrial production because algal pigments are superior to artificial coloring agents (Han et al. 2013).

Protection of the aquatic flora and fauna as well as availability of clean and potable water for human and animal consumption has been a biggest challenge for decades (Hussain et al. 2017). The manufacturing of essential elements by microalgal culture is dependent on the availability of nutrients in the growth media at the optimum concentrations; for example, apart from N and P, other essential vitamins and micronutrients are also needed for the algal growth (Ahmad et al. 2016). WWs contain these crucial nutrients in higher amounts and hence are suitable for the culture of microalgae, but to fetch good results, different microalgal strains are required for the WW treatment; however, studies have reported mixotrophic microalgae results in elevated biomass production and enhanced lipid accumulation (Cheng et al. 2017). MWs generally contain lower levels of N and P but contain relatively higher amounts of toxic metals and therefore require primary, secondary, and tertiary treatment methods, and this troublesome work has also been solved with the help of microalgae (Aslam et al. 2018). Manure leached from livestock and animal wastewaters is rich in N:P ratio in which half of the proportion is covered by ammonia. Animal manure is used in the agricultural crop cultivation, but its release into the water bodies makes it unfit for human and animal consumption (Prandini et al. 2016). Accumulation of elevated P and N concentrations in the WWs results in eutrophication but removal of P from the wastewaters is a very critical step, and in majority of the treatments, chemical are used either for the precipitation of P (solid state) or its transformation into activated sludge fraction by the microbial activity; however, as it is not completely recyclable, it is either dumped in the landfills or used in fertilizer production (Wang et al. 2016). Microalgae can efficiently recover P and N from WWs and therefore microalgal biorefinery is gaining much attraction in the past few decades; for example, O_2 produced by microalgae diminishes the cost of mechanical aeration in the treatment of ponds (Ahluwalia and Goyal 2007). The bacterial species (heterotrophic aerobes) demand oxygen in order to degrade inorganic and organic elements in the WW treatment plants. Wastewater remediation by algal strains is safer and eco-friendly as it does not produce extra pollutants and also offers the possibility of nutrient recycling; for example, biomass produced by microalgae has higher concentration of P and N, so it can serve as a cheap fertilizer as well as an animal food supplement (Munoz and Guieysse 2006). Majority of the research performed on the WW treatment by microalgal species have been conducted either in laboratories or in algal ponds and have suggested that microalgal strains can grow in different types of media ranging from sewage WWs to agricultural WWs as all the WWs have extremely high N, P, and heavy metal concentrations. Many factors influence microalgal growth as researchers have found that microalgae grow faster in the primary settled sewage waters with longer photoperiod and CO_2 addition and increase in the temperature decreases the production of biomass. Wastewaters used as growth media are different from other growth media due to difference in the N and P concentrations. Major fraction of nitrogen in the wastewater comes from ammonia and studies have indicated that elevated levels of ammonia can have a negative effect on the growth of microalgal strains, while heavy metals such as mercury and toxins can also affect algal growth (Wrigley and Toerien 1990). Pathogenic bacteria or predatory zooplanktons are the governing biotic agents that can also negatively impact the growth of microalgae and some other microbes can also compete with microalgae for essential elements. Population growth of microalgae in wastewaters is dependent upon the initial density as it can differ in different wastewaters and the method of treatment selected (Lau et al. 1995). Different microalgal strains tolerate differently the elevated levels of pollutants present in the WWs. Several reports have indicated that unicellular chlorophytic microalgal strains are far more tolerant toward the pollutants and thus can efficiently accumulate nutrients from the wastewaters (Ruiz-Marin et al. 2010). *Scenedesmus* and *Chlorella* strains occur in abundance among the phytoplanktons present in the HRAPs and oxidation ponds; however, they differ in their efficiencies (Masseret et al. 2000). *C. vulgaris* more efficiently accumulates P and N amounts from the wastewater treatment ponds than *C. kessleri*; similarly, growth rate of *C. vulgaris* is slower than *S. obliquus* in the MWs (Ruiz-Marin et al. 2010). *Scenedesmus* and *Chlorella* strains have already been tested for their excellent N, ammonia, and P recovery from the secondary treated WWs, suggesting that they can participate in the tertiary treatment of the wastewaters. Ruiz-Marin et al. (2010) tested *S. obliquus* growth in semicontinuous culture conditions and reported that growth rate is higher than in the batch culturing conditions as nutrients undergo depletion and also chlorophyll content and growth rate of the algal cells decline in the batch culture with the passage of time. P and N recovery from the primary treated sewage WWs by *C. vulgaris* is also excellent (Lau et al. 1995). Wang et al. (2010) tested *Chlorella* growth in the pretreated wastewater from all the three treatment stages by taking into account parameters such as removal of N, P, and metal ions. Their results emphasized that pretreated WWs favor faster microalgal growth as sufficiently high P and N concentrations are available in these WWs. A study conducted with *C. minutissima* reported that this strain can grow well in raw sewage wastewaters, in the dark (heterotrophic form) as well as in the light (mixotrophic form), can utilize different types of nutrients, and can tolerate different pH levels and salts. This strain can completely remove N, P, and ammonia from wastewaters although it grows best in the mixotrophic form as biomass yield is relatively higher in this form. Researchers suggest that this algal strain is best suited for HRAPs as it can thrive well in the untreated wastewater (Bhatnagar et al. 2010).

8.3.1 Nutrients in Wastewater

8.3.1.1 Ammonia (NH_3)

Agriculture wastewaters and other wastewaters are rich sources of ammonia due to the excessive utilization of nitrogen fertilizers that leach into the water sources. Microalgae can utilize this ammonia from the wastewater systems for its metabolic functions (Castro et al. 2017). *S. dimorphus* cultivated at higher concentrations of

NH_3 along with CO_2 in the PBRs can utilize up to 90% of the NH_3 for biomass conversion rich in fatty acids, sugars, and proteins (Kang and Wen 2015). The presence of NH_3 at optimum concentrations is essential as it can affect algal growth; for example, cellular composition and growth of microalgae are affected by dairy wastewater, which has lower NH_3 content. However, dairy wastewater combined with wastewater from the slaughterhouses not only results in better growth and cellular composition of microalgae but also increases lipid content due to higher NH_3 content (Lu et al. 2016). *Chlorella* sp. When tested for NH_3 removal from three different types of wastewaters indicated that relatively higher NH_3 concentrations result in decreased growth of algae; however, increasing the levels of NH_3 can result in decreased lipid content (unsaturated C16 and C18 fatty acid chains) within the cells of microalgae (Wang et al. 2015). Other study has also reported that NH_3 can affect the composition of products of microalgae and it is the ultimate source of nitrogen assimilated by them. No oxidation–reduction reactions are needed for NH_3 assimilation by the microalgae such as nitrites and nitrates (Collos and Harrison 2014). Nitrogen is a crucial element associated with the proper growth and development of plants because it participates in the production of macromolecules (chlorophylls, proteins, enzymes, DNA/RNA, etc.). N (organic form) originates from NO_2^-, NH_4^+, NH_3 (inorganic forms), and microalgae can utilize all these for its metabolic reactions. The enzymes nitrite and nitrate reductases reduce nitrite and nitrate. Electrons are transferred to nitrate from nicotinamide adenine dinucleotide (NADH) in the presence of nitrate reductase to form nitrite, which gets reduced by ferredoxin (Fd) and nitrite reductase to ammonium (NH_4^+). The inorganic N (all forms) are finally transformed to NH_4^+ before amino acids are synthesized from them. Glutamine synthase allows addition of NH_4^+ to glutamine by adenosine triphosphate (ATP) and glutamate (Glu). This amino acid (glutamate) elevates *C. vulgaris* growth up to 70% in the WWs (Wang et al. 2010).

8.3.1.2 Phosphorus

Phosphorus enters the water resources by agricultural fertilizers and composes a fraction of up to 80%. After nitrogen, phosphorus is the second most key ingredient, and it is essential for microalgal growth and metabolism, as it plays an important role in the fatty acid and protein synthesis. Phosphorus can be removed by different methods but needs to be processed before being used for the agricultural practices and microalgae can perform this task very well (Solovchenko et al. 2016). Microalgae utilize phosphorus in two forms, namely HPO_4^- and $H_2PO_4^-$ for the formation of ATP by phosphorylation process (Fig. 8.6). Microalgae can also able to assimilate phosphate from the pond systems as polyphosphates via an uptake method that is controlled by the intensity of light, temperature, and concentration of phosphates (Brown and Shilton 2014). Enhanced pH levels can result in phosphorus precipitation and therefore influence phosphorus uptake by the microalgal cells. *C. vulgaris* and *Bacillus licheniformis* when grown together showed that phosphorus assimilation rate reached to 92% at the pH of 7, while the phosphorus removal rate

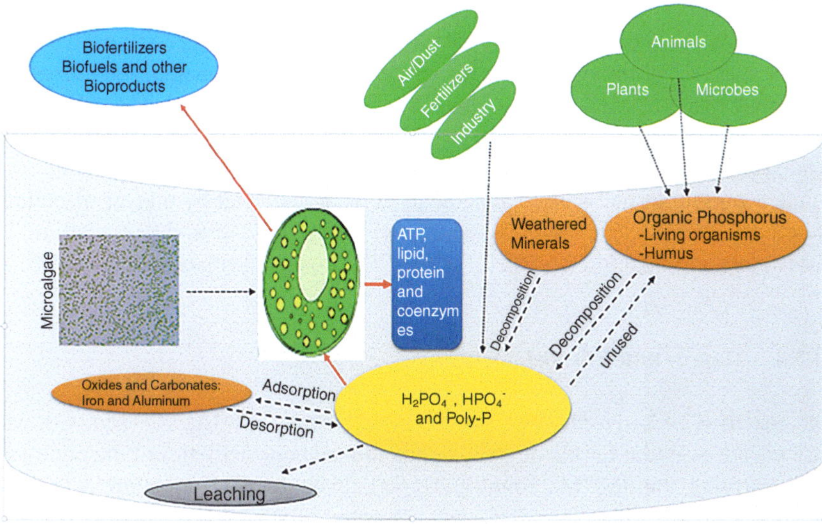

Fig. 8.6 Phosphorus cycle: utilization of $H_2PO_4^-$ and HPO_4^- into useful products by microalgae. (Adapted from Hussain et al. 2021)

declined when pH was dropped to 3, indicating the importance of pH in phosphorus recovery (Liang et al. 2013). Phosphorus participates in fatty acid synthesis, genetic materials, ATP, amino acids, and carbohydrate intermediates in the microalgal cells. Inorganic PO_4^{2-} is essential for metabolism and cellular growth of microalgae as its shortage can affect the rate of photosynthesis (Suganya et al. 2016). HPO_4^{2-}/ H_2PO^{4-} forms are utilized together during the metabolic activities of microalgae to form ATP from adenosine diphosphate (ADP) by phosphorylation. Few microalgal strains are capable of utilizing P as organic esters as well apart from other inorganic P forms for their growth and metabolism. However, orthophosphate has been found in abundance in the freshwater bodies and can cause WW eutrophication. Just like N assimilation, P accumulation by the microalgae is also influenced by environmental factors such as pH and dissolved oxygen (Ji Kabra et al. 2014).

8.3.1.3 Heavy Metals

The major contributors of the wastewaters are the heavy metals, which are responsible for the toxicity of the aquatic plants and animals. Since they do not decompose easily, they very often enter the food chains and cause serious threat to plants, animals, and even humans. Several studies have already been conducted and are still in progress to solve this problem, and all the studies have reported that microalgae can solve this issue (Zeraatkar et al. 2016). *Nitzschia perminuta* and *Nitzschia palea* are two microalgal strains that have the potential to accumulate relatively higher heavy metal proportions within the cells, which can be beneficial in the WW treatment (Chen et al. 2013). Microalgal adsorption potential against the heavy metals has been increased by several methods. Imani et al. (2011) have tested the potential of *Ceramium rubrum* for copper removal from the aqueous substance and found that this strain can tolerate higher levels of copper. The addition of CH_2OH and NaOH further increases the efficiency of this strain against heavy metal accumulation. *Phacus* sp. (Euglenophyta) is resistant against thallium, a pollutant present in the wastewaters, as its cells are covered by a layer of pellicle. Microalgae under the conditions of heavy metal stress can thrive well in the wastewaters and therefore are useful in the WW treatment (Plachno et al. 2015).

8.3.1.4 Macro- and Microelements

Both macro (C, Ca, N, H, P, O, K, Mg) and micro (Mn, Cu, Zn, Mo, Co, Mn) nutrients are essential for the development and metabolic activities of the photosynthesis carrying entities. Macronutrients participate in the formation of cellular structures, while micronutrients are essential for enzyme synthesis. Five micronutrients, namely Ca, Zn, Mn, Fe, and Cu, participate in the microalgal photosynthesis. Mn and Cl in the ionic forms are involved in O_2 production. Macronutrients especially C, P, and N as well as micronutrients such as Cu, Zn, Mg, Mo, Ca, Fe are essential for microalgal growth as a lack of them inhibits their growth WWs (Di Caprio et al. 2015).

8.3.1.5 Carbon Dioxide (CO_2)

Autotrophic algae can utilize CO_2 either from the atmosphere directly or from the gaseous discharge of the industries for their photosynthetic process. CO_2 balances pH of the growth media and researchers have suggested that CO_2 positively affects biomass and lipid production by microalgae. Carbon as CO_3^{2-} can also be taken up by microalgae directly for its growth or it can be converted to CO_2 by the enzyme carbonic anhydrase (Goncalves et al. 2016).

8.3.1.6 Minor Minerals

Minor minerals also play a role in the growth and metabolism of microalgae as they participate in photosynthesis, enzyme synthesis, and storage of energy. Manganese (Mn) is required for microalgae as it is essential for enzyme synthesis. Mn also partakes in the evolution of O_2 in photosynthesis, and studies have shown that deficiency of Mn can inhibit growth of microalgae due to the shortage of the synthesis of PSUs (photosynthesis structure units) (Yang et al. 2016). Iron (Fe) is needed in the respiratory e-transport chain and photosynthetic activity of microalgae, and iron deficiency also alters growth of microalgae. Supplementation of iron in the growth medium of *Lyngbya majuscula* resulted in increased biomass production.

Zinc (Zn) acts as a cofactor of various enzymes and also participates in the synthesis of indole acetic acid. Zn can cause toxicity in smaller concentrations and thus can affect microalgal growth. Wastewaters are rich in different trace elements and thus can be treated by microalgae as they can utilize the nutrients for their growth and metabolic activities (Yruela 2013).

8.4 Pollutant Removal by Microalgae from the Wastewaters

Wastewater treatment by tertiary method for the removal of nitrate, phosphate, and ammonia is more expensive in comparison to the primary treatment. Microalgae serve as a potential candidate for carrying out tertiary treatment of wastewater as a result of symbiosis between the bacteria and microalgae as heavy metals, inorganic, and organic nutrients can be removed from the wastewaters (Whitton et al. 2015). Microalgae liberate O_2 by photosynthesis which in turn is accepted by the bacterial strains (aerobes) resulting in organic nutrient degradation. These aerobic bacteria in turn produce CO_2 as a result of mineralization and this CO_2 is essential for microalgal growth. Numerous studies have indicated that P and N levels in the WWs can be lowered by microalgae as they utilize them for their growth. *Chlorella* Species are best known for their higher tolerance against heavy metals and higher P and N levels in the WWs. Microalgal strains can increase wastewater pH by inorganic carbon utilization and thus can enhance the removal of NH_3 and phosphorus. Excessive phosphorus levels in the WWs can cause eutrophication but microalgae can take up orthophosphate (inorganic P) for its metabolic reactions and cellular growth. Many microalgal species prefer ammonium as a nitrogen source as it requires lesser energy consumption than nitrate and nitrite. Nitrate and nitrite reductase enzymes reduce nitrate and nitrite to ammonium ions so that microalgae can take it up. *Chlorella* sp. can successfully remove 83% of the inorganic nitrogen and 90% inorganic phosphorus from the MWs (Wang et al. 2010). Microalgal strains have the potential of removing heavy metals, namely arsenic, cadmium, mercury, and bromine from the wastewaters and both biotic and abiotic factors influence microalgal growth in the WWs (Abdel-Raouf et al. 2012). Researchers have indicated that higher density of microalgal strains in the wastewaters can result in faster growth and faster pollutant removal but higher density can lead to auto-inhibitor accumulation and reduced photosynthesis. Biochemical constitution and growth features of microalgae depend upon the methods of cultivation, which are classified into four categories, namely (1) photo-heterotrophic, (2) mixotrophic, (3) photoautotrophic, and (4) heterotrophic (Wang et al. 2014). Although microalgal cultivation by photoautotrophic method is rather more popular where microalgal cells utilize solar energy and CO_2, other methods such as mixotrophic and heterotrophic ones result in higher production of biomass along with industrially important materials (Mitra et al. 2012). These two approaches, however, demand external carbon supplements either in the form of acetate or in the form of glucose and this makes the whole process of cultivation expensive (Bhatnagar et al. 2011). *Scenedesmus* and *Chlorella* strains have proved to be successful in the olive oil mill and paper industry WW treatment; however, care must be taken as some microalgal strains are sensitive to certain heavy metals that can limit their growth (Sharma et al. 2012). Cultivation of *S. obliquus* in batch or semicontinuous cultivation methods in the laboratory for the treatment of MW and agricultural wastewaters revealed considerably higher amounts of lipid accumulation (Pittman et al. 2011). Many researchers have tested microalgal cultivation in the open and raceway pond systems, but with the advancement of the technology, photobioreactors (PBRs) have been proved superior to other conventional methods. PBRs can maintain ideal growth conditions for microalgae cultivation and also avoid contamination by other microbes (Salama et al. 2017). However, PBRs have several advantages but still they do not meet the standards of commercial production and therefore certain modifications are needed for enhanced bioenergy production. A two-stage hybrid cultivation system utilizes fast growing microalgal strains transferred from the PBR via open raceway ponds. This system has shown significantly higher yield of microalgal biomass and elevated lipid accumulation when stress conditions were maintained (Narala et al. 2016). This hybrid system can maintain a higher density of microalgae as well as reduces the risk of microbial contamination as the cultivation time in this system is short (Płaczek et al. 2017). Huntley and Redalje (2007) cultivated *H. pluvialis* and *C. vulgaris* in the two-stage hybrid system and found that lipid production by these strains increased significantly, resulting in higher biodiesel yield (Fig. 8.7). Industrial and domestic wastewaters have a high P concentration; for example, domestic urine wastewater and feces have very high P concentrations that can supply 22% of the world's P requirement (Vasconcelos Fernandes et al. 2015). N and P recovery by struvite demands magnesium supplementation along with the maintenance of higher pH, which finally removes only P; however, microalgae can efficiently remove both N and P from the wastewaters. *C. vulgaris* cultivated in a photo-microbial fuel cell with wastewater resulted in significantly higher proportions of N and P (Caporgno et al. 2015). Similarly, microalgae cultivated at the cathode of a membrane-based system resulted in higher removal of P. *Chlamydomonas mexicana* cultivated in the piggery wastewater efficiently removed P, Ca, and C when higher alkaline pH was maintained. Similar results were obtained with *Micractinium reisseri* when cultivated in MWs (Ji Kabra et al. 2014). Emerging organic contaminants (EOCs) and persistent organic pollutants (POPs) cover a wide range of elements that enter the water resources from pharmaceuticals, surfactants, pesticides and cannot be easily removed from the water bodies. EOC removal by activated sludge WWTPs (wastewater treatment plants) has been in practice for years and microalgal efficiency of EOC removal appears to be the best as several studies have confirmed this fact (Kurade et al. 2016). Many researchers have confirmed microalgal efficiency against the recovery of different organic and inorganic compounds by algae-based WWTPs using biodegradation or bioaccumulation methods (Xiong et al. 2016). Microalgal cultivation in HRAPs has shown successful removal of EOCs up to 99% from the

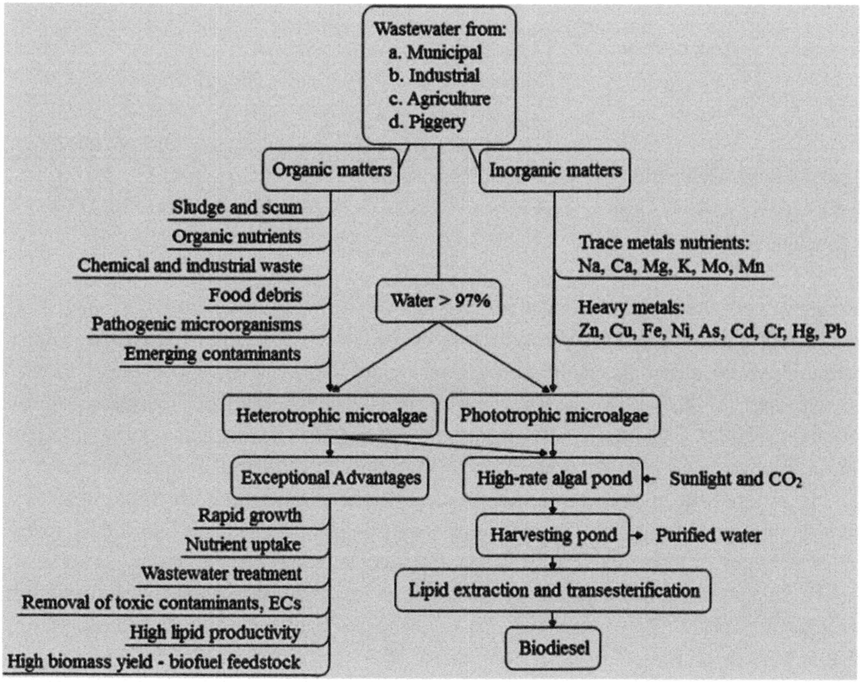

Fig. 8.7 Wastewater treatment for algal biodiesel production. (Adapted from Kadir et al. 2018)

WWs; however, the hydraulic residence time (HRT) has been found to affect the removal efficiency of microalgae (Matamoros et al. 2015).

8.5 Lipid Extraction of Microalgae

Production of bioactive materials from microalgal species such as polyunsaturated fatty acids (PUFA) has gained enough interest by many researchers all over the world. *Dunaliella salina*, *C. vulgaris*, *N. oculata*, *N. gaditana* along with some diatoms are capable of producing fatty acids (Han et al. 2017). Intensive research has been already done in order to enhance PUFA production by microalgae. It has been reported that fatty acid production increases under stress conditions. *N. oculata* when cultivated under carbohydrate and N deficiency resulted in higher lipid accumulation (about 40%) (Hong et al. 2017). Su et al. (2016) also obtained comparable findings with *Porphyridium purpureum* where arachidonic acid and PUFA biosynthesis was found to be increased under P stress as P stress can alter the composition of oleic and palmitoleic acids. *D. salina* produced a higher fraction of fatty acids under higher temperature stress conditions (Castilla Casadiego et al. 2016). Microalgae convert CO_2 into mono- or polysaccharides as the primary source of energy as carbohydrates

Table 8.1 Most common unsaturated fatty acids (Adapted from Allen et al. 2018)

Symbol	Common name	Systematic name	Structure	MP (°C)
12:0	Lauric acid	Dodecanoic acid	$CH_3(CH_2)_{10}COOH$	44.2
14:0	Myristic acid	Tetradecanoic acid	$CH_3(CH_2)_{12}COOH$	52.0
16:0	Palmitic acid	Hexadecanoic acid	$CH_3(CH_2)_{14}COOH$	63.1
18:0	Stearic acid	Octadecanoic acid	$CH_3(CH_2)_{16}COOH$	69.6
20:0	Arachidic acid	Eicosanoic acid	$CH_3(CH_2)_{18}COOH$	75.4

MP melting point temperature

participate in the synthesis of materials required for growth and sustenance of microalgal cells followed by production of bioactive compounds that are highly valuable at commercial scale (Canavate et al. 2017). Two *Chlorella*, namely *C. vulgaris* (UTEX 259) and *C. variabilis* (NC64A), showed increased starch efficiency under CO_2 stress (Cheng et al. 2015). The microalgal protein content can be manipulated for the production of materials of industrial importance such as for the production of protein supplements. Protein extraction from microalgae is a complicated process; however, different methods have been tested so far by using different types of solvents so as to keep the costs as low as possible (Fujisawa et al. 2017). Dry algal biomass pretreatment is a crucial stage of protein extraction from microalgae; for example, microalgal dry biomass when treated at high temperature resulted in an increased protein content of 15% (Eboibi et al. 2015). Insertion of *psbA* gene, which is responsible for the production of the bovine protein milk amyloid A (MAA) (colostrum), by genetic manipulation of *Chlamydomonas reinhardtii* also showed an increased protein content (Gimpel et al. 2015). Hempel and Maier (2016) reported that *D. salina* cultivated in a medium supplemented with glucose and NaCl also showed an increase in the protein content. Lipid accumulation (PUFA, saturated ones, TGA, and glycolipids) is variable among the microalgal strains and cultivation conditions. Fatty acids are classified into two categories: unsaturated (with one mono or several polyunsaturated double bonds as shown in Table 8.1) and saturated (acyl chains without double bonds as shown in Table 8.2). N, P starvation as well as strong radiation can enhance the accumulation of saturated and unsaturated fatty acids and TGA storage in the cells of microalgae. Under lower light conditions, polar lipids (glycolipids and phospholipids) are produced; however, certain microalgal strains can produce lipids under dark or light conditions as well. C, N, and P are essential for the microalgal growth, and lipid content is usually higher in dark (heterotrophic) conditions (Rodolf et al. 2009). Lipids produced under heterotrophic conditions are more suitable to be used as biodiesel as it can reduce the biodiesel cost; however, oil accumulation within the microalgal cells is also dependent upon the carbon composition and the growth conditions. Light, cultivation conditions, and temperature are the key elements of the microalgal biotechnology. Higher lipid accumulation and faster growth of the microalgae reduce the price of biodiesel production and studies have shown that some microalgal strains can maintain higher lipid contents even without N (Eroglu et al. 2015).

Table 8.2 Most common fatty acids (Adapted from Allen et al. 2018)

Symbol	Common name	Systematic name	Structure	MP (°C)
16:1D9	Palmitoleic acid	Hexadecenoic acid	$CH_3(CH_2)_5CH{=}CH{-}(CH_2)_7COOH$	−0.5
18:1D9	Oleic acid	9-Octadecenoic acid	$CH_3(CH_2)_7CH{=}CH{-}(CH_2)_7COOH$	13.4
18:2D9,12	Linoleic acid	9,12-Octadecadienoic acid	$CH_3(CH_2)_4(CH{=}CHCH_2)_2(CH_2)_6COOH$	−9.0
18:3D9,12,15	α-Linolenic acid	9,12,15-Octadecatrienoic acid	$CH_3CH_2(CH{=}CHCH_2)_3(CH_2)_6COOH$	−17.0
20:4D5,8,11,14	Arachidonic acid	5,8,11,14-Eicosatetraenoic acid	$CH_3(CH_2)_4(CH{=}CHCH_2)_4(CH_2)_2COOH$	−49.0
20:5D5,8,11,14,17	EPA	5,8,11,14,17-Eicosapentaenoic acid	$CH_3CH_2(CH{=}CHCH_2)_5(CH_2)_2COOH$	−54.0
22:6D4,7,10,13,16,19	DHA	Docosahexaenoic acid	22:6(n-3)	

MP melting point temperature

The selected strain of microalgae must be either able to grow in the changing cultivation conditions or have a high lipid composition or can accumulate higher levels of lipids under stress conditions. Genetic engineering can increase the lipid accumulation characteristics and decrease photoinhibition and photosaturation and thus can make biofuel cheaper (Rodolf et al. 2009). Rodolf et al. (2009) tested 30 strains of microalgae for lipid and biomass production and reported that three strains of *Nannochloropsis* were dominant. Wastewater ponds have higher populations of *Scenedesmus*, *Actinastrum*, *Euglena*, and *Chlorella* and all of them are able to thrive under stress conditions with higher lipid accumulation (Lyon et al. 2015). *Chlorella*, *Nannochloropsis*, and *Phaeodactylum* strains have the highest biomass production rate and thus are most preferred at the industrial scale (Griffiths and Harrison 2009). Microalgae strains best known for highest lipid production, growth rate, and biofuel production are *C. vulgaris*, *C. minutissima*, *C. protothecoides*, *S. obliquus*, *D. salina*, *D. tertiolecta*, *Chlamydomonas reinhardtii*, and *Chlorococcum* spp. (Bahadar and Bilal Khan 2013). Raeesossadati and Ahmadzadeh (2015) found that biomass production and proportion as well as CO_2 fixation ratio by *S. obliquus*, *C. vulgaris*, and *Chlorococcum littorale* increase under changing concentrations of CO_2. *Chrysotila carterae*, a coccolithophorid microalgae, can also utilize CO_2 for biomass production by generation of $CaCO_3$. A cost-effective and efficient transesterification and lipid extraction method is important for increased biofuel production from microalgal strains (Roberts et al. 2013). The common methods of lipid extraction from microalgae are hexane Soxhlet extraction, microwave-assisted extraction (MAE), Bligh-Dyer method, and ultrasound-assisted extraction (UAE), and algal biomass is filtered from all these methods in order to remove residues followed by filtration for the gravimetric measurement of lipid content. Soxhlet extraction has been proven to be the best as it can recover 100% lipid content from the microalgae strain (Prommuak et al. 2012). Lam and Lee (2012), however, emphasize that methanol–chloroform solvent mixture results in higher lipid yield due to their affinity toward polar and nonpolar lipids. Han et al. (2019) reported that for enhanced biomass production and lipid yield by microalgal strains, iron ions are required in sufficient quantity as they participate in the electron transfer and enzyme activity reactions. Depletion of iron in the culture media can alter photosynthetic activity, protein synthesis as well as growth (Chen et al. 2011). Zn and B decrease the biomass and lipid content of *S. obliquus* cultivated in wastewaters as they are present at higher concentrations. Cultivation of *S. obliquus* in the MW along with the supplementation of the iron ions in the culture medium resulted in elevated lipid yield and increased biomass contents (Li et al. 2010).

Chemical solvents most popularly utilized for microalgal lipid extraction include hexane, chloroform, and 1-butanol as they can easily penetrate the microalgal cell wall and thus can separate lipids into organic phase (Bahadar and Bilal Khan 2013). Both nonpolar and polar solvents were tested by Valdez et al. (2011) and they found that nonpolar solvents have highest lipid extraction rates but the lipids have lower C content than the ones extracted by polar solvents. Solvent extraction method suggests that (a) nonpolar solvents result in the extraction of lipids with lower C content and thus have low density, (b) polar solvents although give lower yields but have higher lipid content, (c) carbon content is influenced by the selected solvent, and (d) N and C contents of the lipids extracted by polar solvents are lower. Pretreatment steps (chemical or biological or physical) can result in increased lipid extraction efficiency of the microalgal strains (Ghasemi Naghdi et al. 2016). Microalgal lipid extraction by microwave also seems promising as heat can easily agitate the algal cell wall rendering lipid extraction easier (Bahadar and Bilal Khan 2013). Ultrasound associated extraction reduces total time of extraction and improves the solvent efficiency, resulting in increased lipid yield (Yoo et al. 2012). Enzyme treatment for the disruption of the cell wall is also advantageous as it does not harm the lipids but requires longer cycle time and strain dependence (Ghasemi Naghdi et al. 2016). Pretreatment with nitrous acid reduces lipid yield as it damages the macromolecules. Aqueous pore formation in the cell wall of microalgae by the electric field application also increases the lipid extraction efficiency (Bai et al. 2014). Adekunle et al. (2016) tested supercritical CO_2 and hexane (solvent) against the total lipid yields as well as composition of *Botryococcus braunii* and found that traditional extraction methods are far more superior. Kanakraj and Dixit (2016) tested the efficiency of supercritical fluid and solvent extraction methods and found that water degumming reduces P content and therefore makes the extraction of lipids much easier by phase separation.

8.6 Algal Biofuel Production

Biofuels can reduce CO_2 emission in the atmosphere and thus can maintain the equilibrium between generation and consumption of CO_2 as microalgae are able to absorb this CO_2 produced by biofuel combustion (Sahar Sadaf et al. 2018). Algal biofuels fall under the category of third-generation biofuels and numerous efforts has been done to increase biofuel yield by microalgae; for example, *Chlamydomonas* cultured in dairy wastewaters showed higher lipid production and is therefore considered suitable for the generation of biofuel (Arora et al. 2016). Microalgal strains are rich in lipid content and therefore they can produce good quality biofuels (Chisti 2007). Several chemical and mechanical methods have been employed for the biofuel extraction from microalgae but ultrasonic, supercritical fluid, and microwave extractions have been found superior to the solvent extraction method. Lipid yield of the microalgal cells is increased by rupturing the cell wall by sonication where solvent-free good quality crude oil is extracted by using supercritical fluids, while microwaves also lead to higher lipid yields from the microalgal cells. Lipid extraction by ultrasonic and critical CO_2 yielded 70% and 86% of the fatty acids, while the majority of them were PUFAs (Drira et al. 2016). Microalgal lipid yield by the dry biomass by ionic liquids (Ils) has also been proved to be promising due to excellent physicochemical properties of these liquids. The cationic and anionic components of the Ils enhance the lipid yield by solvents (Khoo et al. 2020). Lipid extraction from *Nannochloropsis* sp. By cholinium (Ch) amino acid-assisted IL resulted in higher lipid yield as found by Chua et al. (2018) and similar results were obtained by Zhang et al. (2018a), who tested lipid extraction from *Thraustochytrium* sp. By using phosphonium tetrabutylphosphonium propanoate (P_{444}-Prop) and imidazolium-1-ethyl-3-methylimidazolium ethylsulfate (C_2mim-$EtSO_4$). Algal biomass is used in the ethanol production, biodiesel and hydrogen production followed by biogas and compounds of industrial interest by auto-, mixo-, or heterotrophic cultivation modes. Wastewaters are rich in essential nutrients and CO_2, which results in higher yields of lipids, and several microalgal strains that are fast growing and can produce high-quality crude oil are already known. When algal biomass amounts increase in the wastewaters, the lipid contents decrease as slow-growing strains have higher lipid contents in comparison to the fast-growing strains (Sharma et al. 2012). Researchers have proved that growth of microalgae under stress conditions increases the lipid content and different biotic and abiotic factors can cause stress in the microalgae (Ji et al. 2018). *C. sorokiniana* and *Acutodesmus dimorphus* cultivated under BG11 salt stress conditions showed increased lipid content. Algal biomass is rich in TAG, which is composed of C16 and C18 fatty acids just like the vegetable oil and is therefore suitable for biofuel generation. Microalgae (freshwater strain) cultivated in BG11 salt stress can switch its FAME profile by reducing the percentage of saturated fatty acids (SFAs) (Zhang et al. 2018b). Transesterification method reduces the viscosity of crude oil extracted from microalgae by chemical conversion to FAME or biodiesel (Tan et al. 2017). Ethanol or methanol are used in the transesterification of biodiesel as they are cheap and their chemical and physical properties are optimum. Methanol produces FAME, while fatty acid ethyl ester (FAEE) is produced when ethanol is used for transesterification. For the production of FAME and glycerol, catalysts such as acids, bases, or enzymes are used to speed up the reaction as they enhance the yield as well as the rate of reaction (Kim et al. 2013). Bases used as catalysts result in higher FAME yield by reducing the reaction time; for example, biodiesel production increased by *C. vulgaris* when sodium methoxide (CH_3Ona) was used as the catalyst as alkaline catalysts are highly efficient in small concentration (Pragya et al. 2013). Transesterification is affected by many factors such as catalysts, temperature, alcohol/oil molar ratio, percentage of water, and also the stirring rate. The use of heterogeneous catalysts in the generation of biodiesel by microalgae species has appeared to be effective as they are recyclable and can be used continuously without any clearing step till the desired product is obtained. They are eco-friendly, cheap, and resistant to corrosion. The commonly used heterogeneous catalysts are calcium oxide (CaO), magnesium oxide (MgO), and barium oxide (BaO) as they increase biodiesel yield by transesterification but they are susceptible to the free fatty acids (FFA) content and can result in soap formation; therefore, acid catalysts are used for transesterification. Heterogeneous acid catalysts demand higher temperatures, supreme reaction rates, and alcohol/oil molar ratio in order to speed up transesterification (Lam and Lee 2014). Tungstated zirconia, used as a catalyst, increases the biodiesel yield of *S. obliquus*. Recently, a one-step transesterification (direct) is used by researchers as both transesterification and extraction steps can run simultaneously one after the other in a reactor as it reduces the number of solvents, makes separation easy, and reduces reaction time as well (Guldhe et al. 2017). The in situ transesterification is affected by the selected feedstock as dry algal biomass is more preferred because it allows more penetration of chemicals and therefore either acid or alkaline catalysts are used. Biodiesel production from *Nannochloropsis* sp. By Mg–Zr solid base catalyst via single-step transesterification was significantly higher (Pragya et al. 2013). Algal biofuel production is classified into five stages, namely (1) microalgal culture in open ponds or PBRs with continuous supply of wastewaters rich in C, N, P, CO_2, other minor minerals, and algal biomass harvesting, (2) removal of water content from the algal biomass either by oven drying or flocculation or centrifugation, (3) lipid extraction by using solvents from the biomass of microalgae, (4) production of biofuels by transesterification where lipids are converted into biofuel, and (5) biofuel transportation from the production unit to the consumers (Collotta et al. 2018).

Wastewaters require proper treatment before being discharged into the natural water resources as they have very high concentrations of inorganic, organic elements, and heavy metals that can cause eutrophication and pose a threat to aquatic plants and animals. Traditional techniques that use MW treatment can remove significant amounts of N and P but this efficiency can be further increased by using more advanced techniques; for example, by the usage of microalgal species that have a higher nutrient removal rate contribute to tertiary wastewater treatment as biofuels and industrially important chemicals can be produced from the microalgal biomass (Dayananda et al. 2016). *Botryococcus braunii* (Trebouxyophiceae), a provincial green microalga commonly grows in the brackish and freshwaters, has been exploited as a substrate for biofuel production. Rinna et al. (2017) tested the efficiency of two *B. braunii* strains against wastewater treatment for biodiesel production. They tested the N and P removing efficiencies of both the strains and reported that the lipid yield varies among both the two strains (*B. braunii* LB572 and *B. braunii* IBLC116) on the basis of total dry biomass produced by them. This difference can be attributed to the less availability of light due to the tubid nature of wastewater that reduces the lipid accumulation in the microalgal cells. Lipid content by *B. braunii* LB572 was considerably higher than *B. braunii* IBLC116, suggesting that WW casts a conclusive effect on microalgal growth; however, the fatty acid percentages of the two strains showed no comparable difference. Influent wastewaters (IWW) are generally richer in N content in comparison to the effluent wastewaters (EWW), and N stress enhances lipid production by microalgae, as many researchers have also confirmed this fact (Klok et al. 2014). Intracellular microalgal fatty acids are composed of 4–24 carbon atom chains and most common fatty acids are oleic and palmitic acids. Both *B. braunii* LB572 and *B. braunii* IBLC116 strains also contain small amounts of behenic and myristic fatty acids followed by PUFAs (eicosapentaenoic and arachidonic acids). Both *B. braunii* LB572 and *B. braunii* IBLC116 strains showed higher concentration of oleic acid followed by palmitic and stearic acids, while trace amounts of elaidic, eicosenoic, and palmitoleic acids followed by unsaturated C18 (three) and C20 (four) polyenoic acids when they were cultivated in EWWs (Nascimento et al. 2014). *Botryococcus* strains are rich in oleic acid content as well as SFA and MUFAs, a feature different from other microalgal strains and oil yielding crops that have higher content of PUFAs. The peculiarity of biofuel is based on the ratios of SFA, MUFA, and PUFA proportions. The cetane number (CN) as well as saponification of the saturated FAME have been found to be higher for *B. braunii*-IBL cultivated in IWW, and higher SFA content in the biofuels of this strain indicates the higher oxidative stability of the fuel. Higher PUFA and MUFA contents indicate that biofuels are more sensitive toward oxidation. Biofuels with higher CNs indicate that they are obtained from long-chain unsaturated FAMEs (Stansell et al. 2012). FAMEs obtained from *B. braunii* strains have high melting points and have been found apt for the production of biofuels but there are significant differences among the strains in relation to the biodiesel attributes (Nascimento et al. 2013). Pushpakumari Kudahettige et al. (2018) examined the upshot of salt deficiency and nutrient starvation in response to the production of biomass and lipid accumulation followed by FAME profiling of *S. dimorphus* and *Selenastrum minutum* microalgal species cultivated in MWs. They found that under salt stress, neutral lipids accumulate in higher proportion, while chlorophyll (a, b) and carotenoid content decreases in *S. dimorphus* similar to *S. obliquus* and *C. sorokiniana* (Ji et al. 2018; Zhang et al. 2018a). In *P. tricornutum*, N stress causes membrane lipid conversion to TAG (substrate for biodiesel production), while in *Nannochloropsis* sp. N stress results in enhanced levels of MFA and SFAs with a decrease in PUFAs (Yang et al. 2013). Many researchers have confirmed that salt stress casts a negative effect on the chlorophyll content; for example, salt stress results in chlorophyll disintegration in *C. sorokiniana*, while for blue-green alga (*Anabaena* sp.) and yeasts cells, salt stress causes vacuolization and cell death (Kim et al. 2016). *S. dimorphus* grown in 5% salinity as well as nitrogen stress resulted in increased lipid production. Nutrient starvation positively affects lipid accumulation in *S. minutum* and *C. vulgaris* as NADPH (reducing agent) is involved in the synthesis of fatty acids, and stress can cause higher accumulation of lipids during photosynthesis process by upregulation of acetyl CoA carboxylase (ACCase), resulting in increased carbon content during fatty acid synthesis (Guarnieri et al. 2011). Higher salt content can increase intracellular lipid content in microalgae; for example, *Neochloris oleoabundans* cultivated in salt and N stress conditions showed that ascorbate–glutathione and proline cycles have essential role in the activation of glycerol-3-phosphate dehydrogenase and glycerol-3-phosphate acyltransferase genes that are linked with the production of TAG (de Jaeger et al. 2018). *S. dimorphus* cultivated under stress conditions also activates glycerol-3-phosphate dehydrogenase, and glycerol-3-phosphate dehydrogenase genes therefore results in increased lipid accumulation in contrast to carbohydrate content. Microalgal lipids belong to two categories, namely structural lipids and storage lipids; however, under the conditions of nutrient stress, majority of the accumulated lipids belong to the storage category. These storage lipids can be further transformed to biofuels by the process of transesterification. C16 and C18 fatty acids have been found to be the dominant ones in *S. minutum* and *S. dimorphus* under salt stress (5%) and thus can be used for biofuel production. *C. mexicana* and *S. obliquus* show increased accumulation of 18:1 and 18:3 fatty acids under the conditions of salt stress (Salama et al. 2013). Both *S. dimorphus* and *S. minutum* strains have higher ratios of SFA/PUFA and MUFA/PUFA under 5% salinity stress and thus are stable against oxidation and suitable for biofuel production (Singh and Mallick 2014). Several studies have shown that supplementing the wastewaters with plant hormones can increase growth and lipid production of microalgae but this idea is not practical due to higher cost of the plant hormones. MWs are deficient in certain nutrients essential for microalgal growth and metabolism and supplementation of these essential nutrients to the MWs can fasten the algal growth and lipid accumulation (Yu et al. 2017). Increased growth and enhanced accumulation of lipids have been noticed in *Scenedesmus* sp. When BG11 medium was supplemented with metal ions. Increasing the percentage of iron ions in the synthetic media results in increased biomass content and increased lipid accumulation in microalgal cells (Salama et al. 2017). *C. vulgaris* and *C. pyrenoidosa* can excellently adapt in the wastewaters but the lipid accumulation and biodiesel production by these strains are not up to the standards (Deng et al. 2009). Cellular composition of the algae species can differ due to the culture conditions and both abiotic and biotic factors can also alter biomass and lipid production. Li et al. (2011) tested 14 microalgal strains belonging to *Haematococcus*, *Chlorella*, *Chlamydomonas*, *Chloroccum*, and *Scenedesmus* genus for biomass and lipid production when cultivated in centrate wastewater. They found that all the selected 14 strains can thrive well in the autotrophic and heterotrophic BG-11 growth media, indicating that they are capable of taking up inorganic and organic carbon for biomass production. All the strains when cultivated in centrate wastewater showed variations in the biomass content; for example, *C. nocturama* UTEX 1338 had the lowest biomass content, while *C. kessleri* UTEX 398 had the highest biomass content. The FAME content was higher for *C. kessleri* UTEX 398 than *C. protothecoides* UTEX 25 strain. And both strains had 100% ammonia removal efficiency. P removal rate was found to be higher when pH of the cultivation medium was higher. Light period and intensity as well as CO_2 cast a positive effect on microalgal biomass production and on the FAME profiles. Light period casted a more pronounced effect on COD removal than light intensity and CO_2 concentration, while total phosphorus removal was found to be dependent on the CO_2 concentration (Zhou et al. 2011). *C. kessleri* UTEX 398 demanded both CO_2 concentration and light intensity for efficient total nitrogen removal, while *C. protothecoides* UTEX 25 required only efficient lighting periods. However, both strains were successful in the removal of ammonia. Increasing the concentration of CO_2 increases the growth rate of *C. reinhardtii*, *C. pyrenoidosa*, and *S. oblique* (Chiu et al. 2008). Reyimu and Özçimen (2017) found that *Tetraselmis suecica* shows higher biomass production in MWs and results of Sydney et al. (2011) with *B. braunii* were also found to be similar. Amit and Ghosh (2018) tested the biomass-producing efficiency of three freshwater (*Spirulina* sp., *S. abundans*, and *Nostoc muscorum*) and one marine (*Tetraselmis indica*) microalgal strains in wastewaters. Four different types of wastewaters (primary-treated sewage (PTS), secondary-treated sewage (STS), campus wastewater (CWW), and paper mill wastewater (PWW)) were selected for microalgal cultivation. Significantly higher growth rate was observed in *T. indica* in comparison to other strains in STS as well as highest production of biomass in STS than PTS followed by CWW and PWW (Amit et al. 2017). Therefore, STS appears to be the best growth medium for all the selected microalgal strains. PTS casts a negative effect on the metabolic activity of microalgae due to the presence of toxic elements. *T. suecica* has also shown higher biomass production in fish farm wastewaters (Michels et al. 2014). *N. oculata* and *T. suecica* grow faster when cultivated in wastewaters as reported by Reyimu and Özçimen (2017). *S. abundans* also resulted in higher biomass yield with MWs, while lower yields were obtained with PWW due to the presence of toxic elements that hinder growth. Growth was lower in PTS in comparison to STS due to the elevated concentration of macro nutrients (Sharma et al. 2014). Sirin and Sillanpää (2015) investigated the *T. suecica* growth in sewage wastewater after secondary treatment coupled with photon flux and found that *T. indica* grows faster in STS than other treated wastewaters. Nitrogen is essential for the accumulation of lipids in microalgal strains and *T. indica* can efficiently remove significant proportions of P and N from PTS, STS, CWW, and PWW, respectively, while higher removal rates were observed in the STS (Abou-Shanab et al. 2014). Sirakov and Velichkova (2014) found that *T. chuii* removes a higher fraction of N and P from the aquaculture wastewaters but *T. indica* showed the highest nitrate removal rate in CWW. However, *T. indica* showed highest ammonium and total organic carbon (TOC) removal rates in STS followed by *N. muscorum*, *S. abundans*, and *Spirulina* sp. *C. marina* also removes a higher fraction of nitrate from the domestic sewage waters (Kumar et al. 2015). C16–C18 fatty acids were found to be dominant among the selected four microalgal strains and maximum lipid yield was obtained with *T. indica* grown in STS followed by *S. abundans*, *N. muscorum*, and *Spirulina* sp. (Montero et al. 2011).

Tripathi et al. (2019) tested the potential of *Scenedesmus* sp. ISTGA1 (isolated from marble mining site) against biofuel production as well as its potential for the remediation of WW and compared it with the BG-11 medium used as a control. They found that the amount of biomass was higher in the wastewater in comparison to the BG-11, suggesting that wastewaters are suitable for microalgal cultivation. Renuka et al. (2013) also obtained similar results with *Calothrix* sp. When cultivated in a mixture of tap and sewage water (1:1). Alvarez-Díaz et al. (2017) also found that *C. vulgaris*, *C. kessleri*, and *S. obliquus* show higher biomass and lipid production in wastewaters and these results were similar to that obtained for *Scenedesmus* sp. ISTGA1. *Scenedesmus* sp. ISTGA1 shows the abundance of following fatty acids, namely palmitic, stearic, palmitoleic, linolenic, oleic, and linoleic acids, and it produces both saturated and unsaturated fatty acids when grown in wastewaters. Palmitic and stearic acids were dominant when *Scenedesmus* sp. ISTGA1 was grown in wastewaters. This strain favors the production of MFAs in comparison to PUFAs. Biodiesel cost is dependent upon the cultivation method adapted for algal biomass harvesting (Adeniyi et al. 2018). Algae turf scrubber (ATS), a method by Dr. Walter Adey, takes into account algae and WW combinations for the production of biofuel but its usage has been limited only to US regions (Kangas et al. 2017). Algae used as a substrate for the treatment of WWs can also be employed to generate biomethane by the method of anaerobic digestion as algae has sufficient percentage of lipids, carbohydrates, and proteins (Ganesh Saratale et al. 2018). Approximately 75% of wastewater is untreated due to the absence of a proper management system in India and this causes eutrophication (Williams et al. 2019). Marella et al. (2019) tested the biomass content and lipid production followed by nutrient-removing efficiency of an algal flow-way system (simple self-seeding type) with untreated urban WW. They also analyzed the efficacy of algal flow-way (AFW) biomass as a substrate for biodiesel production by FAME profiling in different seasons. They found that total lipid dry cell weight (DCW) was at peak during summer while it was lowest in the rainy season; however, lipid content increased along with the increased growth rate throughout all the seasons. Filamentous green algal and cyanobacterial species are abundant during winter and rainy seasons that increase growth rate but diatoms dominate their populations as they have a lower lipid proportion. Microalgal strains produce lipids enclosed in tiny sacs known as lipid bodies (LBs) and these LBs serve as energy producers during the conditions of stress. LB number was reported to increase three times from initial growth cycle to the final growth cycle except in *Nitzschia palea* and *Gomphonema parvulum* as well as *Navicula lanceolata* had the highest LB diameter. Sixty percent of the total cell volume in *N. amphibia* and *N. umbanota* is contributed by LBs due to their smaller cell size (Lin et al. 2018). C16 and C18 fatty acids were found to be dominant in the microalgae, while eicosapentaenoic acid (EPA) were the dominant PUFAs in diatoms, while diatoms showed the abundance of EPA during winter season. Unsaturation of fatty acid is very much influenced by light and temperature as microalgae produce SFAs under higher temperature and higher light intensity, while MUFA and PUFA are accumulated at lower ranges (Hu et al. 2008). Increase in the proportion of SFA was observed during the summer, while MUFA and PUFA contents increased during the winter. Factors governing the storage potential of biodiesel are CN, iodine value (IV), and saponification value (SV) associated with the degree of unsaturation (DU). Marella et al. (2019) reported that SV is variable throughout the four seasons of the year, while CN is higher during the summer and IV is higher during the winter. DU of microalgal biofuel is rather high due to higher contents of PUFA and MUFA but DU values are lower in summer in comparison to winter where DU values are higher. The H atom saturation to the C chain of the fatty acid is correlated with the CN and cold filter plugging point (CFPP), which in turn are indicated as long-chain saturated factor (LCSF). Highest LCSF values were obtained during the summer season, while these values were found to be lower during winter. CFPP can be defined as the temperature when the crystallization of fuels starts and it hinders the engine's fuel filters. Higher DU values suggest that the biofuel has low melting point and therefore the biofuel is more suitable in the winter. Cloud point (CP) can be defined as the temperature when biodiesel starts to solidify. CP is responsible for the determination of the biodiesel quality as CP values have been found to be lower in both summer and winter (Talebi et al. 2013). Eladel et al. (2019) tested the growth potential, lipid production, FAMEs, and nutrient removal rate by a dominant microalga (*C. sorokiniana*) from a wastewater treatment plant. They found an abundance of C16 and C18 fatty acids in *C. sorokiniana*, indicating that this strain is suitable for biofuel production. *C. sorokiniana* cultivated in wastewater shows a mild increase in the CN similar to the findings of Singh et al. (2017) with *Parachlorella kessleri*. Leong et al. (2020) examined an algae–bacteria associated system in relation to bioremediation of MWs and its efficacy for biofuel production. The amounts of the lipids extracted from the MWs were higher than that obtained from the synthetic wastewater as the biomass content was higher in the MWs. The higher lipid content in the MWs can be attributed to the higher accumulation of nitrate and ammonium, which causes increased activity of acetyl-CoA carboxylase and thus increased lipid synthesis in the microalgae (Sánchez-García et al. 2013). Higher lipid accumulation can also be correlated with the availability of iron in the MWs, resulting in increased biomass production. Liu et al. (2008) also obtained similar results with *C. vulgaris*. C16 and C18 fatty acids were found to be dominant in the FAME profiling, suggesting their suitability for biofuel production (Nayak et al. 2018). Trace amounts of C6, C12, C14, C20, and C22 fatty acids were also obtained with the microalgal–bacterial system when cultivated in the MWs, while these fatty acids were absent when the cultivation was carried out in synthetic wastewater. FAMEs had the highest proportion of SFAs followed by PUFAs and MUFAs; however, *C. vulgaris* when cultivated with a bacterial-activated sludge system showed two-fold increase in the SFA content (Dasan et al. 2020).

Bacteria and microalgae have different growth rates depending upon the biotic and abiotic factors; however, the growth rate of the two can be improved in the algal-activated sludge (AAS) system by (1) external supply of air in order to increase the concentration of CO_2 and (2) solids retention time (SRT) selection as it can inhibit bacterial growth, therefore allows algae to remove more nutrients from the AAS (Huesemann et al. 2016). Katam and Bhattacharyya (2020) investigated the effect of SRT in the removal of C and other nutrients in the AAS reactors as well as on lipid and FAME composition by selecting *Chlorella* strains. They found that N and P stress causes increased accumulation of lipids by the *Chlorella* strains with increased biomass content, which can be attributed to the reduced bacterial growth. Several studies have shown that cultivation of *Chlorella* strains in different types of wastewaters results in higher lipid content. Cho et al. (2011) examined the potential of UV radiation as well as filtration on the biomass content and biofuel-generating efficiency of *Chlorella* sp. 227 when cultured in MWs. They found that C16 and C18 fatty acids constitute majority of the total fatty acids present in the algal cells and FAMEs also has the similar compositions. Leong et al. (2019) examined the nutrient-removing potential of microalgal–bacterial consortia grown in an activated sludge system with MW and synthetic wastewaters in relation to biomass and lipid production. They found that biomass production increased with increasing the microalgal–bacterial concentration; however, the lipid content first increased and then declined with further increase in the biomass. The abundance of microalgae in the freshwaters has already been known and *Cladophora*, *Spirogyra*, and *Oedogonium* species have been isolated from the MWs as well as from wastewaters of aquaculture (Neveux et al. 2016). Ge et al. (2017) used *Spirogyra* sp. against bioremediation of MWs and production of biomass. They found that the lipid accumulation by this strain of microalgae does not fit the biodiesel standards because the carbohydrate ratio was rather higher and it can interfere with the lipid synthesis. Addition of Na_2CO_3 as a catalyst, however, increases the lipid accumulation within the microalgal cells under the hydrothermal liquefaction (HTL) conditions. HTL not only improves the pre- and posttreatment but also increases the faster conversion of biochemical materials into economically useful byproducts; for example, conversion rate of microalgae increases up to 77% during protein extraction just before the HTL (Neveux et al. 2016). Aketo et al. (2019) screened 58 microalgal strains for wastewater bioremediation as well as biofuel generation and found that *P. kessleri* NKG021201 and *Chloroidium saccharophilum* NKH13 are apt for biodiesel production and the selected 58 strains showed an abundance of C18 fatty acids. Biodiesel yields from *P. kessleri* NKG021201 and *Chloroidium saccharophilum* NKH13 meet the requirements of European biodiesel standard EN 14214. Abomohra et al. (2018) investigated the potential of flocculants (inorganic) coupled with cationic starch (organic) for harvesting *S. obliquus* cultivated in MWs for FAME analysis. They found that flocculants positively affect FAME yields as they disrupt the algal cell walls and cell membranes and therefore initiate lipid extraction. Several researchers have supported that pretreatment enhances lipid yield from microalgal cells; for example, lipid extraction from *C. vulgaris* was increased by hot-water mixed with acid pretreatment (Park et al. 2014). Abomohra et al. (2018) also reported that ferric sulphate used as a flocculant increases lipid yield from *S. obliquus* as it is not only able to disrupt the microalgal cell wall but can also penetrate the cell membrane and therefore enhances lipid extraction. Kim et al. (2014) examined response of hydraulic retention time (HRT) in relation to total N, P, COD degradation as well as recovery of total suspended solids (TSS) via an algal consortium isolated from HRAP with MWs. It has been reported that accumulation of lipids takes place in algae during the late growth phase. Increased HRT causes increased lipid accumulation as nutrient stress is established but increasing the HRT results in decreased lipid content after a certain period of time, suggesting that lower HRT is sufficient enough for higher lipid and biomass contents (Kim et al. 2014). Biofuels rich in oleic acid are more stable as they are stable against oxidation and therefore enable expanded storage duration (Knothe 2008). FAMEs derived from the algal consortium were rich in C12, C14, C16, and C18 fatty acids and a very minute change was observed in the FAME proportions with changed HRT. The lipid proportion and FAME analysis of the algal biomass obtained in the HRAPs was found to be suitable for use as biodiesel. Jämsä et al. (2017) for the first time conducted a study with a Finnish isolate UHCC0027 (Scenedesmaceae) for biodiesel production with wastewater under cold climatic conditions. They observed that overall lipid fraction decreased when levels of N and P in the PBRs were decreased. Schwenk et al. (2013) also observed a comparable trend with *Scenedesmus* strain. UHCC0027 strain when transferred to real wastewater from the synthetic culture medium did not really show much difference in the lipid content (Gim et al. 2014). However, transfer of algal cells to wastewater at colder temperatures resulted in higher lipid accumulation as algal cell division slows down at colder temperature ranges (Cea et al. 2015). Palmitate and α-linolenate fatty acids were found to be dominant in the biomass of UHCC0027 strain grown in PBR just like other green microalgal strains (*Scenedesmus* and *Chlorella*) (Teoh et al. 2013). Hexadecatetraenoic acid, which is lodged by the thylakoid membrane glycolipids of chloroplasts, was also found in abundance in UHCC0027 algal cells; however, such PUFAs exert an adverse effect on the CN and oxidative stability of biodiesel and therefore are not preferred for the generation of biofuels. Biomass produced by UHCC0027 algal cells during the mid-exponential growth phase contains fatty acids appropriate for the conversion into biodiesel in comparison to other fatty acids produced in the early- or late-growth phases. Sharma et al. (2020) studied two microalgal consortia for the WW treatment so as to obtain an increased lipid fraction and biofuel yield by cultivating microalgae in WW (different dilutions) as well as in growth medium (BG-11). The selected wastewater dilutions ranged from 25% to 100%. Maximum lipid accumulation and increased chlorophyll content were observed for 75% dilution of wastewater for both the algal consortia but at 50% dilution only increased biomass production was noticed for both consortia and no pronounced differences were observed at 25% and 100% dilutions as well as in BG-11 medium. Similar results have been observed for *C. vulgaris* and *P. kessleri* when used for the MW phyco-remediation (Singh et al. 2017). Relatively increased yield of lipid was obtained with *C. vulgaris* when urban wastewaters were used for cultivation in comparison to the BG-11 medium (Malibari et al. 2018) and increased lipid content has been observed for *C. sorokiniana* when cultivated in aquaculture wastewater than in the commercially available culture media (Guldhe et al. 2017). Algal consortia showed abundance of MUFAs, SFAs, and PUFAs and a mixture of 13 different fatty acids as determined by FAME analysis. SFAs were found to be more dominant in both the algal consortia than PUFAs and MUFAs. Major SFAs were C16 followed by C14 and C12. C22 was in highest proportion followed by C18 among the unsaturated fatty acids. Palmitic, oleic, and linoleic acids determine the properties of biofuels as they provide stability to the biofuel (Anto et al. 2019). PUFAs were abundant for the two algal consortia with BG-11 as the growth medium. Singh et al. (2019) cultivated *P. kessleri*-I in two growth media (wastewater and control) to compare lipid and biomass production. They found that *P. kessleri*-I when cultivated in wastewater resulted in the production of biodiesel composed of SFAs and MUFAs, suggesting that this strain can be used for the operation of engines.

8.7 Conclusion

Biofuels have appeared as renewable and eco-friendly tools of sustainable energy generation that can reduce the dependence on the non-sustainable energy sources, and very intensive research has been done in the past few decades and research is still ongoing. Biofuels of the first and second generations have already faced criticism; however, biofuel generation by microalgae has attracted the interest of researchers in being more promising. Microalgae have the potential to grow in diverse environments and different types of culture media, and the oil produced by them are not only more in amount but also easy to handle in comparison to that produced by the oil-yielding crops. The microalgal efficiency of thriving in WWs along with their highly efficient nutrient recovery potential has opened the doors of treating WWs for producing biofuels and products of industrial and pharmaceutical interest. Tremendous success has been obtained with the culturing of microalgae in municipal wastewaters, industrial wastewaters, dairy wastewaters, and animal manures. They utilize nitrates, phosphates, ammonium, heavy metals, and carbon dioxide present in the wastewaters for their metabolic reactions and therefore release oxygen that can be utilized by bacteria; thus, microalgae maintain a symbiotic relationship with the bacteria present in wastewaters. Several freshwater and few marine microalgae have already been investigated for their potential to generate biofuels when cultivated in different types of wastewaters and positive results have been obtained. Biofuels produced by microalgae emit lower fractions of harmful gases into the environment. Therefore, it can be concluded that microalgae are promising tools of the future for the treatment of wastewaters, biofuel (biodiesel) production along with materials that are important at commercial scale.

Acknowledgments I convey my regards to Gábor Draskovits, Laboratory Researcher and Dr. József Marek, Animal Health Laboratory, Prophyl Kft., Dózsa György út 18, Mohács, Hungary, for motivating me in the completion of this chapter. I thank Dr. Pradeep Kumar Shukla, Assistant Professor, Department of Biological Sciences, Faculty of Science, Sam Higginbottom University of Agriculture, Technology and Sciences, Prayagraj, UP, India, for his guidance. Last but not the least, I also thank Prof. (Dr.) Pramod W. Ramteke (now retired), former Dean PG Studies and Head, Department of Biological Sciences, Faculty of Science, Sam Higginbottom University of Agriculture, Technology and Sciences, Prayagraj, UP, India, for always being an inspiration to me.

References

Abdel-Raouf N, Al-Homaidan A, Ibraheem I (2012) Microalgae and wastewater treatment. Saudi J Biol Sci 19:257–275

Abomohra AE-F, Jin W, Sagar V, Ismail GA (2018) Optimization of chemical flocculation of *Scenedesmus obliquus* grown on municipal wastewater for improved biodiesel recovery. Renew Energy 115:880–886

Abou-Shanab RAI, El-Dalatony MM, El-Sheekh MM, Ji M-K, Salama E-S, Kabra AN, Jeon B-H (2014) Cultivation of a new microalga, Micractinium reisseri, in municipal wastewater for nutrient removal, biomass, lipid, and fatty acid production. Biotechnol Bioprocess Eng. 19: 510–518

Adekunle AS, Oyekunle JAO, Obisesan OR, Ojo OS, Ojo OS (2016) Effects of degumming on biodiesel properties of some non-conventional seed oils. Energy Rep 2:188–193

Adeniyi OM, Azimov U, Burluka A (2018) Algae biofuel: current status and future applications. Renew Sustain Energy Rev 90:316–335

Ahluwalia SS, Goyal D (2007) Microbial and plant derived biomass for removal of heavy metals from wastewater. Bioresour Technol 98:2243–2257

Ahmad A, Buang A, Bhat AH (2016) Renewable and sustainable bioenergy production from microalgal co-cultivation with palm oil mill effluent (POME): a review. Renew Sustain Energy Rev 65:214–234

Ajayan KV, Selvaraju M, Unnikannan P, Sruthi P (2015) Phycoremediation of tannery wastewater using microalgae *Scenedesmus* species. Int J Phytoremediation 17:907–916

Aketo T, Hoshikawa Y, Nojima D, Yabu Y, Maeda Y, Yoshino T, Takano H, Tanaka T (2019) Selection and characterization of microalgae with potential for nutrient removal from municipal wastewater and simultaneous lipid production. J Biosci Bioeng 29:565–572

Alam F, Date A, Rasjidin R, Mobin S, Moria H, Baqui A (2012) Biofuel from algae—is it a viable alternative? Procedia Eng 49:221–227

Allen J, Unlu S, Demirel Y, Black P, Riekhof W (2018) Integration of biology, ecology and engineering for sustainable algal-based biofuel and bioproduct biorefinery. Bioresour Bioprocess 5:47

Allnutt FCT, Kessler BA (2015) Biomass and biofuels from microalgae. Algae 2:289–310

Alvarez-Díaz PD, Ruiz J, Arbib Z, Barragán J, Garrido-Pérez JA, Perales JA (2017) Freshwater microalgae selection for simultaneous wastewater nutrient removal and lipid production. Algal Res 24:477–485

Amit, Ghosh UK (2018) An approach for phycoremediation of different wastewaters and biodiesel production using microalgae. Environ Sci Pollut Res 25:18673–18681

Amit, Chandra R, Ghosh UK, Nayak JK (2017) Phycoremediation potential of marine microalga *Tetraselmis indica* on secondary treated domestic sewage for nutrient removal and biodiesel production. Environ Sci Pollut Res 24:20868–20875

Anto S, Pugazhendhi A, Mathimani T (2019) Lipid enhancement through nutrient starvation in *Chlorella* sp. and its fatty acid profiling for appropriate bioenergy feedstock. Biocatal Agric Biotechnol 20:101–179

Arashiro LT, Ferrer I, Rousseau DP, Van Hulle SW, Garfí M (2019) The effect of primary treatment of wastewater in high rate algal pond systems: biomass and bioenergy recovery. Bioresour Technol 280:27–36

Arora N, Patel A, Sartaj K, Pruthi PA, Pruthi V (2016) Bioremediation of domestic and industrial wastewaters integrated with enhanced biodiesel production using novel oleaginous microalgae. Environ Sci Pollut Res 23:20997–21007

Aslam A, Thomas-Hall SR, Manzoor M, Jabeen F, Iqbal M, uz Zaman Q, Schenk PM, Tahir MA (2018) Mixed microalgae consortia growth under higher concentration of CO_2 from unfiltered coal fired flue gas: fatty acid profiling and biodiesel production. J Photochem Photobiol B 179: 126–133

Aslan S, Kapdan IK (2006) Batch kinetics of nitrogen and phosphorus removal from synthetic wastewater by algae. Ecol Eng 28:64–70

Bahadar A, Bilal Khan M (2013) Progress in energy from microalgae: a review. Renew Sustain Energy Rev 27:128–148

Bai X, Ghasemi Naghdi F, Ye L, Lant P, Pratt S (2014) Enhanced lipid extraction from algae using free nitrous acid pretreatment. Bioresour Technol 159:36–40

Barros AI, Gonçalves AL, Simões M, Pires JCM (2015) Harvesting techniques applied to microalgae: a review. Renew Sustain Energy Rev 41:1489–1500

Bhatnagar A, Bhatnagar M, Chinnasamy S, Das KC (2010) *Chlorella minutissima*—a promising fuel alga for cultivation in municipal wastewaters. Appl Biochem Biotechnol 161:523–536

Bhatnagar A, Chinnasamy S, Singh M, Das K (2011) Renewable biomass production by mixotrophic algae in the presence of various carbon sources and wastewaters. Appl Energy 88:3425–3431

Branyikova I, Prochazkova G, Potocar T, Jezkova Z, Branyik T (2018) Harvesting of microalgae by flocculation. Fermentation 4:93

Brennan L, Owende P (2010) Biofuels from microalgae—a review of technologies for production, processing, and extractions of biofuels and co-products. Renew Sustain Energy Rev 14:557–577

Brown N, Shilton A (2014) Luxury uptake of phosphorus by microalgae in waste stabilisation ponds: current understanding and future direction. Rev Environ Sci Biotechnol 13:321–328

Bux F (2013) Biotechnological applications of microalgae: biodiesel and value added products. CRC Press, Taylor & Francis Group, Boca Raton

Canavate JP, Armada I, Hachero-Cruzado I (2017) Common and species-specific effects of phosphate on marine microalgae fatty acids shape their function in phytoplankton trophic ecology. Microb Ecol 74:623–639

Caporgno MP, Taleb A, Olkiewicz M, Font J, Pruvost J, Legrand J, Bengoa C (2015) Microalgae cultivation in urban wastewater: nutrient removal and biomass production for biodiesel and methane. Algal Res. 10:232–239

Castilla Casadiego DA, Albis Arrieta AR, Angulo Mercado ER, Cervera Cahuana SJ, Baquero Noriega KS, Suarez Escobar AF, Morales Avendaño ED (2016) Evaluation of culture conditions to obtain fatty acids from saline microalgae species: *Dunaliella salina*, *Sinecosyfis* sp., and *Chroomonas* sp. Biomed Res Int 2016:1–7

Castro JS, Calijuri ML, Assemany PP, Cecon PR, de Assis IR, Ribeiro VJ (2017) Microalgae biofilm in soil: greenhouse gas emissions, ammonia volatilization and plant growth. Sci Total Environ 574:1640–1648

Cea M, Sangaletti-Gerhard N, Acuña P, Fuentes I, Jorquera M, Godoy K, Osses F, Navia R (2015) Screening transesterifiable lipid accumulating bacteria from sewage sludge for biodiesel production. Biotechnol Rep 8:116–123

Chen M, Tang H, Ma H, Holland TC, Ng KYS, Salley SO (2011) Effect of nutrients on growth and lipid accumulation in the green algae *Dunaliella tertiolecta*. Bioresour Technol 102:1649–1655

Chen X, Mao X, Cao Y, Yang X (2013) Use of siliceous algae as biological monitors of heavy metal pollution in three lakes in a mining city, southeast China. Oceanol Hydrobiol Stud 42:233–242

Cheng YS, Labavitch JM, Vander Gheynst JS (2015) Elevated CO_2 concentration impacts cell wall polysaccharide composition of green microalgae of the genus *Chlorella*. Lett Appl Microbiol 60:1–7

Cheng J, Ye Q, Yang Z, Yang W, Zhou J, Cen K (2017) Microstructure and antioxidative capacity of the microalgae mutant *Chlorella* PY-ZU1 during tilmicosin removal from wastewater under 15% CO_2. J Hazard Mater 324:414–419

Chew KW, Yap JY, Show PL, Suan NH, Juan JC, Ling TC, Lee DJ, Chang JS (2017) Microalgae biorefinery: high value products perspectives. Bioresour Technol 229:53–62

Chisti Y (2007) Biodiesel from microalgae. Biotechnol Adv 25:294–306

Chiu S, Kao C, Chen C, Kuan T, Ong S, Lin C (2008) Reduction of CO_2 by a high-density culture of *Chlorella* sp. in semicontinuous photobioreactor. Bioresour Technol 99:3389–3396

Cho S, Luong TT, Lee D, Oh Y-K, Lee T (2011) Reuse of effluent water from a municipal wastewater treatment plant in microalgae cultivation for biofuel production. Bioresour Technol 102:8639–8645

Christenson L, Sims R (2011) Production and harvesting of microalgae for wastewater treatment, biofuels, and bioproducts. Biotechnol Adv 29:686–702

Chua ET, Brunner M, Atkin R, Eltanahy E, Thomas-Hall SR, Schenk PM (2018) The ionic liquid cholinium arginate is an efficient solvent for extracting high-value *Nannochloropsis* sp. lipids. ACS Sustain Chem Eng 7:2538–2544

Collos Y, Harrison PJ (2014) Acclimation and toxicity of high ammonium concentrations to unicellular algae. Mar Pollut Bull 80:8–23

Collotta M, Champagne P, Mabee W, Tomasoni G (2018) Wastewater and waste CO_2 for sustainable biofuels from microalgae. Algal Res 29:12–21

Craggs RJ, Lundquist TJ, Benemann JR (2013) Wastewater treatment and algal biofuel production. In: Algae for biofuels and energy. Springer, Dordrecht, pp 153–163

Dasan YK, Lam MK, Yusup S, Lim JW, Show PL, Tan IS, Lee KT (2020) Cultivation of *Chlorella vulgaris* using sequential-flow bubble column photobioreactor: a stress-inducing strategy for lipid accumulation and carbon dioxide fixation. J CO_2 Util 41:101226

Dasgupta CN, Suseela M, Mandotra S, Kumar P, Pandey MK, Toppo K, Lone JA (2015) Dual uses of microalgal biomass: an integrative approach for biohydrogen and biodiesel production. Appl Energy 146:202–208

Daverey A, Pandey D, Verma P, Verma S, Shah V, Dutta K, Arunachalam K (2019) Recent advances in energy efficient biological treatment of municipal wastewater. Bioresour Technol Rep 7:100252

Davis R, Markham J, Kinchin C, Grundl N, Tan ECD, Humbird D (2016) Process design and economics for the production of algal biomass algal biomass production in open pond systems and processing through dewatering for downstream conversion. Nat Renew. Energy Lab, Golden

Dayananda C, Sarada R, Usha Rani M, Shamala TR, Ravishankar GA (2016) Autotrophic cultivation of *Botryococcus braunii* for the production of hydrocarbons in exosaccharides in various media. Biomass Bioenergy 31:87–93

de Jaeger L, Carreres BM, Springer J, Schaap PJ, Eggink G, Martins dos Santos VAP, Wijffels RH, Martens DE (2018) *Neochloris oleoabundans* is worth its salt: transcriptomic analysis under salt and nitrogen stress. PLoS One 13:e0194834

Dean AP, Sigee DC, Estrada B, Pittman JK (2010) Using FTIR spectroscopy for rapid determination of lipid accumulation in response to nitrogen limitation in freshwater microalgae. Bioresour Technol 101:4499–4507

de-Bashan LE, Bashan Y (2010) Immobilized microalgae for removing pollutants: review of practical aspects. Bioresour Technol 101:1611–1627

Demirel Y (2016) Energy production, conversion, storage, conservation, and coupling, 2nd edn. Springer, London

Deng X, Li Y, Fei X (2009) Microalgae: a promising feedstock for biodiesel. Afr J Microbiol Res 3: 1008–1014

Di Caprio F, Altimari P, Pagnanelli F (2015) Integrated biomass production and biodegradation of olive mill wastewater by cultivation of *Scenedesmus* sp. Algal Res 9:306–311

Ding J, Zhao F, Cao Y, Xing L, Liu W, Mei S, Li S (2015) Cultivation of microalgae in dairy farm wastewater without sterilization. Int J Phytoremediation 17:222–227

Dong T, Knoshaug EP, Pienkos PT, Laurens LM (2016) Lipid recovery from wet oleaginous microbial biomass for biofuel production: a critical review. Appl Energy 177:879–895

Drira N, Piras A, Rosa A, Porcedda S, Dhaouadi H (2016) Microalgae from domestic wastewater facility's high rate algal pond: lipids extraction, characterization and biodiesel production. Bioresour Technol 206:239–244

Eboibi BE, Lewis DM, Ashman PJ, Chinnasamy S (2015) Influence of process conditions on pretreatment of microalgae for protein extraction and production of biocrude during hydrothermal liquefaction of pretreated *Tetraselmis* sp. RSC Adv 5:20193–20207

Eladel H, Abomohra AE-F, Battah M, Mohmmed S, Radwan A, Abdelrahim H (2019) Evaluation of *Chlorella sorokiniana* isolated from local municipal wastewater for dual application in nutrient removal and biodiesel production. Bioprocess Biosyst Eng 42:425–433

Eroglu E, Smith SM, Raston CL (2015) Biomass and biofuels from microalgae, vol 2. Springer, Cham

Fujisawa T, Leverenz RL, Nagarnine M, Kerfeld CA, Unno M (2017) Raman optical activity reveals carotenoid photoactivation events in the orange carotenoid protein in solution. J Am Chem Soc 139:10456–10460

Ganesh Saratale R, Kumar G, Banu R, Xia A, Periyasamy S, Dattatraya Saratale G (2018) A critical review on anaerobic digestion of microalgae and macroalgae and co-digestion of biomass for enhanced methane generation. Bioresour Technol 262:319–332

Gao F, Li C, Yang Z-H, Zeng G-M, Mu J, Liu M, Cui W (2016) Removal of nutrients, organic matter, and metal from domestic secondary effluent through microalgae cultivation in a membrane photobioreactor. J Chem Technol Biotechnol 91:10

Ge S, Madill M, Champagne P (2017) Use of freshwater macroalgae Spirogyra sp. for the treatment of municipal wastewaters and biomass production for biofuel applications. Biomass Bioenergy 111:212–223

Ghasemi Naghdi F, González LM, Chan W, Schenk PM (2016) Progress on lipid extraction from wet algal biomass for biodiesel production. Microb Biotechnol 9:718–726

Gikas P (2017) Towards energy positive wastewater treatment plants. J Environ Manag 203:621–629

Gim GH, Kim JK, Kim HS, Kathiravan MN, Yang H, Jeong SH, Kim SW (2014) Comparison of biomass production and total lipid content of freshwater green microalgae cultivated under various culture conditions. Bioprocess Biosyst Eng 37:99–106

Gimpel JA, Hyun JS, Schoepp NG, Mayfield SP (2015) Production of recombinant proteins in microalgae at pilot greenhouse scale. Biotechnol Bioeng 112:339–345

Goncalves AL, Rodrigues CM, Pires JCM, Simoes M (2016) The effect of increasing CO_2 concentrations on its capture, biomass production and wastewater bioremediation by microalgae and cyanobacteria. Algal Res 14:127–136

Gonçalves AL, Pires JC, Simões M (2017) A review on the use of microalgal consortia for wastewater treatment. Algal Res 24:403–415

Gouveia L, Graça S, Sousa C, Ambrosano L, Ribeiro B, Botrel EP, Neto PC, Ferreira AF, Silva CM (2016) Microalgae biomass production using wastewater: treatment and costs: scale-up considerations. Algal Res 16:167–176

Green FB, Lundquist T, Oswald W (1995) Energetics of advanced integrated wastewater pond systems. Water Sci Technol 31:9–20

Griffiths MJ, Harrison ST (2009) Lipid productivity as a key characteristic for choosing algal species for biodiesel production. J Appl Phycol 21:493–507

Guarnieri MT, Nag A, Smolinski SL, Darzins A, Seibert M, Pienkos PT (2011) Examination of triacylglycerol biosynthetic pathways via *de novo* transcriptomic and proteomic analyses in an unsequenced microalga. PLoS One 6:e25851

Guldhe A, Singh P, Ansari FA, Singh B, Bux F (2017) Biodiesel synthesis from microalgal lipids using tungstated zirconia as a heterogeneous acid catalyst and its comparison with homogeneous acid and enzyme catalysts. Fuel 187:180–188

Hall CAS, Lambert JG, Balogh SB (2014) EROI of different fuels and the implications for society. Energy Policy 64:141–152

Han D, Li Y, Hu Q (2013) Astaxanthin in microalgae: pathways, functions and biotechnological implications. Algae 28:131–147

Han TH, Zhang S, Cho MH, Hwang S-J (2017) Enhancement of volatile fatty acids removal by a co-culture of microalgae and activated sludge. KSCE J Civil Eng 21:2106–2112

Han S-F, Jin W, Abomohra AE-F, Tu R, Zhou X, He Z, Chen C, Xie G-J (2019) Municipal wastewater enriched with trace metals for enhanced lipid production of the biodiesel-promising microalga *Scenedesmus obliquus*. Bioenergy Res 12:1127–1133

Harun R, Doyle M, Gopiraj R, Davidson M, Forde GM, Danquah MK (2013) Process economics and greenhouse gas audit for microalgal biodiesel production. In: Advanced biofuels and bioproducts. Springer, New York, pp 709–744

Hempel F, Maier UG (2016) Microalgae as solar-powered protein factories. In: Vega MC (ed) Advanced technologies for protein complex production and characterization, pp 241–262

Hena S, Fatimah S, Tabassum S (2015) Cultivation of algae consortium in a dairy farm wastewater for biodiesel production. Water Res Ind 10:1–14

Hong S-J, Park YS, Han M-A, Kim ZH, Cho B-K, Lee H, Choi H-K, Lee C-G (2017) Enhanced production of fatty acids in three strains of microalgae using a combination of nitrogen starvation and chemical inhibitors of carbohydrate synthesis. Biotechnol Bioprocess Eng 22: 60–67

Hu Q, Sommerfeld M, Jarvis E, Ghirardi M, Posewitz M, Seibert M, Darzins A (2008) Microalgal triacylglycerols as feedstocks for biofuel production: perspectives and advances. Plant J 54: 621–639

Huesemann M, Crowe B, Waller P, Chavis A, Hobbs S, Edmundson S, Wigmosta M (2016) A validated model to predict microalgae growth in outdoor pond cultures subjected to fluctuating light intensities and water temperatures. Algal Res 13:195–206

Huntley ME, Redalje DG (2007) CO_2 mitigation and renewable oil from photosynthetic microbes: a new appraisal. Mitig Adapt Strat Glob Change 12:573–608

Hussain F, Shah SZ, Zhou W, Iqbal M (2017) Microalgae screening under CO_2 stress: growth and micro-nutrients removal efficiency. J Photochem Photobiol B 170:91–98

Hussain F, Shah SZ, Ahmad H, Abubshait SA, Abubshait HA, Laref A, Manikandan A, Kusuma HS, Iqbal M (2021) Microalgae an ecofriendly and sustainable wastewater treatment option:

biomass application in biofuel and bio-fertilizer production. A review. Renew Sustain Energy Rev 137:110603

Imani S, Rezaei-Zarchi S, Hashemi M, Borna H, Javid A, Zand AM, Abarghouei HB (2011) Hg, Cd and Pb heavy metal bioremediation by *Dunaliella* alga. J Med Plants Res 5:2775–2780

Jämsä M, Lynch F, Santana-Sánchez A, Laaksonen P, Zaitsev G, Solovchenko A, Allahverdiyeva Y (2017) Nutrient removal and biodiesel feedstock potential of green alga UHCC00027 grown in municipal wastewater under Nordic conditions. Algal Res 26:65–73

Ji Kabra AN, Salama e-S, Roh HS, Kim JR, Lee DS, Jeon BH (2014) Effect of mine wastewater on nutrient removal and lipid production by a green microalga *Micratinium reisseri* from concentrated municipal wastewater. Bioresour Technol 157:84–90

Ji X, Cheng J, Gong D, Zhao X, Qi Y, Su Y, Ma W (2018) The effect of NaCl stress on photosynthetic efficiency and lipid production in freshwater microalga—*Scenedesmus obliquus* XJ002. Sci Total Environ 633:593–599

Kadir WNA, Lam MK, Uemura Y, Limb JW, Leed KT (2018) Harvesting and pre-treatment of microalgae cultivated in wastewater for biodiesel production: a review. Energy Convers Manag 171:1416–1429

Kanakraj S, Dixit S (2016) A comprehensive review on degummed biodiesel. Biofuels 7:537–548

Kang J, Wen Z (2015) Use of microalgae for mitigating ammonia and CO_2 emissions from animal production operations—evaluation of gas removal efficiency and algal biomass composition. Algal Res. 11:204–210

Kangas P, Mulbry W, Klavon P, Laughinghouse HD, Adey W (2017) High diversity within the periphyton community of an algal turf scrubber on the Susquehanna River. Ecol Eng 108:564–572

Katam K, Bhattacharyya D (2020) Effect of solids retention time on the performance of alga-activated sludge association in municipal wastewater treatment and biofuel production. J Appl Phycol 32:1803–1812

Kesaano M, Sims RC (2014) Algal biofilm based technology for wastewater treatment. Algal Res 5: 231–240

Khaldi H, Maatoug M, Dube CS, Ncube M, Tandlich R, Heilmeier H, Laubscher RK, Dellal A (2017) Efficiency of wastewater treatment by a mixture of sludge and microalgae. J Fundam Appl Sci 9:1454–1472

Khoo KS, Chew KW, Yew GY, Leong WH, Chai YH, Show PL, Chen W-H (2020) Recent advances in downstream processing of microalgae lipid recovery for biofuel production. Bioresour Technol 304:122996

Kim J, Yoo G, Lee H, Lim J, Kim K, Kim CW, Park MS, Ji-Won Y (2013) Methods of downstream processing for the production of biodiesel from microalgae. Biotechnol Adv 31:862–876

Kim B-H, Kang Z, Ramanan R, Choi J-E, Cho D-H, Oh H-E, Kim H-S (2014) Nutrient removal and biofuel production in high rate algal pond using real municipal wastewater. J Microbiol Biotechnol 24:1123–1132

Kim BH, Ramanan R, Kang Z, Cho DH, Oh HM, Kim HS (2016) *Chlorella sorokiniana* HS1, a novel freshwater green algal strain, grows and hyperaccumulates lipid droplets in seawater salinity. Biomass Bioenergy 85:300–305

Klok PA, Lamers JP, Martens DE, Draaisma RB, Wijffels RH (2014) Edible oils from microalgae: insights in TAG accumulation. Trends Biotechnol 32:521–528

Knothe G (2008) Optimizing fatty ester composition to improve fuel properties. Energy Fuels 22: 1358–1364

Kumar D, Santhanam P, Jayalakshmi T, Nandakumar R, Ananth S, Shenbaga Devi A, Balaji Prasath B (2015) Excessive nutrients and heavy metals removal from diverse wastewaters using marine microalga *Chlorella marina* (Butcher). Indian J Geomarine Sci 44:97–103

Kurade MB, Kim JR, Govindwar SP, Jeon B-H (2016) Insights into microalgae mediated biodegradation of diazinon by *Chlorella vulgaris*: microalgal tolerance to xenobiotic pollutants and metabolism. Algal Res 20:126–134

Lam MK, Lee KT (2012) Immobilization as a feasible method to simplify the separation of microalgae from water for biodiesel production. Chem Eng J 191:263–268

Lam MK, Lee KT (2014) Scale-up and commercialization of algal cultivation and biofuel production. In: Biofuels from algae. Elsevier, pp 261–286

Langley NM, Harrison STL, Van Hille RP (2012) A critical evaluation of CO_2 supplementation to algal systems by direct injection. Biochem Eng J 68:70–75

Lardon L, Helias A, Sialve B, Steyer J-P, Bernard O (2009) Life-cycle assessment of biodiesel production from microalgae. ACS Publications

Lau PS, Tam NFY, Wong YS (1995) Effect of algal density on nutrient removal from primary settled wastewater. Environ Pollut 89:59–66

Leong WH, Zaine SNA, Ho YC, Uemura Y, Lam MK, Khoo KS, Kiatkittipong W, Cheng CK, Show PL, Lim JW (2019) Impact of various microalgal-bacterial populations on municipal wastewater bioremediation and its energy feasibility for lipid-based biofuel production. J Environ Manag 249:109384

Leong WH, Kiatkittipong K, Kiatkittipong W, Cheng YW, Lam MK, Shamsuddin R, Mohamad M, Lim JW (2020) Comparative performances of microalgal-bacterial co-cultivation to bioremediate synthetic and municipal wastewaters whilst producing biodiesel sustainably. Processes 8:1427

Li X, Hu HY, Gan K, Sun YX (2010) Effects of different nitrogen and phosphorus concentrations on the growth, nutrient uptake, and lipid accumulation of a freshwater microalga *Scenedesmus* sp. Bioresour Technol 101:5494–5500

Li Y, Zhou W, Hu B, Min M, Chen P, Ruan RR (2011) Integration of algae cultivation as biodiesel production feedstock with municipal wastewater treatment: Strains screening and significance evaluation of environmental factors. Bioresour Technol 102:10861–10867

Liang Z, Liu Y, Ge F, Xu Y, Tao N, Peng F, Wong M (2013) Efficiency assessment and pH effect in removing nitrogen and phosphorus by algae-bacteria combined system of *Chlorella vulgaris* and *Bacillus licheniformis*. Chemosphere 92:1383–1389

Lin Q, Zhuo W-H, Wang X-W, Chen C-P, Gao Y-H, Liang J-R (2018) Effects of fundamental nutrient stresses on the lipid accumulation profiles in two diatom species *Thalassiosira weissflogii* and *Chaetoceros muelleri*. Bioprocess Biosyst Eng 41:1213–1224

Liu ZY, Wang GC, Zhou BC (2008) Effect of iron on growth and lipid accumulation in *Chlorella vulgaris*. Bioresour Technol 99:4717–4722

Liu Y-J, Gu J, Liu Y (2018) Energy self-sufficient biological municipal wastewater reclamation: present status, challenges and solutions forward. Bioresour Technol 269:513–519

Lu Q, Zhou W, Min M, Ma X, Ma Y, Chen P, Urriola PE, Shurson GC, Ruan R (2016) Mitigating ammonia nitrogen deficiency in dairy wastewaters for algae cultivation. Bioresour Technol 201: 33–40

Lyon SR, Ahmadzadeh H, Murry MA (2015) Biomass and biofuels from microalgae, vol 2. Springer, Berlin, pp 95–115

Malibari R, Sayegh F, Elazzazy AM, Baeshen MN, Dourou M, Aggelis G (2018) Reuse of shrimp farm wastewater as growth medium for marine microalgae isolated from Red Sea–Jeddah. J Clen Prod 198:160–169

Marella TK, Datta A, Patil MD, Dixit S, Tiwari A (2019) Biodiesel production through algal cultivation in urban wastewater using algal floway. Bioresour Technol 280:222–228

Masseret E, Amblard C, Bourdier G, Sargos D (2000) Effects of a waste stabilization lagoon discharge on bacterial and phytoplanktonic communities of a stream. Water Environ Res 72: 285–294

Mata TM, Martins AA, Caetano NS (2010) Microalgae for biodiesel production and other applications: a review. Renew Sustain Energy Rev 14:217–232

Matamoros V, Gutierrez R, Ferrer I, Garcia J, Bayona JM (2015) Capability of microalgae-based wastewater treatment systems to remove emerging organic contaminants: a pilot-scale study. J Hazard Mater 288:34–42

Mehrabadi A, Craggs R, Farid MM (2015) Wastewater treatment high-rate algal ponds (WWT HRAP) for low-cost biofuel production. Bioresour Technol 184:202–214
Michels MHA, Vaskoska M, Vermuë MH, Wijffels RH (2014) Growth of *Tetraselmis suecica* in a tubular photobioreactor on wastewater from a fish farm. Water Res. 65:290–296
Mitra D, van Leeuwen JH, Lamsal B (2012) Heterotrophic/mixotrophic cultivation of oleaginous *Chlorella vulgaris* on industrial co-products. Algal Res 1:40–48
Montero MF, Manuela A, Guillermo GR (2011) Isolation of high-lipid content strains of the marine microalga Tetraselmis suecica for biodiesel production by flow cytometry and single-cell sorting. J Appl Phycol 23:1053–1057
Mulbry W, Kondrad S, Pizarro C, Kebede-Westhead E (2008) Treatment of dairy manure effluent using freshwater algae: algal productivity and recovery of manure nutrients using pilot-scale algal turf scrubbers. Bioresour Technol 99:8137–8142
Munoz R, Guieysse B (2006) Algal–bacterial processes for the treatment of hazardous contaminants: a review. Water Res 40:2799–2815
Muylaert K, Beuckels A, Depraetere O (2015) Biomass and biofuels from microalgae, vol 2. Springer, pp 75–94
Narala RR, Garg S, Sharma KK, Thomas-Hall SR, Deme M, Li Y, Schenk PM (2016) Comparison of microalgae cultivation in photobioreactor, open raceway pond, and a two-stage hybrid system. Front Energy Res 4:29
Nascimento IA, Marques SSI, Cabanelas ITD, Periera SA, Druzian JI, Souza CO, Vich DV, Carvalho GC (2013) Screening microalgae strains for biodiesel production: lipid productivity and estimation of fuel quality based on fatty acids profiles as selective criteria. Bioenergy Res 6: 1–13
Nascimento IA, Marques SSI, Cabanelas ITD, de Carvalho GC, Nascimento MA, de Souza CO, Druzian JI, Hussain J, Liao W (2014) Microalgae versus land crops as feedstock for biodiesel: productivity, quality, and standard compliance. Bioenergy Res. 7:1002–1013
Nayak M, Suh WI, Lee B, Chang YK (2018) Enhanced carbon utilization efficiency and FAME production of *Chlorella* sp. HS2 through combined supplementation of bicarbonate and carbon dioxide. Energy Convers Manag 156:45–52
Neveux N, Magnusson M, Mata L, Whelan A, de Nys R, Paul N (2016) The treatment of municipal wastewater by the macroalga *Oedogonium* sp. and its potential for the production of biocrude. Algal Res 13:284–292
Olguí EJ (2003) Phycoremediation: key issues for cost-effective nutrient removal processes. Biotechnol Adv 22:81–91
Olguín EJ (2012) Dual purpose microalgae–bacteria-based systems that treat wastewater and produce biodiesel and chemical products within a Biorefinery. Biotechnol Adv 30:1031–1046
Palmer CM (1969) A composite rating of algae tolerating organic pollution. J Phycol 5:78–82
Park JBK, Craggs RJ, Shilton AN (2011) Wastewater treatment high-rate algal ponds for biofuel production. Bioresour Technol 102:35–42
Park J, Oh Y, Lee J, Lee K, Jeong M, Choi S (2014) Acid-catalyzed hot-water extraction of lipids from *Chlorella vulgaris*. Bioresour Technol 153:408–412
Passos F, Ferrer I (2014) Microalgae conversion to biogas: thermal pretreatment contribution on net energy production. Environ Sci Technol 48:7171–7178
Passos F, Uggetti E, Carrère H, Ferrer I (2014) Pretreatment of microalgae to improve biogas production: a review. Bioresour Technol 172:403–412
Pienkos PT, Darzins A (2009) The promise and challenges of microalgal-derived biofuels. Biofuels Bioprod Biorefining 3:431–440
Pittman JK, Dean AP, Osundeko O (2011) The potential of sustainable algal biofuel production using wastewater resources. Bioresour Technol 102:17–25
Plachno BJ, Wolowski K, Augustynowicz J, Lukaszek M (2015) Diversity of algae in a thallium and other heavy metals-polluted environment. Int J Limnol 51:139–146
Płaczek M, Patyna A, Witczak S (2017) Technical evaluation of photobioreactors for microalgae cultivation. E3S Web of Conferences. EDP Sciences

Pragya N, Pandey KK, Sahoo P (2013) A review on harvesting, oil extraction and biofuels production technologies from microalgae. Renew Sustain Energy Rev 24:159–171
Prandini JM, da Silva ML, Mezzari MP, Pirolli M, Michelon W, Soares HM (2016) Enhancement of nutrient removal from swine wastewater digestate coupled to biogas purification by microalgae *Scenedesmus* spp. Bioresour Technol 202:67–75
Prommuak C, Pavasant P, Quitain AT, Goto M, Shotipruk A (2012) Microalgal lipid extraction and evaluation of single-step biodiesel production. Eng J 16
Pushpakumari Kudahettige N, Pickova J, Gentili FG (2018) Stressing algae for biofuel production: biomass and biochemical composition of *Scenedesmus dimorphus* and *Selenastrum minutum* grown in municipal untreated wastewater. Front Energy Res 6:132
Raeesossadati MJ, Ahmadzadeh H (2015) Biomass and biofuels from microalgae 2:117–136
Ranjith Kumar R, Hanumantha Rao P, Arumugam M (2015) Lipid extraction methods from microalgae: a comprehensive review. Front Energy Res 2:61
Rawat I, Kumar RR, Mutanda T, Bux F (2011) Dual role of microalgae: phycoremediation of domestic wastewater and biomass production for sustainable biofuels production. Appl Energy 88:3411–3424
Razzak SA, Ali SAM, Hossain MM (2017) Biological CO_2 fixation with production of microalgae in wastewater—a review. Renew Sustain Energy Rev 76:379–390
Renuka N, Sood A, Ratha S, Prasanna R, Ahluwalia A (2013) Nutrient sequestration, biomass production by microalgae and phytoremediation of sewage water. Int J Phytoremediation 15: 789–800
Reyimu Z, Özçimen D (2017) Batch cultivation of marine microalgae *Nannochloropsis oculata* and *Tetraselmis suecica* in treated municipal wastewater toward bioethanol production. J Clean Prod 150:40–46
Rincón L, Jaramillo J, Cardona C (2014) Comparison of feedstocks and technologies for biodiesel production: an environmental and techno-economic evaluation. Renew Energy 69:479–487
Rinna F, Buono S, Cabanelas ITD, Nascimento IA, Sansone G, Barone CMA (2017) Wastewater treatment by microalgae can generate high quality biodiesel feedstock. J Water Process Eng 18: 144–149
Roberts GW, Fortier MOP, Sturm BSM, Stagg-Williams SM (2013) Promising pathway for algal biofuels through wastewater cultivation and hydrothermal conversion. Energy Fuels 27:857–867
Rodolf L, Zittelli GC, Bassi N, Padovani G, Biondi N, Bonini G, Tredici MR (2009) Microalgae for oil: strain selection, induction of lipid synthesis and outdoor mass cultivation in a low-cost photobioreactor. Biotechnol Bioeng 102:100–112
Rodolfi L, Chini Zittelli G, Bassi N, Padovani G, Biondi N, Bonini G, Tredici MR (2009) Microalgae for oil: strain selection, induction of lipid synthesis and outdoor mass cultivation in a low-cost photobioreactor. Biotechnol Bioeng 102:100–112
Ruiz-Marin A, Mendoza-Espinosa LG, Stephenson T (2010) Growth and nutrient removal in free and immobilized green algae in batch and semi-continuous cultures treating real wastewater. Bioresour Technol 101:58–64
Sahar Sadaf S, Iqbal J, Ullah I, Bhatti HN, Nouren S, ur-Rehman H, Nisar J, Iqbal M (2018) Biodiesel production from waste cooking oil: an efficient technique to convert waste into biodiesel. Sustain Cities Soc 41:220–226
Salama ES, Kim HC, Abou-Shanab RAI, Ji MK, Oh YK, Kim SH, Jeon B-H (2013) Biomass, lipid content, and fatty acid composition of freshwater *Chlamydomonas mexicana* and *Scenedesmus obliquus* grown under salt stress. Bioprocess Biosyst Eng 36:827–833
Salama E-S, Kurade MB, Abou-Shanab RA, El-Dalatony MM, Yang I-S, Min B, Byong-Hun J (2017) Recent progress in microalgal biomass production coupled with wastewater treatment for biofuel generation. Renew Sustain Energy Rev 79:1189–1211
Sánchez-García D, Resendiz-Isidro A, Villegas-Garrido TL, Flores-Ortiz CM, Chávez-Gómez B, Cristiani-Urbina E (2013) Effect of nitrate on lipid production by *T. suecica*, *M. contortum* and *C. minutissima*. Cent Eur J Biol 8:578–590

Schwenk D, Seppälä J, Spilling K, Virkki A, Tamminen T, Oksman-Caldentey KM, Rischer H (2013) Lipid content in 19 brackish and marine microalgae: influence of growth phase, salinity and temperature. Aquat Ecol 47:415–424

Sharma KK, Schuhmann H, Schenk PM (2012) High lipid induction in microalgae for biodiesel production. Energies 5:1532–1553

Sharma GK, Khan S, Ahmad F, Gupta N (2014) Nutrient sequestration and phycoremediation of sewage waste water by selective microalgae. Green Farming 5:1–4

Sharma J, Kumar V, Kumar SS, Malyan SK, Mathimani T, Bishnoia NR, Pugazhendhi A (2020) Microalgal consortia for municipal wastewater treatment—lipid augmentation and fatty acid profiling for biodiesel production. J Photochem Photobiol B Biol 202:111638

Singh DK, Mallick N (2014) Accumulation potential of lipids and analysis of fatty acid profile of few microalgal species for biodiesel feedstock. J Microbiol Biotechnol Res 4:37–44

Singh P, Kansal A, Carliell-Marquet C (2016) Energy and carbon footprints of sewage treatment methods. J Environ Manag 165:22–30

Singh Y, Singla A, Upadhyay A, Singh AK (2017) Sustainability of moringa-oil-based biodiesel blended lubricant. Energy Sources Part A Recov Util Environ Eff 39:313–319

Singh AK, Farooqi H, Abdin MZ, Kumar S (2019) Bioremediation of municipal wastewater Bioremediation of municipal wastewater and biodiesel production by cultivation of *Parachlorella kessleri*-I. In: Sukla LB et al (eds) The role of microalgae in wastewater treatment. Springer, Singapore

Sirakov IN, Velichkova KN (2014) Bioremediation of wastewater originate from aquaculture and biomass production from microalgae species-*Nannochloropsis oculata* and *Tetraselmis chuii*. Bulg J Agric Sci 20:66–72

Sirin S, Sillanpää M (2015) Cultivating and harvesting of marine alga *Nannochloropsis oculata* in local municipal wastewater for biodiesel. Bioresour Technol 191:79–87

Solovchenko A, Verschoor AM, Jablonowski ND, Nedbal L (2016) Phosphorus from wastewater to crops: an alternative path involving microalgae. Biotechnol Adv 34:550–564

Sonune A, Ghate R (2004) Developments in wastewater treatment methods. Desalination 167:55–63

Stansell GR, Gray SVM, Sym SD (2012) Microalgal fatty acid composition: implications for biodiesel quality. J Appl Phycol 24:791–801

Stephens E, Ross IL, King Z, Mussgnug JH, Kruse O, Poston C, Borowitzka MA, Hankamer B (2010) An economic and technical evaluation of microalgal biofuels. Nat Biotechnol 28:126–128

Su G, Jiao K, Li Z, Guo X, Chang J, Ndikubwimana T, Sun Y, Zeng X, Lu Y, Lin L (2016) Phosphate limitation promotes unsaturated fatty acids and arachidonic acid biosynthesis by microalgae *Porphyridium purpureum*. Bioprocess Biosyst Eng 39:1129–1136

Suganya T, Varman M, Masjuki HH, Renganathan S (2016) Macroalgae and microalgae as a potential source for commercial applications along with biofuels production: a biorefinery approach. Renew Sustain Energy Rev 55:909–941

Sydney EB, da Silva TE, Tokarski A, Novak AC, de Carvalho JC, Woiciecohwski AL, Larroche C, Soccol CR (2011) Screening of microalgae with potential for biodiesel production and nutrient removal from treated domestic sewage. Appl Energy 88:3291–3294

Talebi AF, Mohtashami SK, Tabatabaei M, Tohidfar M, Bagheri A, Zeinalabedini M, Hadavand Mirzaei H, Mirzajanzadeh M, Malekzadeh Shafaroudi S, Bakhtiari S (2013) Fatty acids profiling: a selective criterion for screening microalgae strains for biodiesel production. Algal Res 2:258–267

Tan X, Lam MK, Uemura Y, Lim JW, Wong CY, Lee KT (2017) Cultivation of microalgae for biodiesel production: a review on upstream and downstream processing. Chin J Chem Eng 26: 17–30

Teoh ML, Phang SM, Chu WL (2013) Response of Antarctic, temperate, and tropical microalgae to temperature stress. J Appl Phycol 25:285–297

Tripathi R, Gupta A, Thakur IS (2019) An integrated approach for phycoremediation of wastewater and sustainable biodiesel production by green microalgae, *Scenedesmus* sp. ISTGA1. Renew Energy 135:617–625

Trivedi J, Aila M, Bangwal DP, Kaul S, Garg MO (2015) Algae based biorefinery—how to make sense? Renew Sustain Energy Rev 47:295–307

Valdez PJ, Dickinson JG, Savage PE (2011) Characterization of product fractions from hydrothermal liquefaction of *Nannochloropsis* sp. and the influence of solvents. Energy Fuels 25:3235–3243

van Beilen JB (2010) Why microalgal biofuels won't save the internal combustion machine. Biofuels Bioprod Biorefining 4:41–52

Vargas e Silva F, Monteggia LO (2015) Pyrolysis of algal biomass obtained from high-rate algae ponds applied to wastewater treatment. Front Energy Res 3:31

Vasconcelos Fernandes T, Shrestha R, Sui Y, Papini G, Zeeman G, Vet LE, Wijffels RH, Lamers P (2015) Closing domestic nutrient cycles using microalgae. Environ Sci Technol 49:12450–12456

Wang L, Min M, Li Y, Chen P, Chen Y, Liu Y, Wang Y, Ruan R (2010) Cultivation of green algae *Chlorella* sp. in different wastewaters from municipal wastewater treatment plant. Appl Biochem Biotechnol 162:1174–1186

Wang J, Yang H, Wang F (2014) Mixotrophic cultivation of microalgae for biodiesel production: status and prospects. Appl Biochem Biotechnol 172:3307–3329

Wang J, Zhou W, Yang H, Wang F, Ruan R (2015) Trophic mode conversion and nitrogen deprivation of microalgae for high ammonium removal from synthetic wastewater. Bioresour Technol 196:668–676

Wang L, Liu J, Zhao Q, Wei W, Sun Y (2016) Comparative study of wastewater treatment and nutrient recycle via activated sludge, microalgae and combination systems. Bioresour Technol 211:1–5

Whitton R, Ometto F, Pidou M, Jarvis P, Villa R, Jefferson B (2015) Microalgae for municipal wastewater nutrient remediation: mechanisms, reactors and outlook for tertiary treatment. Environ Technol Rev 4:133–148

Williams BA, Grantham HS, Watson J, Alvarez SJ, Simmonds JS, Rogeliz CA, da Silva MA, Forero-Medina G, Etter A, Nogales J (2019) Minimising the loss of biodiversity and ecosystem services in an intact landscape under risk of rapid agricultural development. Environ Res Lett 15:014001

Woertz I, Fefer A, Lundquist T, Nelson Y (2009) Algae grown on dairy and municipal wastewater for simultaneous nutrient removal and lipid production for biofuel feedstock. J Environ Eng 135:1115–1122

Wrigley TJ, Toerien DF (1990) Limnological aspects of small sewage ponds. Water Res. 24:83–90

Xiong J-Q, Kurade MB, Abou-Shanab RA, Ji M-K, Choi J, Kim JO, Jeon B-H (2016) Biodegradation of carbamazepine using freshwater microalgae *Chlamydomonas mexicana* and *Scenedesmus obliquus* and the determination of its metabolic fate. Bioresour Technol 205: 183–190

Yang ZK, Niu YF, Ma YH, Xue J, Zhang MH, Yang WD, Liu J-S, Lu S-H, Guan Y, Li H-Y (2013) Molecular and cellular mechanisms of neutral lipid accumulation in diatom following nitrogen deprivation. Biotechnol Biofuels 6:67

Yang I-S, Salama E-S, Kim J-O, Govindwar SP, Kurade MB, Lee M, Roh H-S, Jeon B-H (2016) Cultivation and harvesting of microalgae in photobioreactor for biodiesel production and simultaneous nutrient removal. Energy Convers Manag 117:54–62

Yoo G, Park WK, Kim CW, Choi YE, Yang JW (2012) Direct lipid extraction from wet *Chlamydomonas reinhardtii* biomass using osmotic shock. Bioresour Technol 123:717–722

Yruela I (2013) Transition metals in plant photosynthesis. Metallomics 5:1090–1109

Yu Z, Song M, Pei H, Jiang L, Hou Q, Nie C, Zhang L (2017) The effects of combined agricultural phytohormones on the growth, carbon partitioning and cell morphology of two screened algae. Bioresour Technol 239:87–96

Zeraatkar AK, Ahmadzadeh H, Talebi AF, Moheimani NR, McHenry MP (2016) Potential use of algae for heavy metal bioremediation, a critical review. J Environ Manag 181:817–831
Zhang Q, Hu J, Lee DJ, Chang Y, Lee YJ (2017) Sludge treatment: current research trends. Bioresour Technol 243:1159–1172
Zhang Y, Ward V, Dennis D, Plechkova NV, Armenta R, Rehmann L (2018a) Efficient extraction of a docosahexaenoic acid (DHA)-rich lipid fraction from *Thraustochytrium* sp. using ionic liquids. Materials 11:1986
Zhang L, Pei H, Chen S, Jiang L, Hou Q, Yang Z, Yu Z (2018b) Salinity-induced cellular cross-talk in carbon partitioning reveals starch-to lipid biosynthesis switching in low-starch freshwater algae. Bioresour Technol 250:449–456
Zhou W, Li Y, Min M, Hu B, Chen P, Ruan R (2011) Local bioprospecting for high-lipid producing microalgal strains to be grown on concentrated municipal wastewater for biofuel production. Bioresour Technol 102:6909–6919
Zhu S, Qin L, Feng P, Shang C, Wang Z, Yuan Z (2019) Treatment of low C/N ratio wastewater and biomass production using co-culture of *Chlorella vulgaris* and activated sludge in a batch photobioreactor. Bioresour Technol 274:313–320

Chapter 9
Positive Influence and Future Perspective of Marine Alga on Biofuel Production

Sivasankari Sekar

Abstract This chapter intended to elaborate the uniqueness of marine algae consisting of micro- and macroalgae and its positive impact on biofuel production. In the present scenario, algae magnetize the attention across the globe owing to their sustainable biomass production. Greenhouse gas emissions have created numerous negative impacts on the Earth and its ecosystem. Among them, the most prevalent one is CO_2 with its emission around 31.5 Gt. Algae play a dual role as a regulator by utilizing CO_2 from the atmosphere as the main source of carbon and in a similar mode by making the biomass sustainable for biofuel production. The bio-oil produced in the form of lipids has been considered as the primary source of biofuel, while the energy sources such as ethanol, butanol, methane, and hydrogen are considered as secondary energy sources produced from the waste biomass after bio-oil extraction. A biorefinery based on algae is worthy of note in order to contribute to the SDGs. Moreover, the chapter concludes with review of the commercial analysis of algal biofuels.

Keywords Marine algae · Biofuel · Biomass · Waste management · CO_2 mitigation

Abbreviations

AD	Anaerobic digestion
CDR	Carbon dioxide removal
CO_2	Carbon dioxide
CPME	Cyclopentyl methyl ether
DAB	Defatted algal biomass
FAME	Fatty acid methyl ester
GHG	Greenhouse gas

S. Sekar (✉)
Department of Green Energy Technology, Mandanjeet School of Green Energy Technologies, Pondicherry University, Puducherry, India

N. Srivastava, P. K. Mishra (eds.), *Technological Advancement in Algal Biofuels Production*, Clean Energy Production Technologies,
https://doi.org/10.1007/978-981-19-6806-8_9

GM	Genetically modified
HTL	Hydrothermal liquefaction
OPS	Open pond system
PBR	Photobioreactor
SCF	Supercritical fluid
SDGs	Sustainable Development Goals

9.1 Introduction

Each successive day, our fuel demand grows enormously; however, the amount of fossil fuel available is diminishing. As a result of the continuous use of fossil fuels, there is an accumulation of GHG in our atmosphere, which causes global warming (Gaurav et al. 2017). In order to deal with this situation, various research projects have been conducted globally with full vigor to explore alternative fuels for more than two decades. It is unfortunate that despite the vast amount of research on alternative fuels, a proper alternative energy fuel has not yet been commercialized. There have been many generations of alternative fuels ranged from the first to the fourth. Though it has been revealed that each generation has some distinct limitations (Adeniyi et al. 2018). The use of genetically modified algae as a fuel is considered to be the fourth generation. The GM algae fuel has undisputed implications for fuel research.

The marine ecosystem consists of micro and macro forms of algae; macroalgae or seaweeds are seagrasses that do not have real roots. Algal community consists of both prokaryotes and eukaryotes has tolerance to drought, salinity, osmotic pressure, temperature stress, and anoxygenic and radiations (Guedes et al. 2011). Marine algae have special features, including sustainable biomass, reduced complexity, dual performance in cultivation, etc. (Abdullah et al. 2019). Marine algae also have recently attracted considerable attention as a source of sustainable biofuels, fueled by their ability to produce a maximum of tenfolds biomass per year than any terrestrial crop (Mata et al. 2010; Wigmosta et al. 2011). They produce biofuels on a primary and secondary basis. An in-depth discussion of marine algae is presented in this chapter, reviewing the progress of microalgal biorefineries for the production of renewable biofuels and feeds and discussing the potential of the algae biorefinery developments, that could be applied in multi-sectors.

9.2 Positive Influence of Marine Algae

9.2.1 Potential Alternative Resource

In recent years, several studies have been looking for an alternative to lignocellulosic biomass. Moreover, much attention has been given to microalgae and macroalgae found in the marine ecosystem especially with genetically engineered strain, which would create a promising source for third- and fourth-generation biofuels.

Algal biofuel products include lipid, biodiesel, and other biofuels, including ethanol, butanol, hydrogen, and methane, which can be produced from algal biomass. The versatility of this algal biomass has made it popular over the years. Even though algae produce significantly less photosynthesis than terrestrial plants, they have an excellent growth rates and oil production than the higher plants (Rodolfi et al. 2009; Ahmed et al. 2012). Biodiesel can be produced from the accumulated lipid fraction of the algae by using methods such as extraction and conversion. The leftover biomass after the extraction of the bio-oil was termed as defatted algal biomass (DAB). Which has a multivalent application as an aquaculture feed, bioresource for bioenergy, bioabsorbent, biofertilizer, etc., (Chandra et al. 2015; Kumar et al. 2018).

Algae grow on incompatible environments from polluted water sources, soda lakes, and wasteland to moderate environment like hygienic water, seawater or backwater, Arctic environments, desert. Even occur in association with other organisms such as corals, plants, and fungi (lichens) with unfavorable conditions, which will not be possible for higher plants (Norton et al. 1996; Ahmed et al. 2012; Wooldridge 2013; Raja et al. 2014). Additionally, algae minimize direct and indirect use of the land for cultivation. Algae have long been considered a possible source of biofuel because of these characteristics and their high productivity potential.

9.2.2 Microalgae

In various freshwater and marine environments, microalgae naturally grow as single-celled microscopic prokaryotic organisms with a massive diversity of species estimated at around 300,000-350,000 (Alam et al. 2015; Sydney et al. 2019). As an alternative to superior plants, microalgae have a number of advantages that make them an attractive candidate, and a promising technology for biofuel production, which include the following: (1) they have rapid biomass production with huge CO_2 fixation, (2) they have reduced complexity of cell structure; (3) can cope with large temperature fluctuations and environmental conditions, (4) need low nutrient intake, (5) have the capability of producing valuable by-products with primary product, (6) have intensive production cycle with excellent photosynthetic efficiency, (7) can be both autotrophic and heterotrophic (consuming residual nutrients) and (8) yield high lipid content (Sydney et al. 2019; Makareviciene et al. 2020).

9.2.3 Macroalgae

Evolutionarily, macroalgae are highly diverse. Over 40,000 species have been identified, but many are still unidentified (Suganya et al. 2016). Similar in structure to microalgae, macroalgae are multicellular eukaryotic cells that generally contain carbohydrates as a component of their morphological structure. The macroalgae or seaweeds seem to be valuable biomass and a perfect candidate for biofuel production because of their positive influencing factors such as cost-effective cultivation without much contamination and less effort to harvest the biomass. Essentially, marine seaweed can reach a length of 50 m and has low lignin content with moderate cellulose, and this content makes the substrate sustainable (Yahmed et al. 2016; Gaurav et al. 2017).

Macroalgae or seaweeds possess several advantages as a substrate for biofuel research. Seaweeds grow faster to produce more biomass, are easily degradable materials and require no significant land (deforestation or preparing of land), freshwater, pesticides, or fertilizers for cultivation like terrestrial energy plants. Each hectare will yield 20 tons of biomass DW (dry weight). Moreover, they have been proposed as an ideal tool for mitigating CO_2. Since there are so many species, there is no conflict between food and fuel. As water is used as the cultivation medium, maintaining a constant temperature should not be a difficult issue. The culture area can also be moved easily (Gaurav et al. 2017; Kumar et al. 2021; Christensen 2020).

It has been attempted to use the algal biomass for anaerobic digestion to produce biodiesel, methane, and hydrogen (Table 9.1) and for the fermentation of ethanol and butanol, using various microorganisms to produce sugars derived from the enzyme hydrolysis of the algae. As a final note, algae fix over 40% of Earth's carbon, providing the world with a significant amount of oxygen (Farzanah et al. 2019).

9.3 Sustainable Biomass

9.3.1 Imperative Selection of Species

A wide variety of algae exist, from photosynthetic to heterotrophic and unicellular and multicellular to larger and more complex structures, such as prokaryotes (cyanobacteria) and eukaryotes (diatoms).

A few marine microalgal species, such as *Isochrysis*, *Nannochloropsis*, *Skeletonema*, and *Chaetoceros*, have been used widely as shellfish food or for aquaculture purposes. Green algae or diatoms are the most appropriate microalgae for energy production from alternative sources. Microalgae have so far primarily been produced on specific species with a high tolerance to extreme environmental conditions, which allowed them to be grown outdoors in raceways or ponds. Microalgae cultivation for alternative energy has expected to be based on more sophisticated technology that cultivates pure monocultures of species elected with

Table 9.1 Marine algal resource for renewable and sustainable algal biorefinery

Marine microalgae	Biofuel production	References
Chlorella protothecoides	Bio-oil, 31.17 wt%	Zou et al. (2010)
U. prolifera	Bio-oil, 41.3 wt %	Changwei et al. (2010)
Chlorella ovalis *Chlorella* sp. *Dunaliella tertiolecta* *Tetraselmis chui* *Tetraselmis* sp. *Thalassiosira pseudonana* *Nannochloropsis oculata* *Rhinomonas reticulata* *Isochrysis galbana* *Porphyridium purpureum*	51.0% FAME yields (DW) 59.8% FAME yields (DW) 9.9% FAME yields (DW) 5.3% FAME yields (DW) 12.6% FAME yields (DW) 15.2% FAME yields (DW) 21.0% FAME yields (DW) 45.5% FAME yields (DW) 9.1% FAME yields (DW) 7.6% FAME yields (DW)	Slocombe et al. (2013)
Nannochloropsis salina	21.8 ± 5.8% FAME content by mass	Bartley et al. (2014)
Nannochloropsis oculata	Bio-oil, 18 MJ/kg	Choi (2014)
Tetraselmis sp.	Bio-oil, 25 wt %	Choi (2014)
Nannochloropsis sp.	Bio-oil, 19.5 wt %	Gong et al. (2014)
Dunaliella salina	Bio-oil, 55.4 wt %	Ceylan and Kazan (2015)
Dunaliella tertiolecta	Bio-oil,36.9 wt %	Gao et al. (2017)
Cryptococcus curvatus	Bio-oil, 39 MJ/kg	Jafarian and Tavasoli (2018)
Closterium sp., *Chlorella* sp., *Oscillatoria* sp., *Spirulina* sp., *Navicula* sp., *Pinnularia* sp., *Spirogyra* sp., *Gomphonema* sp., *Scenedesmus* sp., *Zygnema* sp., *Frustulia* sp.	Biodiesel (89.7%, 0.25% free fatty acid content)	Karmakar et al. (2018)
Chlorella vulgaris	Bio-oil, 47.7 wt %	Ansaha et al. (2018)
Spirulina platensis	Bio-oil, 18.65 wt %	Zainan et al. (2018)
Saccharina japonica	Bio-oil, 47 wt %	Gautam and Vinu (2018)
Spirulina (Arthrospira) platensis	Bio-oil, 37.77 wt %	Ly et al. (2019)
Desmodesmus armatus	Cassava wastewater Biodiesel (21.91%)	Okpozua et al. (2019)

(continued)

Table 9.1 (continued)

Marine microalgae	Biofuel production	References
Cyanobacteria	Bio-oil, 41.1 wt%	Sotoudehniakarani et al. (2019)
Chlamydomonas debaryana	Bio-oil, 43.8 wt%	Aramkitphotha et al. (2019)
Isochrysis sp.	Bio-oil, 39.05 wt%	Ma et al. (2020)
Saccharina japonica	Bio-oil, 39.05 wt%	Rahman et al. (2020)
Marine macroalgae		
Ulva reticulata	87.5 mL H_2/g biohydrogen	Kumar et al. (2021)
Padina boergesenii, *Colpomenia sinuosa*, and *Ulva* sp.	Biogas (501 mL/5% volatile solids of *Ulva* sp.)	Farzanah et al. (2019)

definite physiological potential for sugar, hydrogen production, and lipid biosynthesis (Lyons et al. 2009).

As opposed to seagrass, macroalgae do not have roots, and they transport their food (nutrition) through their tissues by diffusion. In macroalgae, color or chlorophyll determines their classification, such as blue-green algae, green algae, brown algae, and red algae (Kim 2015). Seaweed is used as an energy source either through anaerobic digestion (AD), which creates biogas, or fermentation to produce ethanol. Fermentation inhibitors such as sulfated polysaccharides, polyphenols, and salt should be handled cautiously in order to reach fruitful fermentation. Energy potentials are most evident in *Laminaria* and *Ulva* species among macroalgae. Macroalgae have been reported as an effective candidate for energy source because of their enormous growth pattern of about 20% of their total biomass of *Ulva lactuca* (Fortes and Luning 1980; Waite et al. 1972; Farzanah et al. 2019). As the temperature gets cold outside, the macroalgae growth dominates an extensive part of the interior portion; however, in summer, they occupy a confined portion of the middle (Lee 2008).

9.3.2 Cultivation

In order to generate sustainable biomass from micro and macroalgae, cultivation is the main method. For a long time, industrial-scale production has been started in this way. Biomass cultivation systems typically fall into two categories: open ponds and closed photobioreactors (PBRs) (Behera and Varma 2016; Ganesana et al. 2020). However, the best microalgae cultivation system is still debated. Various cultivation systems have been studied, and scientists still have conflict over the most suitable one.

Compared to closed systems, open ponds like raceways have relatively low capital and operational costs, but they yield low productivity and a lack of control over the cultivation. The photobioreactor (PBR), a closed cultivation system, yields better productivity, although it is uneconomical.

9.3.3 Merits and Demerits Vs OPS Cultivation

The OPS cultivation of microalgae for large or industrial-scale production has been proposed as a feasible method by which any open land, pond, and lake of natural to artificial origin can be adopted. Preferably, algae are cultivated in liquid suspension, which consistently enhances the addition of nutrients and other required elements. Since it is an open system, light and airflow could be utilized from the sunlight and atmosphere. The Raceway system has been reported as a highly productive open pond system. The salient features of the raceway system are as follows:

1. The pond has a shallow depth of 30–50 cm with the shape of an ovoid.
2. Paddle wheels are specially designed to mix the pond mechanically.
3. The paddle wheel mixing enhances the vertical mixing to minimize biomass flocculation and enhance proper aeration.
4. It requires minimum capital investment.
5. The cost of operation was minimized due to less frequent observation of biomass and nutritional update.

This successive model is currently being used in all major industries. Even though it has its own limitations, energy consumption in mixing and artificial light sources are not appropriate for mass cultivation. In the case of natural sunlight, too much light causes cell death. Biomass densities cannot be achieved with this system, hence, it is often contaminated by microorganisms that grow quickly. In a study, the *Tetraselmis* sp. M8 was cultured in the OPS system for a period of 32 days, with an effective biodiesel extraction from the harvested mass (Narala et al. 2016).

9.3.4 Merits and Demerits Vs "PBR" Cultivation

The closed PBR is referred to as another form of photobioreactor system for culturing microalgae. A suspension method of cultivation has been followed the same as an "open pond"; hence, the system was controlled and closed. In this PBR technology, the water is circulated by pumps and relies on unnatural light and heat. While biomass cultivation is intended for energy or biofuel production, natural light and waste heat will be considered an economical route. In addition, an automated system is used to monitor nutrient and gas levels continuously. Compared to the open ponds, closed PBR offers a high level of productivity, minimal contamination, competent CO_2 consumption, uninterrupted runs, and optimal growth conditions.

Microalgae production with PBRs is subject to many designs and operational challenges that could be rectified before being implemented as a commercial process. Many systems are being tested worldwide, but here some of them are evaluated:

1. The costs associated with capital and operational expenses are the main disadvantages.
2. Both external and internal walls of the system become fouled during the continuous culture and need to be cleaned.
3. As the dirt accumulation on the external side and algae growth climb up on the internal part, the light will no longer enter the PBR.
4. It is being a challenge to blend the suspension to make photosynthetic efficiency as high as possible.
5. Maintaining the anti-flocculent state of the suspension requires energy, usually either from a pumping system or from gas sparging.
6. It is imperative to minimize parasitic energy loads in order to maintain good energy balances on an overall basis and embodied energy application in the components of the production system.

The developed intermediate system of OPS under greenhouses was considered a more controlled environment than OPS. The photobioreactor has to be improved by the manufacturers to be economical by constructing the system using materials such as transparent pipes, plates, solar for light, and additional nutrients without external energy (gravity feeding), etc. The CO_2 bubbling method has a potential to act together as a source of CO_2 and as a blender, thereby reducing the cost of production. In Israel, a number of projects have been started to produce algal biomass for biofuel production.

9.3.5 Combined Cultivation

Algae are designed to be cultivated in two steps, raceways and photobioreactor, combined to accomplish this cultivation system. A biomass rapid cultivation system is used in the PBR, while a stress cultivation system is used in OPS. In the former step of the process, a photoreactor provides an excellent defense to the growing biomass. Efforts are made to maximize CO_2 capture. In ponds with low nitrogen but high carbon dioxide levels, the suspended microalgae are transferred to nutrient-free ponds. Due to the higher algal biomass density in the second step, and the nutrient depletion in this phase, the open raceway in Step 2 poses fewer problems due to a lower risk of external contamination. Huntley and Redalje (2007) have found that photoreactor cultivation in conjunction with open pond cultivation provides the highest levels of astaxanthin production. Biofuel companies are currently testing the technology. A significant amount of research has been conducted on this topic at the University of Florence (Rodolfi et al. 2009).

9.3.6 Demand for Algal Growth

A number of environmental factors, such as light, water temperature, pH, salinity, and turbidity, also play an important role in algae's growth in OPS and PBR. In addition, algae require nutrients (nitrogen, phosphorus, silicate for diatoms) and carbon for their growth and all metabolic activities. The carbon can be sourced from power plant exhaust gas which contains substantial quantities of cost-effective CO_2. In many biofuel projects, CO_2 is recycled in order to generate energy, and this is a common part of the industries. A number of commercial products may be extracted and separated from algal biomass if biorefinery-type processes are applied. Biomass from algae offers an additional commercial openings for ethanol, butanol, methane, and hydrogen in addition to lipids.

In order to allow sufficient light to penetrate the PBR, the form, structure, and orientation of PBR specifically, the depth of the system, have to be taken into account. Light access can be restricted by poor design, but excessive sunlight can also inhibit algal growth. As with inadequate design, reduced sunlight may lead to reduced productivity, but algae may also be photoinhibited by overexposure to sunlight. Designed systems should allow the proficient assimilation of CO_2 and other nutrients. Throughout the world, microalgae thrive in different ecological conditions and are exposed to different climatic conditions, which makes PBR design problematic, as it may not be one design that would fit in all related applications.

Key factors that have been proven to influence microbial growth are light and temperature. Unless an artificial light source is free of charge, only natural sunlight will provide a viable commercial working environment. Therefore, growing high biomass should take place primarily in the tropical regions in the middle of 35° south and north of the Earth's equatorial plane. It is a topic that has been widely debated and will be extended in the productivity section.

9.4 Composition Vs Complexity

Microalgal biomass has a unique chemical composition based on the type of algae that will selected for biofuel production. Although microalgae coverings are an important aquatic resource, they are still poorly understood, especially considering the vast diversity of organisms in the water. Hence, selecting the species based on the preferred means of biofuel production is important. In microalgae, biomass composition is directly related to cultural conditions; it has the capability of accruing significant amounts of carbon as starch or lipids under firm culturing conditions (Hu 2013). The algal biomass may have a balanced composition of cellulose, starch, lipids, sugars, peptidoglycan, and proteins, or it can be exclusively rich in proteins and lipids (Rudolf et al. 2015; Cavalier-Smith 2017). At the time of cell wall

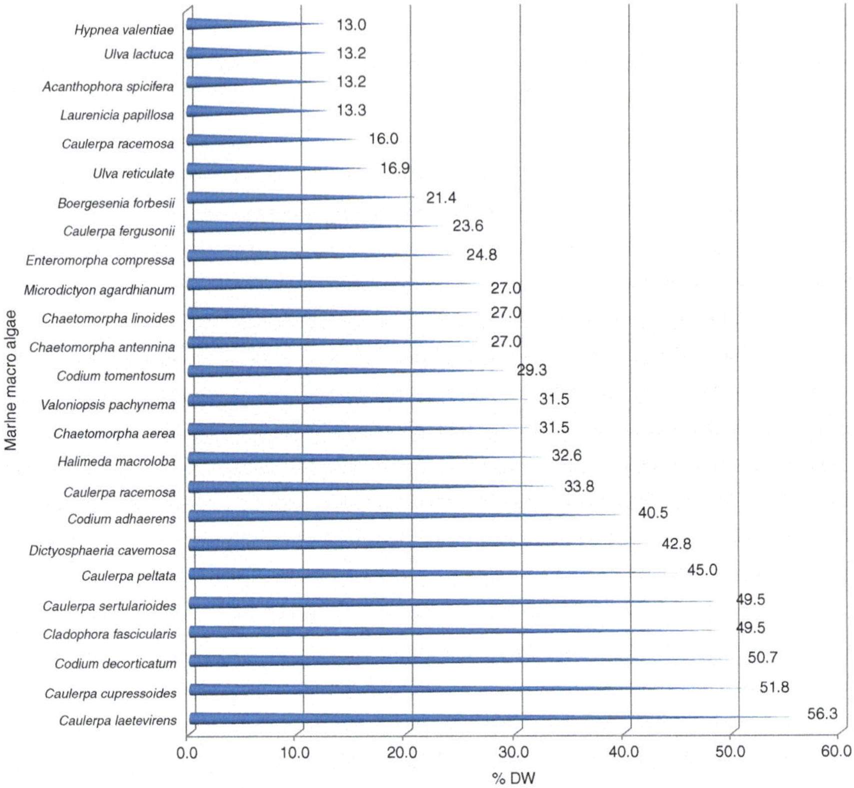

Fig. 9.1 Structural carbohydrates of macroalgae

formation of algae, cellulose is habitually deposited on the extracellular matrix (Chan et al. 2019).

Biomass from macro- and microalgae contains no or low lignin or lignin-like materials, so the complexity of the biomass reported is less than higher plants' lignocellulosic biomass. This marine biomass consists of a significant amount of carbohydrates, which has been illustrated graphically in Figs. 9.1 and 9.2. This illustration clearly shows that the algal biomass contains the maximum amount of structural carbohydrates at around 40–50%. Biomass from algae contains proteins, lipids, vitamins, carotenoids, and antioxidants in addition to carbohydrates, so it can be used in the production of fuel as primary energy and value-added biomolecules as secondary by-products (Rosenberg et al. 2008; Farzanah et al. 2019). When living under certain stress conditions, microalgae are susceptible to excessive lipid accumulation.

Microalgae have been extensively studied in terms of their chemical composition, but they have been studied less, despite the fact that they have been consumed as a food in Japan and China. The structural composition of the selected macroalgae of *Padina boergesenii*, *Colpomenia sinuosa*, and *Ulva* sp. revealed the presence of

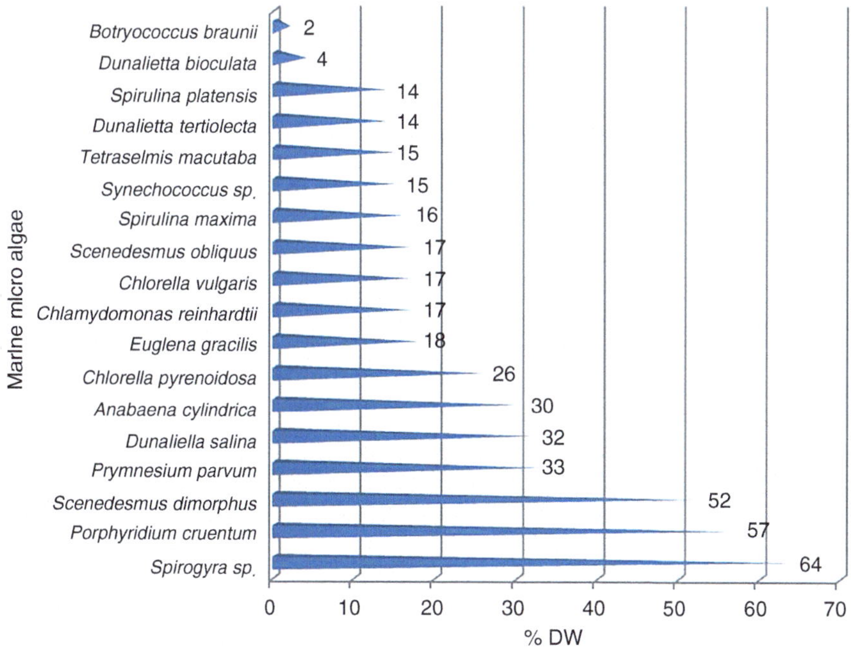

Fig. 9.2 Structural carbohydrates of microalgae

carbohydrates such as glucan, xylan, and arabinan (Farzanah et al. 2019). The species *Sargassum natans* and *Sargassum fluitans* have been reported first time for the presence of lignin-like materials by FTIR analysis. Biological compounds like cellulose, hemicellulose, and lignin have strong absorption of hydroxyl group (–OH) bands at 3406 cm^{-1} (Gaviria et al. 2021).

Although efforts to convert macroalgal biomass to biofuels seem to be increasing, this is still at an early stage. It is agreed by bioenergy feasibility experts that *Ulva* sp., which has a high rate of carbohydrates, reaching 67%, and protein content of 27%, can be a significant feedstock for biorefineries producing platform chemicals (Reich 2011; Fleurence 1999).

9.5 Dual Performance on Cultivation

Inappropriate management of excess wastewater occurs when it is supplied to crops even when they do not require it. Although the efficient source with the significant amount of nutrients creates unexpected pollution on environments, including water bodies, air, and cultivation land (Webb et al. 2013), the algae have the ability to extract nutrients from wastewater and slurries, thereby renovating the issue into a resource (Ledda et al. 2015).

There has been a description of the dual performance of algae when they grow on wastewater as a substrate while utilizing all the nutrients in the wastewater for energy and nutrients that allow them to produce massive amounts of biomass. It is possible that wastewater could be used to satisfy nutrient demands, leading to reduced cost of production of biomass and favorable for water treatment (Leite et al. 2019). As a general rule, the quality of biomass generated varies with the conditions during production. In optimistic conditions, around 90–95% of biomass will be in the form of pure algae (Acien et al. 2016).

A dual performance system must be implemented via a fully validated evaluation approach; it should be used to identify an integrated sustainable production of products in a new supply chain that depends on a standardized approach. This aid of evaluation has been referred to as life cycle assessment (LCA). During this process, each step of the life cycle of the study shall be computed for its energy and resources as input and production as output (Herrera et al. 2021).

Nitrogen and phosphorus in wastewaters of fish aquaculture, amino acids, enzymes of food production industries, and sewage effluent can stimulate algal growth, as well as cleanse the environment through biological methods. This systematic approach has been proven to reduce the amount of eutrophication in the aquatic environment.

A nitrogen reduction efficiency of 72% and a phosphorus reduction efficiency of 28% have been achieved using *C. vulgaris* from wastewater. Other microalgae that have been reported for nutrient removal are *Chlorella vulgaris*, *Scenedesmus dimorphus*, *Scenedesmus obliquus*, *Spirulina*, *Nannochloris*, *Botryococcus braunii*, and *Phormidium bohneri.* Algal species such as *Chlorella vulgaris*, *Scenedesmus bijugatus*, *Oscillatoria amphibia*, and *Lyngbya majuscula* had significant support to heavier metal levels, especially iron and manganese (Bansod and Nandkar 2015). Several studies have demonstrated that algae can perform a dual role in producing energy and other value-added products as well as to treat wastewater without causing harm to the environment (Gonzalez et al. 1997; Laliberte et al. 1997; Martınez et al. 2000; Olguın et al. 2003; Lee and Lee 2001; Aslan and Kapdan 2006; Mostafa 2012).

9.6 CO_2 Sequestration

Numerous sources of CO_2 emissions exist, including emissions from man-made activities (deforestation, fossil fuel combustion, incineration of solid waste, household waste) and commercial/industrial sources like electric power generating units, cement, refineries, fertilizer, and natural gas outlets and decomposition of biomass (Shaikh et al. 2018).

The global CO_2 emission level of the top 18 countries has been illustrated in Fig. 9.3. China emits one-third (29.1%) of the total global CO_2 emission and occupies first place, followed by the USA, India, Russia, Japan, and Germany. In order to reduce CO_2 emissions to the atmosphere significantly, biodegradable,

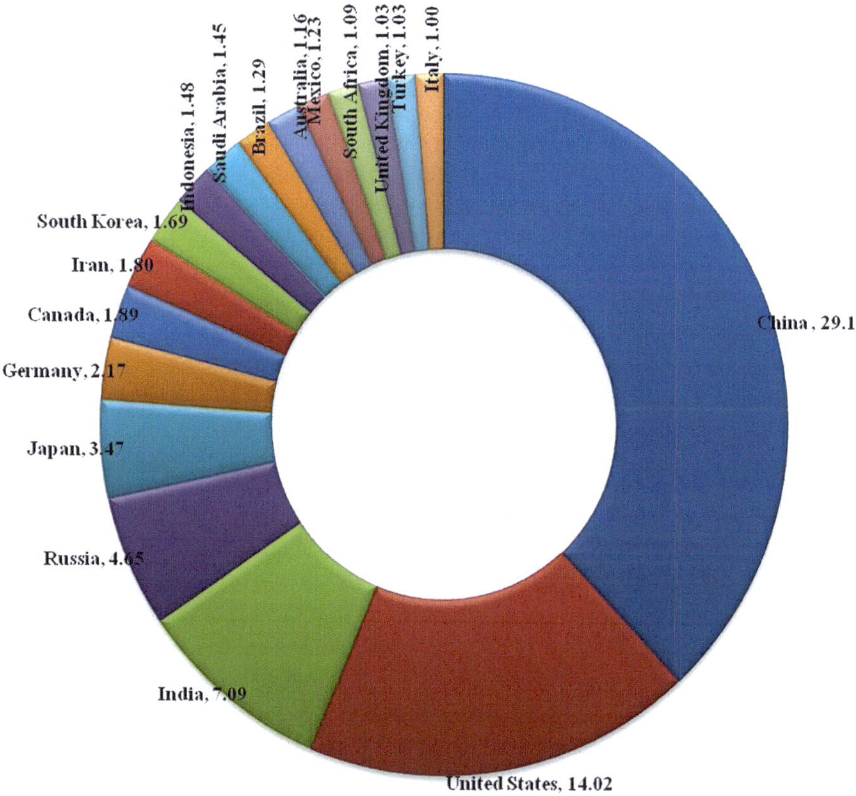

Fig. 9.3 CO_2 emission share (%), 2020

nuclear, and renewable energy sources have to be used more extensively. In our view, many options proposed and used for capturing CO_2 are inherently impermanent from the economic, social, and environmental perspectives. Commonly, mitigation measures have attempted to counteract immediate adverse effects, often by transmitting emissions to other areas. A typical example of such "end-of-pipe technologies" remediation is releasing flue gases into oceans or terrestrial repositories (Packer 2009).

Microalgae are ideal converters of CO_2 and associated nutrient sources into biomass through photosynthesis significantly faster than terrestrial plants since they have acquired considerable recognition for CO_2 sequestration and biofuel production. As microalgae are composed of 45% carbon, it is considered one of the predominant sources algae needs and this has to be provided through CO_2. According to a mass balance, each kg of microalgae needs at least 1.65–1.8 kg of CO_2 (Nilsen 2006; Herrera et al. 2021).

Other beneficial products can also be produced by algae cultivation besides biofuels. Algal biomass can store CO_2 with high economic benefit by producing value-added products. Algal products like pigments, proteins, lipids, antioxidants, animal feed, food supplements, vitamins, amino acids, biomolecules, bioabsorbents, biostimulants, biopesticides, etc., have been extracted from several species of algae. A biorefinery based on algae can produce a broad spectrum of products that will benefit a wide range of industries and contribute to the expansion of numerous divisions (Ahmed et al. 2012; Kumar et al. 2018). In this way, algal biofuels are being seen as "biofactories" for imminent resources due to their synergistic progress with additional business (Brennan and Owende 2010; Mostafa 2012). The production of microalgal biomass in association with available energy and wastewater treatment technologies could likely support CO_2 fixation and biofuel production more sustainably and thereby more affordably (Mostafa 2012).

CDR has been reported as one of the best ways to control climate change. A report suggests that CDR could limit global warming to 1.5 °C once it is effectively implemented. The positive correlation of CDR contributes to the sustainability of ecosystems, environment, and biodiversity (Mehmood and Faisal 2020).

9.7 Primary and Secondary Biofuel Production

(a) **Microalgae**

The conversion of microalgal biomass into biofuels can be achieved by a combination of physical, enzymatic, chemical, thermochemical, and biochemical processes. Economic factors, biomass composition, species selection, availability, end and co-product, and processing ease are all considered when deciding which method to use.

Bringing in artificial eutrophication to water through human intervention can lead to alterations of ecosystems that may result in large blooms of microalgae. The principal source of marine algal biomass is the natural or man-made bloom in ponds and lakes with backwaters. It is possible to harvest biomass from a natural source since it has been overgrown. The man-made eutrophication arrived with a huge amount of organic biomass, and the continuous biomass accumulation that leads to microbial degradation concurrently decreases the dissolved oxygen of the ecosystem. The reduced oxygen level in the ecosystem has negatively influenced the other aquatic habitats. This algal biomass could be a significant resource for biofuel production.

(b) **Macroalgae**

It has been known that seaweeds or macroalgae adapt well to stress factors in marine environments like coastal backwater lakes and ponds. Many brown and green varieties of seaweed have been used for industrial applications for a long time, and now attention is being paid to their use in producing energy, particularly in areas with seaweed resources. In particular, *Ulva* sp. is being investigated as potential feedstock for sustainable biofuels (Lyons et al. 2009).

9.7.1 Bio-oil Extraction from Algal Biomass

It is not a natural occurrence for algae to contain high levels of oil. It occurs when the algae are deficient in nitrogen due to the cultivation under stress. As a result, carbon is stored in lipids within the cell. However, it is quite difficult to change operating conditions in large ponds. It is possible to grow and produce algae with 60% triglycerides and 40% carbohydrates and protein using carbon dioxide and sunlight (Pienkos and Darzins 2009). An overview of micro and macroalgal species producing lipids is given in Table 9.2. The species and cultivation techniques of these algae have to be optimized for the production of lipids (Griffiths and Harrison 2009). Numerous articles describe oil-rich species and how lipids are extracted and converted through different techniques. Many articles explain the growth and lipid conversion processes involved in these microbes (Ganesan et al. 2020). This multi-layered microbe consists of polysaccharides and cellulose as well as an outer envelope made up of lipids and fatty acids. This can be extracted as oil from biomass.

9.7.2 Transesterification

It is important to develop downstream processes, particularly for converting algae whole cells into esters (biodiesel). Microalgae are high in lipids, ranging from 1% to 60%, and converted into C18 range carbon fuels via the process known as transesterification.

This type of process has been identified and is proposed for the extraction of bio-oil from biomass in Fig. 9.4. It is considered one of the most desirable challenges of the biofuel production process. During transesterification, a triglyceride molecule that is a complex ester reacts with a molecule of alcohol to produce a simple ester.

Transesterification is commonly catalyzed by several acid catalysts (sulfonic acid, sulfuric acid, HCl) and base catalysts (KOH, NaOH, sodium methoxide, sodium ethoxide, potassium carbonate). It is popular industrially to use the base-catalyzed reaction since it is less corrosive than the acid-catalyzed reaction. The carbonyl group gains a proton in the acid, whereas the alcohol loses a proton, rendering a nucleophilic compound. A methanol and ethanol solvent has been proposed to extract methyl and ethyl esters. Later the homogenous catalysts were replaced by heterogeneous catalysts:

1. Methanol and chloroform have been widely used in a commercial method to extract lipids specifically.
2. There has also been a great deal of work on using mechanical methods for oil extraction, such as microwave, milling, bead beating, and ultrasound. More importantly, these methods require no chemical treatment to be used.
3. The large amount of beads (ceramics or glass) are typically used in the beating process to disrupt the cells of microbes on a small scale. The vigorous shaking of

Table 9.2 Lipid/oil content of marine algae

Microalgae	Lipid/oil %	Macroalgae	Lipid/oil %
Dunaliella primolecta	23	*Ulva lactuca*	9.6–11.4
Dunaliella salina	6.0–25	*Enteromorpha compressa*	11.45
Dunaliella tertiolecta	18–71	*Caulerpa peltata*	11.42
Nannochloropsis oculata	22.7–29.7	*Valoniopsis pachynema*	9.09
Chaetoceros calcitrans	14.6–16.4	*Caulerpa racemosa*	9.0–10.5
Scenedesmus obliquus	11–55	*Caulerpa sertularioides*	6.99
Tetraselmis suecica	8.5–23	*Hypnea valentiae*	9.6–11.6
Neochloris oleoabundans	29–65	*Acanthophora spicifera*	10.0–12.0
Chlorella protothecoides	14.6–57	*Laurencia papillosa*	8.9–10.8
Porphyridium cruentum	9–14	*Ulva reticulate*	8.50
Spirulina platensis	4.0–16	*Chaetomorpha aerea*	8.50
Schizochytrium sp.	50–77	*Chaetomorpha antennina*	11.45
Phaeodactylum tricornutum	18–57	*Chaetomorpha linoides*	12.00
Isochrysis galbana	7.0–40.0	*Cladophora fascicularis*	15.70
Botryococcus braunii	25–75	*Microdictyon agardhianum*	9.40
Spirogyra sp.	11–21	*Boergesenia forbesii*	11.42
Scenedesmus quadricauda	1.9–18	*Dictyosphaeria cavernosa*	10.51
Scenedesmus dimorphus	16–40	*Caulerpa cupressoides*	10.97
Chlamydomonas reinhardtii	21	*Caulerpa laetevirens*	8.80
Chlorella vulgaris	5.0–58.0	*Caulerpa fergusonii*	7.15
Chlorella pyrenoidosa	2	*Halimeda macroloba*	9.89
Dunaliella bioculata	8	*Codium adhaerens*	7.40
Euglena gracilis	14–20	*Codium decorticatum*	9.00
Prymnesium parvum	22–39	*Codium tomentosum*	7.15
Tetraselmis maculate	3		
Spirulina maxima	4.0–9.0		
Synechococcus sp.	11		
Anabaena cylindrica	4–7		
Desmodesmus armatus	21.91		
Botryococcus braunii	25–75		
Chlorella emersonii	25–63		

the beads with the cell suspension allows the lipids to leak from walls when mixed with a large surface area.

4. The process of crushing algal biomass is comparable to wheat milling, which ensures that maximum nutrients remain intact.
5. In order to meet the demand, the SCF (cyclopentyl methyl ether (CPME) and ethanol (EtOH)) technique produces safe and high-quality end products. A species with complex characteristics can be effectively derived from specific components by this method.
6. *n*-Heptane used in the Soxhlet method of extraction extracts much less oil than the SCF method.

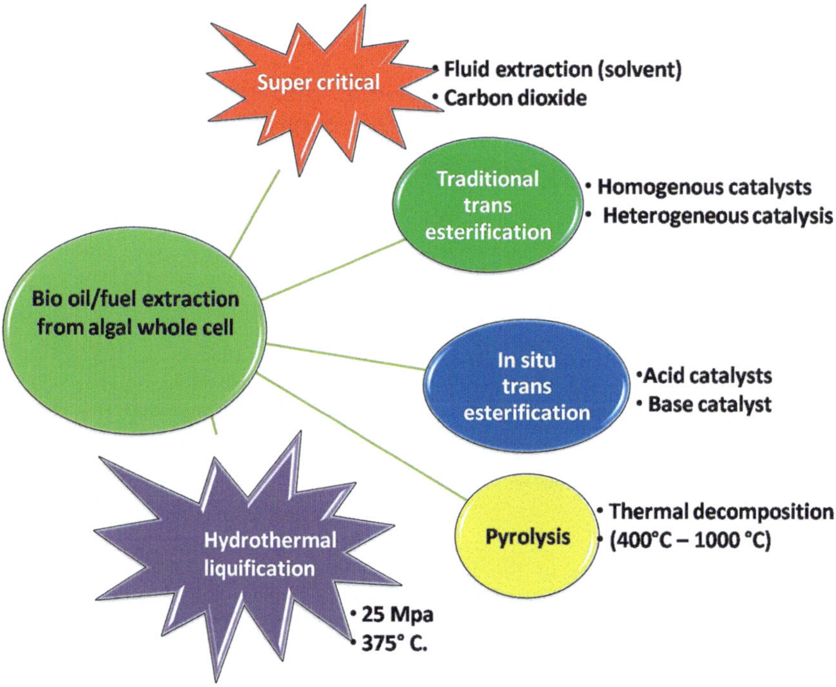

Fig. 9.4 Bio-oil extraction from whole algae cell

7. Despite hexane being the more expensive solvent, no significant difference was reported between algal biomass extracted with hexane (9%) and acetone (8%).
8. The biological enzymes can contribute as an alternative to the biofuel industry, and some of the enzyme candidates like lipase and cellulase are actually extracting oil more effectively. It is proving to be a very productive industry in a green way.

The oil extracted from algal biomass can be used as a feedstock for biodiesel and as a supplement for refinery diesel. Research has recently focused on oil extraction without solvents (Sengupta and Pike 2016). In addition to microalgae, the conversion of seaweed to biodiesel has also received attention (Chen et al. 2015; Demirbas and Demirbas 2011). In order to create liquid fuel/biodiesel from algal oil, the acquired triglyceride/fatty acids must be converted via reduction catalysts. After that, only then will it be a viable substitute for fossil fuels (Karmakar et al. 2018).

9.7.3 HTL of Algal Biomass

In comparison to combustion, hydrothermal liquefaction (HTL) can produce bio-oil that is viscous and thick in composition. Algae nutrients are converted easily using HTL, including proteins, carbohydrates, vitamins, amino acids, etc. Though there is no separate lipid extraction used for this method and there are high operating costs are involved with this technique, it will be less economical to replace it. This method may reduce the cost to some extent because it is able to process algae with a high water content (90%) without drying and pretreatment of algae. HTL has a low conversion rate without a catalyst, though with catalyst, it becomes more efficient (Kumar et al. 2019). It has been shown that the HTL system with nanoparticles significantly improves the conversion ratio (Ganesan et al. 2020).

9.7.4 Pyrolysis

The term pyrolysis refers to the high temperature between 400 and 1000 °C combustion process without oxygen or catalysis, in which biomass gets converted into solid, liquid, and gaseous forms of biofuel. Two distinct conditions are required for pyrolysis: high temperatures for short durations and low temperatures for longer periods. In an aqueous phase, various low molecular weight compounds were observed, primarily alcohols, acids, and ketones. Similar to this, aliphatics, alcohols, carbonyls, acids, phenols, and cresols, among others, were produced in the non-aqueous phase. Because of the high nitrogen content of algae, complete combustion of algae releases more nitrogen into the atmosphere (Bridgwater 1994; Demirbas 2001, 2003; Wang et al. 2018; Obeid et al. 2019; Ganesan et al. 2020).

9.7.5 Derived Biomass of Primary Production to Biofuel

It is anticipated that the leftover biomass will be a good source of carbohydrates after oil extraction, making it competitive with energy plants used for biofuel production. This biomass generated from the primary energy production process has been referred to as defatted algal biomass. When compared to the raw algal biomass, the DAB has less complexity; hence, it will bypass the pretreatment process, thereby reducing biofuel production costs. This DAB biomass will be used to produce bioethanol, methanol, butanol, and hydrogen.

9.8 Marine Alga in Bioethanol Production

The following microalgal species have been studied for bioethanol production: *Chlorococcum infusionum*, *Chlamydomonas reinhardtii*, *Chlorella vulgaris* (Schroeder and Michalak 2018), *Dunaliella* sp., *Nannochloropsis oculata*, *Tetraselmis tetrathele* (Shirai et al. 1998), *Undaria pinnatifida* (Lee et al. 2011), and *Saccharina japonica* (Ji et al. 2016). In recent years, macroalgae have gained more attention as a bioresource for bioethanol production. The species studied were green algae, *Ulva lactuca* L. and *Ulva pertusa* and *Chaetomorpha linum*; red algae, *Gracilaria chilensis*, *Kappaphycus alvarezii*, *Gelidium amansii*, *Gelidium elegans*, and *Gracilaria salicornia*; and brown algae, *Macrocystis pyrifera*, *Laminaria japonica*, *Laminaria hyperborean*, *Saccharina latissima*, *Sargassum fulvellum*, *Undaria pinnatifida*, and *Alaria crassifolia* (Aitken et al. 2014; Yahmed et al. 2016; Schroeder and Michalak 2018).

Algal cells, being the richest source of carbohydrates, are believed to be a cell factory for producing bioethanol. Those specific parts of the nucleus of the cell contain cellulose and hemicellulose. Initially, a complex form of carbohydrate present in algal biomass needs to be broken down into sugar monomers through a process called hydrolysis. It can be accomplished chemically by using acid (sulfuric, hydrochloric, oxalic, etc.) or enzymes such as cellulase and xylanase to produce rhamnose, fructose, mannose, xylose, galactose, and glucose as fermentable monomers. The sugars are then fermented by the action of ethanologenic microbes such as bacteria (*Zymomonas mobilis*, *Clostridium acetobutylicum*, *Clostridium thermocellum*, *Candida tropicalis*, etc.) and yeast (*Saccharomyces cerevisiae*, *Schizosaccharomyces* sp., *Issatchenkia orientalis*, etc.) under oxygen-free conditions (Banuselvi et al. 2008; Sivasankari 2010).

Scenedesmus dimorphus contains 54% wt of carbohydrate and is considered an efficient ethnaol producer. The sugar from *Scenedesmus* sp. was reported to be 94% by enzymatic hydrolysis. Fermenting these sugars requires the right fermentation microbe (Chng et al. 2017). An attempt has been made to produce bioethanol from red macroalgae *Eucheuma denticulatum (Spinosum)* residue from industrial agar extraction. In the first step, the pretreated was hydrolyzed with an acid, followed by variables that will be optimized to obtain the maximum quantity of reducing sugars. The production of 0.12 g bioethanol/g reducing sugars was obtained by a yeast strain *Saccharomyces cerevisiae* (Alfonsin et al. 2019).

There was the highest reported ethanol production from *Saccharina japonica*, in which *Vibrio splendidus* efficiently utilized the alginate content of the biomass and produced around 0.3% ethanol w/w biomass (Wargacki et al. 2012).

The red algae *K. alvarezii* and *G. amansii* have been tried for ethanol production. The fungal enzyme pretreatment enhanced the complete conversion of the carbohydrates into bioethanol and reported the maximum fermentation efficiency of 92.2 and 80.4 in *K. alvarezii* and *G. amansii*, respectively, using *S. cerevisiae*. The maximum production was achieved by the supplementation of residual fungal biomass. The yield was not significant, while the organic or inorganic nitrogen supplements were given in various algal biomass *Ulva prolifera* (61.7), *Gracilaria* sp. (78.4), *Porphyridium cruentum* (70.3), *Gelidium amansii* (84.9), *Ulva fasciata* (88.2), and *Sargassum angustifolium* (73.0) (Sulfahri et al. 2020). As bioethanol oxygenated fuel produces less CO, hydrocarbons, and nitrogen oxides than gasoline and diesel, during combustion, it often produces aldehyde, which contributes to photochemical smog.

Algenol has developed a metabolically engineered algae specie to produce alcohol in a single step by the direct method of closed bioreactor. The metabolically engineered algae have the potential to produce 100 million gallons of ethanol from 1.5 million tons of carbon dioxide. Though the process of making ethanol by the engineered algae, the physical factors (pH, temperature, environmental conditions) and chemical parameters (CO_2, nutrients, water, salinity) of the fermentation should be advised to follow strictly. This technology has been developed by Dow Chemical Company and the Department of Energy at Dow's Freeport, Texas (Voith 2009; Sengupta and Pike 2016). US Renewable Fuels Standard fixed the target of 36 billion gallons of liquid biofuel production before 2022 (Qari et al. 2017; Ganesan et al. 2020).

9.9 Marine Alga in Biohydrogen Production

Since hydrogen fuel is such a reliable energy source, unlike conventional fossil fuels, its only by-product when it burns is water. Fuels derived from fossil fuels are not sustainable and can be produced conventionally through nonrenewable energy sources and produced through the thermochemical process. An alternative to fossil fuels, biohydrogen has the potential to be a sustainable and clean energy source. This energy-efficient process requires ambient temperature and pressure, unlike the chemical process.

A number of prokaryotic and eukaryotic algae produce hydrogen through the hydrogen-metabolism enzymes called hydrogenase or nitrogenase. Various microbial species are used to produce algal hydrogen through photosynthesis and biophotolysis. The production of hydrogen by photosynthetic microorganisms depends on sunlight. The carbon and energy sources for photoautotrophic microalgae and cyanobacteria are carbon dioxide and sunlight, respectively. In a photosynthesis or photolysis reaction, the reducing power is provided by water oxidation under light irradiation (Schutz et al. 2004).

Biological processes like biophotolysis and catabolism of metabolic substrate have been reported as the main production processes of biohydrogen in the microalgae and cyanobacteria. The water is the main source for the biophotolysis mechanism that occurs in vivo induced with illumination and results in the breakdown of water molecules into hydrogen and oxygen. The in vivo biophotolysis occurs in two forms: direct and indirect photolysis. In direct photolysis, the enzymes such as hydrogenase, water-plastoquinone oxidoreductase (photosystem II), and ferredoxin oxidoreductase (photosystem I) were involved for hydrogen production.

The thylakoid membrane of the algae has been reported as the accessible part of the above-stated enzymes.

A direct photolysis process has the advantage that it relies primarily on the water as a feedstock and sunlight as an energy source. Though these two elements are abundant in nature, this technology has significant challenges. Hydrogen production under aerobic conditions has several challenges, including incompatibility in production, oxygen as a strong suppressor of hydrogenase, and huge production area requirements (Show et al. 2019).

In the biological system, hydrogen production occurs depending upon the source of the electron if it is derived from water and the production pathway starts with the photosynthetic hydrogen production of biophotolysis. On the other hand, the electron derived from the catabolism of in vivo substrate leads to oxidative carbon metabolism and produces hydrogen. In the absence of oxygen and light, dark fermentation has been taking place in some of the green algae, where it utilizes the carbohydrate as a source and enters the metabolic pathway of heterotrophic fermentation. The direct photolysis will simultaneously produce oxygen and hydrogen gases in *Chlorella* sp. (Spruit 1958).

The marine macroalgae *Gelidium amansii* has been tried for biological hydrogen production. This richest source of carbohydrates (67.3%) has been used as a substrate for biohydrogen production by sludge seed culture. The hydrogen production rate was higher (52 mL H_2/g-dry biomass) with dilute sulfuric acid than with other acid pretreatments (Sivagurunathan et al. 2017).

The biomass of *Ulva reticulata* (marine macroalgae) was taken as a substrate for biohydrogen production. During the experiment, the biomass was treated with two conditions: microwave treatment with hydrogen peroxide and microwave treatment with hydrogen peroxide at alkaline conditions. A later proposed method clearly indicated that the hydrogen production reached a maximum of 87.5 mL H_2/g COD (Kumar et al. 2021).

Recently, sulfur-deprived anaerobic hydrogen production has been studied in green algae under the sulfur deprivation method. A logistic model was used to evaluate the kinetics of the reaction. Through this model, the growth and production factors such as time, sulfur concentration, and pH were optimized and achieved, improving hydrogen yield of 54% (Bechara et al. 2021).

Microalgal species *Scenedesmus obliquus*, *Chlorella fusca*, *Chlorella vulgaris*, *Dunaliella tertiolecta*, *Chlamydomonas moewusii*, *Chlamydomonas reinhardtii*, *Lobochlamys culleus*, *Chlorococcum littorale*, *Tetraselmis subcordiformis*, *Synechocystis* sp., *Anabaena* sp., *Nostoc* sp., and *Tetraspora* sp. and macroalgal species *Laminaria japonica*, *Chaetomorpha antennina*, and *Ulva reticulata* have been tried for hydrogen production. It has been reported that genetic and metabolic engineering of hydrogenase and nitrogenase plays a significant role in improving hydrogen production. Future biohydrogen production will benefit from genetically modified strains of microalgae. Researchers are studying the probability of microRNAs for hydrogen production (Sivagurunathan et al. 2017; Anwar et al. 2019). The hydrogen production of 170 mL H_2/g volatile solids was obtained from *Chlorella* biomass with a maximum of 70% carbohydrates and reported as an efficient producer of hydrogen (Giang et al. 2019).

9.10 Marine Alga in Biomethane Production

(a) **Methane Production from Microalgae**

The methane production process like hydrothermal liquefaction or pyrolysis has not been suggested for commercial or largescale production of methane because it is an energy-consuming process that requires higher temperature and pressure and sophisticated technologies. However, the possibility of using wet biomass as such in the anaerobic digestion process makes it a feasible technology for methane production.

The conversion of whole algae to biogas is possible through anaerobic digestion (AD) and gasification for syngas-derived fuels and chemicals. According to some reports, the energy requirements of biogas production from algal biomass are comparable to sewage sludge digestion (Razon and Tan 2011).

Studies of marine microalgae (*Nannochloropsis* sp., *Chlorella pyrenoidosa*, *Scenedesmus obliquus*, *Chlorella sorokiniana*, *Dunaliella salina*, and *Nostoc* sp.) that have been conducted on biomethane production are significantly fewer than studies on macroalgae (Razon and Tan 2011; Fermoso et al. 2016; Xiao et al. 2019).

The size of the *Nannochloropsis* sp. is reported to be 2–4 μm; hence, the flocculent aluminum sulfate is used to obtain 90% of biomass for biogas production. The resulting biomass concentration was improved 357-fold, and then the raw biomass, made as de-oiled algal cake and used for biogas production, was 1.5 m^3 (Razon and Tan 2011).

Despite its rigid cell wall, raw microalgae did not produce significant amounts of methane through anaerobic digestion. According to the researchers, pretreatment of algal biomass before methane production greatly enhances production. The researcher utilized various strategies, including thermochemical (acid, alkali, and solvent), ultrasound, enzymatic, and hydrothermal strategy. Considering the above, the (HTL) hydrothermal pretreatment method significantly improved methane production. The evaluation of HTL for biomass has been detailed by Gollakota et al. (2018). The conditions ranged from 523 to 647 K (temperature) and from 4 to 22 MPa (pressure), though according to the substrate, the conditions vary for microalgae ≤108 °C with ≤2 MPa (Xiao et al. 2019).

The use of hydrothermal pretreatment can be an ideal way to enhance anaerobic digestion in methane production from microalgal biomass. However, the expense of hydrothermal pretreatment simultaneously increases the cost of production. As the energy required for the pretreatment was substituted by solar energy and improved the carbohydrate content by 7.4 times than the raw biomass, methane production was also recorded high (57%) in anaerobic digestion (Xiao et al. 2019).

A research study reported the utilization of an existing biogas production facility (digester) for the production of algal biogas without any modification since the biogas plant has a huge amount of nutrients and CO_2 that can be effectively used to cultivate the algal biomass (Wang et al. 2013).

(b) **Methane Production from Macroalgae**

The species such as *Padina boergesenii*, *Colpomenia sinuosa*, and *Ulva* sp. *Gracilaria chilensis* and *Macrocystis pyrifera*, *Saccharina latissima*, *Laminaria digitata*, and *Enteromorpha* sp. showed the potential as significant substrates for methane production (Aitken et al. 2013; Farzanah et al. 2019).

In contrast to microalgae, has a proven research base and provided evidence of the practical feasibility of AD. Though the biogas production from macroalgae is initiated with some obstacles due to the complex nature of the cell structure with enormous sulfur and nitrogen content, the efficiency of the biogas production makes the situation uneconomical (Montingelli et al. 2015).

The macroalgae *G. chilensis* and *M. pyrifera* were cultivated based on three processes reported as "bottom planted," "Thalli tied long-line," and "spore inoculated long line." The species contributed to a new life cycle assessment method, which included methane production in the selected macroalgae. The production of bioethanol and biomethane from a single biomass source rather than a single by-product significantly enhanced methane production. According to the life cycle assessment, it has been recommended that the bottom cultivation method for *G. chilensis* and the long-line cultivation method for *M. pyrifera* may be ideal methods for culturing macroalgae and cumulative energy production (Aitken et al. 2014).

In the same way, the macroalgae *Padina boergesenii*, *Colpomenia sinuosa*, and *Ulva* sp. were also studied for their ability to increase biogas production by extractive-free material using commercial cellulase HTec2 and CTec2. It was possible to achieve 100% glucose conversion rates and to generate biogas with *Ulva* sp. and 501ML in 5% VS, respectively (Farzanah et al. 2019). The synergistic effect of cobalt nanoparticle and microwave pretreatment on *Enteromorpha* sp. showed a 42% increment in biogas production (Zaidi et al. 2019).

Hythane, a mixture of hydrogen and methane, is among the most promising biogases for industrial applications and is a highly efficient and ultraclean burning fuel (Cavinato et al. 2009).

9.11 Marine Alga in Biobutanol Production

Scientists denoted the algae as an evolutionary model for the origin of life on Earth (Falkowski et al. 2004). The potential nature of the algae to withstand adverse conditions like intense temperature, pH, salinity, etc., and mass availability of the carbohydrate content, making it a beneficial resource for biobutanol production (Hirayama et al. 1996). Biobutanol, aliphatic alcohol, has four carbons produced by microbes through the process called solventogenesis. Biobutanol has been promoted as a suitable alternative to gasoline since its fuel quality, like octane, energy content, low volatility, and blending nature, make it a remarkable than ethanol (Cascone 2008).

The primary biobutanol producers reported are *Clostridium acetobutylicum*, *Clostridium beijerinckii*, *Clostridium saccharobutylicum*, *Clostridium saccharoperbutylacetonicum*, and *Clostridium tyrobutyricum* (Maiti et al. 2016a, b; Narchonai et al. 2020; Onay 2020).

9.11.1 Biobutanol from Microalgae

Algal species such as *Laminaria hyperborean*, *Dunaliella tertiolecta*, *Chlamydomonas reinhardtii*, *Saccharina latissima*, *Spirulina fusiformis*, *C. reinhardtii*, *Spirogyra* sp., *Chlorella* sp., *Scenedesmus* sp., *Ankistrodesmus* sp., *Micromonas* sp., *Tetraselmis suecica*, *Saccharina japonica*, *Chlorococcum humicola*, *Rhizoclonium sp.*, and *Nannochloropsis gaditana are effectively used for biobutanol production* (Lakaniemi et al. 2012; Maiti et al. 2016a, b; Narchonai et al. 2020; Salaeh et al. 2019; Onay 2020; Fu et al. 2021).

As with terrestrial crops, microalgae are typically handled easier and grow faster in industrial applications. Because of their high cell division rate, the asexual reproduction of *Chlamydomonas* species typically takes 6 h under ideal conditions (Packer 2009; Chen et al. 2015). In a Logan Lagoon study, algal biomass was harvested and converted into sugar by acid and base methods followed by ABE fermentation of *Clostridium saccharoperbutylacetonicum* with a yield of 0.2% (Ellis et al. 2012).

The study of butanol production by *C. tyrobutyricum* from *S. japonica* hydrolyzate was conducted. A pretreatment method was investigated for its effect on butanol production. Among the methods tested, ultrasonic-assisted acid hydrolyzate yielded the highest butanol production of 0.26% by wt biomass with a productivity of 0.19 g/L/h. The strain *C. tyrobutyricum* has been used with genetic enhancement at heat shock protein overexpression. The strain *C. tyrobutyricum* Ct-pMA12G was found to produce the highest amount of 12 g/45 g mannitol (Fu et al. 2021).

9.11.2 Biobutanol from Macroalgae

Regardless of the type of macroalgae, brown and red macroalgae have been described as suitable resources for biobutanol production. The composition of red algae *Gelidium amansii* with cellulose, glucan, and galactan makes them an excellent feedstock for butanol production. Moreover, some brown algae like *Laminaria* sp., *Saccorhiza* sp., and *Alaria* sp. have the potential to grow up to several meters in length because of the presence of laminarin and mannitol as the main carbon sources for their growth (Horn et al. 2000; Adams et al. 2011; Nobe et al. 2003; Kim 2009; Wi et al. 2009; Yoon et al. 2010).

The reported unique features of the macroalgae to be significant feedstocks are (1) enormous growth rate with huge biomass production with minimum energy supply and (2) potential to utilize the essential nutrients throughout the culture area.

The biomass of macroalgae *Ulva lactuca* was used for butanol production since it has high carbohydrate content and growth rate. The strains *Clostridium beijerinckii* and *C. saccharoperbutylacetonicum* used in the bioconversion of biomass to butanol were reported to have a butanol production of 0.26 g/g of reducing sugars in the lab scale and 0.29 g/g reducing sugar at the large scale (Potts et al. 2012).

The author Malik detailed the use of genetically engineered *Clostridium acetobutylicum* in hyperbutanol production by an oil refining industry in Korea, Bio Fuel Chem, with the production of 585 g of butanol from 1.8 kg of glucose (Malik 2014).

9.12 Commercialization of Algal Biofuel

Increasingly, alternative fuels are replacing fossil fuels because fossil fuels are becoming scarcer. Scientists are still working hard on developing algal biofuel as an alternative fuel, and there is a lot of disputes about which is the best alternative. Popularization would largely depend on the commercialization of biofuel. The commercialization of the biofuel depends on how it can be produced by simple and rabbit technology. The following factors must be considered when commercializing algal biofuel:

1. In general, the cultivation method for algal biomass can be adapted by the industries effectively and without much difficulty.
2. Algae have a tendency to utilize CO_2 as the primary source of carbon for their growth; it has been proposed as a potential tool to reduce greenhouse gases and preserve the environment.
3. Marine algae exhibit a special feature that allows them to adapt to the stress factors of the marine environment.
4. Algae can produce substantial amounts of raw material in a shorter period of time and at a lower cost compared to the raw material obtained from terrestrial plants.

The demand for biodiesel has increased since its wider application. Biodiesel has the advantage of blending easily with fossil fuels. Biofuel blends with at least 5% biodiesel (EN590) were considered acceptable. Biodiesel can also be used in higher blends or at its cleanest level (B100) in engines. Biofuels derived from algae are proposed as unique among the diverse types of biofuels that have recently been developed (Ching et al. 2021). The researchers often recommend that to maximize economies of scale, algal-based biorefineries can generate other value-added products and synergistic bioenergies simultaneously (Solis et al. 2020). Biofuel production has focused mainly on finding a way to use biomass effectively and efficiently so that it can be utilized for commercial production. The pretreatments suggested will differ depending on the substrate for biofuel production. For commercialization of algal biofuels, new innovative technologies have been proposed by researchers to minimize technical difficulties.

Algal biodiesel possesses a variety of different lipid combinations than conventional nonrenewable diesel. The feature adds value to the product to be commercialized.

In order to prove the concept, the algal feedstock that is likely to be used in existing biodiesel refineries is one of the saline features of the algal biodiesel to be commercialized. The distinctive properties of algal biodiesel are reported as energy density, storage stability, and cold temperature performance, which make it a great candidate for the aviation industry (Brennan and Owende 2010).

Biomass from algae is used for a wide variety of fuel applications, such as biodiesel, bioethanol, biohydrogen, biomethane, and biobutanol. As a result, researchers have begun to explore fourth-generation biofuels. The production of biofuels from genetically enhanced algae leads to the other three generations of biofuels. The advanced technologies have been used in the design and implementation of biodiesel to make it an economic fuel (Botero et al. 2017).

Algae have huge, diverse species with different ranges of carbohydrates that made the biomass a suitable feedstock for biofuel production including, macroalgae such as *Cladophora fascicularis*, *Dictyosphaeria cavernosa*, *Caulerpa cupressoides*, *Caulerpa peltata*, *Caulerpa laetevirens*, *Caulerpa sertularioides*, *Codium adhaerens*, and *Codium decorticatum* and microalgae such as *Porphyridium cruentum*, *Spirogyra* sp., and *Scenedesmus dimorphus* (Suganya et al. 2016).

Initially, micro- and macroalgae or seaweed resources can be introduced into established anaerobic digestion plants in relatively small quantities of algal biomass collected from natural sources according to the season. For conversion of macroalgae to energy, this may be the most direct route to commercialization, but even though there is a reduction of raw material costs due to transportation or cultivation methods, energy production technologies still need to be addressed (Lam et al. 2019).

In order to improve the economics of the process and maximize biofuel production, an integrated approach needs to be explored with one suitable process for cultivation, harvesting, the process of biomass, pretreatment, lipid extraction, downstream process, etc.

9.13 Conclusion and Future Perspective

This chapter has elaborated on the positive influence and future perspective of marine algae on biofuel production and the state of the art of this title in 14 major angles as the positive influence of marine alga's biomass, sustainable cultivation methods of both macro- and microalgae, the influence of the composition of algal biomass with its complexity, and dual performance of algae as a resource and preserving the environment through CO_2 sequestration. In addition, the primary and secondary biofuel production from algal biomass with various biofuels such as biodiesel, bioethanol, biobutanol, biohydrogen, and biogas were also specified. The commercialization of algal biofuels is also discussed in terms of advantages and disadvantages.

Algae, exhibit rapid growth and demand for minimal land for cultivation when it comes to significant resources. Furthermore, wastewater can also be used as a growth medium, and CO_2, a greenhouse gas, can be used as a sole carbon source because of its adaptability to multi stress conditions. It is possible to avoid contamination by using marine or seawater culture media, and industrial wastewater also contains the potential to produce biomass. This leads to a reduction in the process of wastewater treatment required.

Developing a sustainable cultivation method for algae is proven to be a challenge for industries that want to produce a lot of biomass for biofuel production. The contents of this chapter combine the merits and demerits of possible cultivation methods to evaluate and select those suitable for commercial cultivation.

Algae are primarily composed of carbohydrates; therefore, the carbohydrate content of species varies significantly. A pictorial representation of carbohydrate analysis would help to select species for biomass production and biofuels.

This is an astonishing opportunity for mankind to develop hydrogen, biodiesel, butanol, ethanol, and jet fuel from algal biomass. Despite this, significant economic and technological improvements are needed in order to make microalgal biofuels viable on a large scale. To meet industrial needs, scientists need to move beyond laboratory scales to large-scale trials. As of now, current technology estimates that the cost of a barrel of algae-based fuel is five- to sevenfold more compared with petroleum-based fuel per barrel.

There are many research articles available on technical point of view about the biofuel production of algae since the existence of diverse species. The economic analysis of biofuel and the technology used in that study has not been projected as well. One of the challenges involved selecting the best cultivation method, extracting oil, and other biofuel production from a single source for different micro- and macroalgae species.

Developing microRNAs for microalgal biohydrogen production could greatly impact its future. In recent years, genetic modification and metabolic engineering have improved biohydrogen production efficiency, environmental friendliness, and sustainability. Future studies on hydrogenases and their ability to absorb light, mass production, and economic evaluation would eliminate the barriers to practical applications and sustainable biohydrogen production.

A newer technology adds nanoparticles, such as gold, silver, and cobalt, to microwave-treated biomass to enhance biofuel production though the cost-effectiveness and economical feasibility of the stated technology should be evaluated further for successful commercial application.

References

Abdullah B, Syed Muhammad SAF, Shokravi Z, Ismail S, Kassim KA, Mahmood AN, Aziz MMA (2019) Fourth generation biofuel: a review on risks and mitigation strategies. Renew Sust Energ Rev 107:37–50. https://doi.org/10.1016/j.rser.2019.02.018

Acien FG, Gomez-Serrano C, Morales-Amaral MM, Fernandez-Sevilla JM, Molina-Grima E (2016) Wastewater treatment using microalgae: how realistic a contribution might it be to significant urban wastewater treatment? Appl Microbiol Biotechnol 100:9013–9022. https://doi.org/10.1007/s00253-016-7835-7

Adams JMM, Toop TA, Donnison IS, Gallagher JA (2011) Seasonal variation in *Laminaria digitata* and its impact on biochemical conversion routes to biofuels. Bioresour Technol 102: 9976–9984

Adeniyi OM, Azimov U, Burluka A (2018) Algae biofuel: current status and future applications. Renew Sust Energ Rev 90:316–335. https://doi.org/10.1016/j.rser.2018.03.067

Ahmed F, Li Y, Schenk PM (2012) Cellular origin, life in extreme habitats and astrobiology. In: Gordon R, Seckbach J (eds) The science of algal fuels. Springer, Netherlands, pp 21–41. https://doi.org/10.1007/978-94-007-5110-1_2

Aitken CM, Jones DM, Maguire MJ, Gray ND, Sherry A, Bowler BFJ et al (2013) Evidence that crude oil alkane activation proceeds by different mechanisms under sulfate-reducing and methanogenic conditions. Geochim Cosmochim Acta 109:162–174. https://doi.org/10.1016/j.gca.2013.01.031

Aitken D, Bulboa C, Godoy-Faundez A, Turrion-Gomez JL, Antizar-Ladislao B (2014) Life cycle assessment of macroalgae cultivation and processing for biofuel production. J Clean Prod 75: 45–56

Alam F, Saleh M, Chowdhury H (2015) Third generation biofuel from algae. Procedia Eng 105: 763–768. https://doi.org/10.1016/j.proeng.2015.05.068

Alfonsin V, Maceiras R, Gutierrez C (2019) Bioethanol production from industrial algae waste. Waste Manag 87:791–797. https://doi.org/10.1016/j.wasman.2019.03.019

Ansaha E, Wang L, Zhang B, Shahbazi B (2018) Catalytic pyrolysis of raw and hydrothermally carbonized *Chlamydomonas debaryana* microalgae for denitrogenation and production of aromatic hydrocarbons. Fuel 228:234–242

Anwar M, Lou S, Chen L, Li H, Hu Z (2019) Recent advancement and strategy on bio-hydrogen production from photosynthetic microalgae. Bioresour Technol 292:121972

Aramkitphotha S, Tanatavikorn H, Yenyuak C, Vitidsant T (2019) Low sulfur fuel oil from blends of microalgae pyrolysis oil and used lubricating oil: properties and economic evaluation. Sustain Energy Technol Assess 31:339–346

Aslan S, Kapdan IK (2006) Batch kinetics of nitrogen and phosphorus removal from synthetic wastewater by algae. Ecol Eng 28(1):64–70

Bansod SR, Nandkar PB (2015) Physiological effects of mining contaminants on algae with special reference to heavy metal toxicity. JIIS 2(1):43–60

Banuselvi R, Sivasankari S, Vinoj G, Raju PN (2008) Bioethanol from the green alga *Valoniopsis pachynema*. In: Prasad BN, Matheuuw L (eds) Recent advances in biotechnology. Excel India Publishers, New Delhi, pp 106–112

Bartley ML, Boeing WJ, Dungan BN, Holguin FO, Schaub T (2014) pH effects on growth and lipid accumulation of the biofuel microalgae *Nannochloropsis salina* and invading organisms. J Appl Phycol 26:1431–1437

Bechara R, Azizi F, Boyadjian C (2021) Process simulation and optimization for enhanced biophotolytic hydrogen production from green algae using the sulfur deprivation method. Int J Hydrog Energy 46(27):14096–14108

Behera KB, Varma A (2016) From algae to liquid fuels. In: Microbial resources for sustainable energy. Springer, Cham. https://doi.org/10.1007/978-3-319-33778-4_3

Botero G, Restrepo S, Cardona A (2017) A comprehensive review on the implementation of the biorefinery concept in biodiesel production plants. Biofuels Res J 4(3):691–703

Brennan L, Owende P (2010) Biofuels from microalgae—a review of technologies for production, processing, and extractions of biofuels and co-products. Renew Sust Energ Rev 14:557–577. https://doi.org/10.1016/j.rser.2009.10.009
Bridgwater AV (1994) Catalysis in thermal biomass conversion. Appl Catal A Gen 116:5–47
Cascone R (2008) Biobutanol: a replacement for bioethanol? Chem Eng Prog 104(8)
Cavalier-Smith T (2017) Euglenoid pellicle morphogenesis and evolution in light of comparative ultrastructure and trypanosomatid biology: semi-conservative microtubule strip duplication, strip shaping and transformation. Eur J Protistol 61:137–179
Cavinato C, Bolzonella D, Eusebi AL, Pavan P (2009) Bio-hythane production by thermophilic two-phase anaerobic digestion of organic fraction of municipal solid waste: preliminary results. In: AIDIC conference series, pp 61–66. https://doi.org/10.3303/ACOS0909008
Ceylan S, Kazan D (2015) Pyrolysis kinetics and thermal characteristics of microalgae *Nannochloropsis oculata* and *Tetraselmis* sp. Bioresour Technol 187:1–5
Chan WS, Kwok ACM, Wong JTY (2019) Knockdown of dinoflagellate cellulose synthase CesA1 resulted in malformed intracellular cellulosic thecal plates and severely impeded cyst-to-swarmer transition. Front Microbiol 10:546. https://doi.org/10.3389/fmicb.2019.00546
Chandra ST, Mudliar SN, Vidyashankar S, Mukherji S, Sarada R, Krishnamurthi K, Chauhan VS (2015) Defatted algal biomass as a non-conventional low-cost adsorbent: surface characterization and methylene blue adsorption characteristics. Bioresour Technol 184:395–404. https://doi.org/10.1016/j.biortech.2014.10.018
Changwei H, Yang W, Li Y, Dong L, Zhu L, Tong D, Qing R, Fan Y (2010) The direct pyrolysis and catalytic pyrolysis of *Nannochloropsis* sp. Residue for renewable bio-oils. Bioresour Technol 101(12):4593–4599
Chen CL, Chang JS, Lee DJ (2015) Dewatering and drying methods for microalgae. Dry Technol 33(4):443–454
Ching PML, Mayol AP, Juan JLGS, Calapatia AM, So RHY, Sy CL, Ubando AT, Culaba AB (2021) AI methods for modeling the vacuum drying characteristics of *Chlorococcum infusionum* for algal biofuel production. Process Integr Optim Sustain 5:247–256
Chng LM, Lee KT, Chan DCJ (2017) Evaluation on microalgae biomass for bioethanol production. IOP Conf Ser Mater Sci Eng 206:012018. 29th Symposium of Malaysian Chemical Engineers (SOMChE), 1–3 December 2016, Miri, Sarawak, Malaysia
Choi JH (2014) Pyrolysis of seaweeds for bio-oil and bio-char production. Chem Eng Trans 37: 121–126
Christensen DL (2020) Seaweed cultivation in the Faroe Islands: analyzing the potential for forward and fiscal linkages. Mar Policy 119:104015
Demirbas A (2001) Biomass resource facilities and biomass conversion processing for fuels and chemicals. Energy Convers Manag 42:1357–1378
Demirbas A (2003) Biodiesel fuels from vegetable oils via catalytic and non-catalytic supercritical alcohol transesterifications and other methods: a survey. Energy Convers Manag 44(13):2093–2109
Demirbas A, Demirbas MF (2011) Importance of algae oil as a source of biodiesel. Energy Convers Manag 52:163–170
Ellis JT, Hengge NN, Sims RC, Miller CD (2012) Acetone butanol and ethanol production from wastewater algae. Bioresour Technol 111:491–495
Falkowski PG, Katz ME, Knoll AH, Quigg A, Raven JA, Schofield O et al (2004) The evolution of modern eukaryotic phytoplankton. Science 305:354–360
Farzanah RH, Brudecki GP, Cybulska I, Oyanedel JRB, Schmidt JE, Thomsen MH (2019) Screening and production of biogas from macro algae biomass of *Padina boergesenii*, *Colpomenia sinuosa*, and *Ulva* sp. In: Bastidas-Oyanedel JR, Schmidt J (eds) Biorefinery. Springer, Cham, pp 727–740. https://doi.org/10.1007/978-3-030-10961-5_33
Fermoso FG, Beltran C, Jimenez A, Fernandez MJ, Rincon B, Borja R, Jeison D (2016) Screening of biomethane production potential from dominant microalgae. J Environ Sci Health Part A 51(12):1062–1067

Fleurence J (1999) Seaweed proteins: biochemical, nutritional aspects and potential uses. Food Sci Technol 10:25–28. https://doi.org/10.1002/ppap.201100070

Fortes MD, Luning K (1980) Growth rates of North Sea macroalgae in relation to temperature, irradiance and photoperiod. Helgola Meeresun 34:15–29

Fu H, Hu J, Guo X, Feng J, Yang ST, Wang J (2021) Butanol production from Saccharina japonica hydrolysate by engineered *Clostridium tyrobutyricum*: the effects of pretreatment method and heat shock protein overexpression. Bioresour Technol 335:125290

Ganesan N, Chitirai A, Felix LO, Nooruddin T (2020) Enhancing starch accumulation/production in *Chlorococcum humicola* through sulphur limitation and 2,4-D treatment for butanol production. Biotechnol Rep 28:e00528

Ganesana R, Manigandanb S, Samuelc MS, Shanmuganathan R, Brindhadevie K, Chie NTL, Ducf PA, Pugazhendhie A (2020) A review on prospective production of biofuel from microalgae. Biotech Rep 27:e00509

Gao L, Sun J, Wei X, Xiao G (2017) Catalytic pyrolysis of natural algae over Mg-Al layered double oxides/ZSM-5 (MgAl-LDO/ZSM-5) for producing bio-oil with low nitrogen content. Bioresour Technol 225:293–298

Gaurav N, Sivasankari S, Kira G, Ninawe A, Selvin J (2017) Utilization of bioresources for sustainable biofuels: a review. Renew Sust Energ Rev 73:205–214. https://doi.org/10.1016/j.rser.2017.01.070

Gautam R, Vinu R (2018) Non-catalytic fast pyrolysis and catalytic fast pyrolysis of *Nannochloropsis oculata* using Co-Mo/g-Al_2O_3 catalyst for valuable chemicals. Algal Res 34:12–24

Gaviria AL, Maldonado JD, Villacis RC, Maciel EO, Bautista RML, Escamilla GC, Vazquez AC, Zepeda CH, Pool FAB, Tussell RT (2021) Presence of polyphenols complex aromatic "lignin" in *Sargassum* spp. from Mexican Caribbean. J Mar Sci Eng 9:6. https://doi.org/10.3390/jmse9010006

Giang TT, Lunprom S, Liao Q, Reungsang A, Salakkam A (2019) Enhancing hydrogen production from *Chlorella* sp. biomass by pre-hydrolysis with simultaneous saccharification and fermentation (PSSF). Energies 12(5):90

Gollakota ARK, Kishore N, Gu S (2018) A review on hydrothermal liquefaction of biomass. Renew Sust Energ Rev 81(1):1378–1392. https://doi.org/10.1016/j.rser.2017.05.178

Gong X, Zhang B, Zhang Y, Huang Y, Xu M (2014) Investigation on pyrolysis of low lipid microalgae *Chlorella vulgaris* and *Dunaliella salina*. Energy Fuel 28(1):95–103

Gonzalez LE, Canizares RO, Baena S (1997) Efficiency of ammonia and phosphorus removal from a colombian agroindustrial wastewater by the microalgae *Chlorella vulgaris* and *Scenedesmus dimorphus*. Bioresour Technol 60(3):259–262

Griffiths MJ, Harrison STL (2009) Lipid productivity as a key characteristic for choosing algal species for biodiesel production. J Appl Phycol 21:493–507. https://doi.org/10.1007/s10811-008-9392-7

Guedes AC, Amaro HM, Malcata FX (2011) Microalgae as sources of carotenoids. Mar Drugs 9: 625–644. https://doi.org/10.3390/md9040625

Herrera A, Imporzano GD, Fernandez FGA, Adani F (2021) Sustainable production of microalgae in raceways: nutrients and water management as key factors influencing environmental impacts. J Clean Prod 287:125005. https://doi.org/10.1016/j.jclepro.2020.125005

Hirayama S, Nakayama H, Sugata K, Ueda R (1996) Process for the production of ethanol from microalgae. Google Patents

Horn SJ, Aasen IM, Ostgaard K (2000) Ethanol production from seaweed extract. J Ind Microbiol Biotechnol 25:249–254

Hu Q (2013) Environmental effects on cell composition. In: Richmond A, Hu Q (eds) Handbook of microalgal culture: J Appl Phycol Biotech, 2nd edn, pp 114–122

Huntley ME, Redalje DG (2007) CO_2 mitigation and renewable oil from photosynthetic microbes: a new appraisal. Mitig Adapt Strat Glob Change 12:573–608

Jafarian S, Tavasoli A (2018) A comparative study on the quality of bioproducts derived from catalytic pyrolysis of green microalgae *Spirulina (Arthrospira) platensis* over transition metals supported on HMS-ZSM5 composite. Int J Hydrog Energy 43(43):19902–19917

Ji SQ, Wang B, Lu M, Li FL (2016) Direct bioconversion of brown algae into ethanol by thermophilic bacterium *Defluviitalea phaphyphila*. Biotechnol Biofuels 9(1):81. https://doi.org/10.1186/s13068-016-0494-1

Karmakar R, Kundu K, Rajor A (2018) Fuel properties and emission characteristics of biodiesel produced from unused algae grown in India. Pet Sci 15:85–395. https://doi.org/10.1007/s12182-017-0209-7

Kim YS (2009) Pretreatment of *Gelidium amansii* for the production of bioethanol. In: The 31st symposium on biotechnology for fuels and chemicals

Kim SK (2015) Handbook of marine biotechnology. Springer, Berlin. ISBN 9783642539701

Kumar AN, Min B, Mohan SV (2018) Defatted algal biomass as feedstock for short chain carboxylic acids and biohydrogen production in the biorefinery format. Bioresour Technol 269:408–416

Kumar MV, Babu AV, Kumar PR (2019) Influence of metal-based cerium oxide nanoparticle additive on performance, combustion, and emissions with biodiesel in diesel engine. Environ Sci Pollut Res 26:7651–7664

Kumar AN, Yoon JJ, Kumar G, Hyoun KS (2021) Biotechnological valorization of algal biomass: an overview. SMAB 1:31–141. https://doi.org/10.1007/s43393-020-00012-w

Lakaniemi AM, Tuovinen OH, Puhakka JA (2012) Production of electricity and butanol from microalgal biomass in microbial fuel cells. Bioenergy Res 5:481–491. https://doi.org/10.1007/s12155-012-9186-2

Laliberte G, Lessard P, Noue J, Sylvestre S (1997) Effect of phosphorus addition on nutrient removal from wastewater with the cyanobacterium *Phormidium bohneri*. Bioresour Technol 59: 227–233

Lam AK, Khoo CG, Lee KT (2019) Scale-up and commercialization of algal cultivation and biofuels production. In: Pandey A, Chang JS, Soccol CR, Lee DJ, Chisti Y (eds) Biomass, biofuels, biochemicals, biofuels from algae, 2nd edn. Elsevier, pp 475–506

Ledda C, Ida A, Allemand D, Mariani P, Adani F (2015) Production of wild *Chlorella* sp. cultivated in digested and membrane-pretreated swine manure derived from a full-scale operation plant. Algal Res 12:68–73

Lee RE (2008) Phycology, vol 91, 4th edn. Cambridge University Press, Cambridge. ISBN 9780521864084

Lee K, Lee CG (2001) Effect of light/dark cycles on wastewater treatments bymicroalgae. Biotechnol Bioprocess Eng 6:194–199

Lee S, Oh Y, Kim D, Kwon D, Lee C, Lee J (2011) Converting carbohydrates extracted from marine algae into ethanol using various ethanolic *Escherichia coli* strains. Appl Biochem Biotechnol 164(6):878–888. https://doi.org/10.1007/s12010-011-9181-7

Leite LDS, Teresa M, Daniel LA (2019) Microalgae cultivation for municipal and piggery wastewater treatment in Brazil. J Water Process Eng 31:1–7. https://doi.org/10.1016/j.jwpe.2019.100821

Ly HV, Choi JH, Woo HC, Kim SS, Kim J (2019) Upgrading bio-oil by catalytic fast pyrolysis of acid-washed *Saccharina japonica* alga in a fluidized-bed reactor. Renew Energy 133:11–22

Lyons H, Lerat Y, Stanley M, Rasmussen MB (2009) A review of the potential of marine algae as a source of biofuel in Ireland, pp 1–88

Ma C, Geng J, Zhang D, Ning X (2020) Non-catalytic and catalytic pyrolysis of *Ulva prolifera* macroalgae for production of quality bio-oil. J Energy Inst 93(1):303–311

Maiti S, Maiti DC, Verma M, Brar SK (2016a) Biobutanol—"a renewable green alternative of liquid fuel" from algae. In: Soccol CR, Brar SK, Faulds CR, Luiz P (eds) Green energy and technology biobutanol—"a renewable green alternative of liquid fuel" from algae, pp 445–465. https://doi.org/10.1007/978-3-319-30205-8

Maiti S, Maiti DC, Verma M, Brar SK (2016b) Biobutanol—"a renewable green alternative of liquid fuel" from algae. In: Soccol C, Brar S, Faulds C, Ramos L (eds) Green fuels technology. Green energy and technology. Springer, Cham. https://doi.org/10.1007/978-3-319-30205-8_18

Makareviciene V, Sendzikiene E (2020) Application of microalgae for the production of biodiesel fuel. In: Konur O (ed) Handbook of algal science, Technology and Medicine. Academic Press, pp 353–365

Malik VS (2014) Editorial biofuel: the butanol perspective and algal biofuel. J Plant Biochem Biotechnol 23(4):337–338. https://doi.org/10.1007/s13562-014-0283-5

Martınez ME, Sanchez S, Jimenez JM, El Yousfi F, Munoz L (2000) Nitrogen and phosphorus removal from urban wastewater by the microalga *Scenedesmus obliquus*. Bioresour Technol 73(3):263–272

Mata TM, Martins AA, Caetano NS (2010) Microalgae for biodiesel production and other applications: a review. Renew Sust Energ Rev 14:217–232

Mehmood U, Faisal M (2020) Sustainable approaches toward the production of bioethanol from biomass. In: Arshad M (ed) Sustainable ethanol and climate change, pp 15–38. https://doi.org/10.1007/978-3-030-59280-6_2.

Montingelli ME, Tedesco S, Olabi AG (2015) Biogas production from algal biomass: a review. Renew Sust Energ Rev 43:961–972

Mostafa SSM (2012) Microalgal biotechnology: prospects and applications. In: Dhal NK, Sahu SC (eds) Plant science. IntechOpen. https://doi.org/10.5772/53694. https://www.intechopen.com/chapters/41642

Narala RR, Garg S, Sharma KK, Thomas-Hall SR, Deme M, Li Y, Peer MS (2016) Comparison of microalgae cultivation in Photobioreactor, open raceway pond, and a two-stage hybrid system. Front Energy Res 4(29)

Narchonai G, Arutselvan C, Lewis Oscar F, Thajuddin N (2020) Enhancing starch accumulation/production in *Chlorococcum humicola* through sulphur limitation and 2,4- D treatment for butanol production. Biotechnol Rep 28:e00528

Nilsen BJ (2006) Production of micro-algae based products. Nordic Innovation Centre, Oslo

Nobe R, Sakakibara Y, Fukuda N, Yoshida N, Ogawa K, Suiko M (2003) Purification and characterization of laminaran hydrolases from *Trichoderma viride*. Biosci Biotechnol Biochem 67:1349–1357

Norton TA, Melkonian N, Andersen R (1996) Algal biodiversity. Phycology 35:308–326

Obeid F, Van TC, Brown R, Rainey T (2019) Nitrogen and sulphur in algal biocrude: a review of the HTL process, upgrading, engine performance and emissions. Energy Convers Manag 181:105–119

Okpozua OO, Ogbonnaa IO, Ikwebea J, Ogbonna JC (2019) Phycoremediation of cassava wastewater by *Desmodesmus armatus* and the concomitant accumulation of lipids for biodiesel production. Bioresour Technol Rep 7:100255

Olguın EJ, Galicia S, Mercado G, Perez T (2003) Annual productivity of *Spirulina* (Arthrospira) and nutrient removal in a pig wastewater recycle process under tropical conditions. J Appl Phycol 15:249–257

Onay M (2020) Enhancing carbohydrate productivity from *Nannochloropsis gaditana* for bio-butanol production. Energy Rep 6(1):63–67

Packer K (2009) Algal capture of carbon dioxide; biomass generation as a tool for greenhouse gas mitigation with reference to New Zealand energy strategy and policy. Energy Policy 37(9):3428–3437

Pienkos PT, Darzins A (2009) The promise and challenges of microalgal-derived biofuels. Biofuels Bioprod Biorefin 3(4):431–440

Potts T, Du J, Paul M, May P, Beitle R, Hestekin J (2012) The production of butanol from Jamaica bay macro algae. AIChE Fall 2010 Annual Meeting 31(1):29–36

Qari H, Rehan M, Nizami AS (2017) Key issues in microalgae biofuels: a short review. Energy Procedia 142:898–903

Rahman NAA, Fermoso J, Sanna A (2020) Stability of Li-LSX zeolite in the catalytic pyrolysis of non-treated and acid pre-treated *Isochrysis* sp. Microalgae Energies 13(4):959

Raja R, Shanmugam H, Ganesan V, Carvalho IS (2014) Biomass from microalgae: an overview. J Oceanogr Mar Sci 2(1):1–7

Razon LF, Tan RR (2011) Net energy analysis of the production of biodiesel and biogas from the microalgae: *Haematococcus pluvialis* and *Nannochloropsis*. Appl Energy 88(10):3507–3514

Reich B (2011) Bioenergy potential of *Ulva lactuca*: biomass yield, methane production and combustion. Bioresour Technol 102:2595–2604. https://doi.org/10.1016/j.biortech.2010.10.010

Rodolfi L, Zitelli GC, Bassi N, Padovani G, Biondi N, Bonini G, Tredici MR (2009) Microalgae for oil: strain selection, induction of lipid synthesis and outdoor mass cultivation in a low-cost photobioreactor. Biotechnol Bioeng 102:100–112

Rosenberg JN, Oyler GA, Wilkinson L, Betenbaugh MJ (2008) A green light for engineered algae: redirecting metabolism to fuel a biotechnology revolution. Curr Opin Biotechnol 19:430–436. https://doi.org/10.1016/j.copbio.2008.07.008

Rudolf M, Tetik N, Ramos-Leon F et al (2015) The peptidoglycan-binding protein SjcF1 influences septal junction function and channel formation in the filamentous cyanobacterium *Anabaena*. mBio 6:4. https://doi.org/10.1128/mBio.00376-15

Salaeh S, Kongjan P, Panphon S, Hemmanee S, Alissara R, Rattana J (2019) Feasibility of ABE fermentation from *Rhizoclonium* spp. hydrolysate with low nutrient supplementation. Biomass Bioenergy 127:105269

Schroeder G, Michalak I (2018) Algae biomass: characteristics and applications, economic aspects of algae biomass harvesting for industrial purposes. In: Chojnacka K, Wieczorek PP (eds) The life-cycle assessment of the product, pp 131–143. https://doi.org/10.1007/978-3-319-74703-3

Schutz K, Happe T, Troshina O, Lindblad P, Leitao E, Oliveira P, Tamagnini P (2004) Cyanobacterial hydrogen production—a comparative analysis. Planta 218:350–359

Sengupta D, Pike RW (2016) Biomass as feedstock. In: Chen WY et al (eds) Handbook of climate change mitigation and adaptation. Springer, Cham, pp 1723–1773. https://doi.org/10.1007/978-3-319-14409-2_25

Shaikh MS, Shaikh PH, Qureshi K, Bhatti I (2018) Green house effect and carbon foot print. In: Reference module in materials science and materials engineering. https://doi.org/10.1016/b978-0-12-803581-8.10456-4

Shirai F, Kunii K, Sato C et al (1998) Cultivation of microalgae in the solution from the desalting process of soy sauce waste treatment and utilization of the algal biomass for ethanol fermentation. World J Microbiol Biotechnol 14:839–842

Show K-Y, Yan Y-G, Lee D-J (2019) Biohydrogen production from algae: perspectives, challenges, and prospects. In: Show K-Y, Yan Y-G, Lee D-J (eds) Biofuels from algae. Biomass, biofuels, biochemicals, 2nd edn, pp 325–343

Sivagurunathan P, Kumar G, Kobayashi T, Xu K, Kim SH (2017) Effects of various dilute acid pretreatments on the biochemical hydrogen production potential of marine macroalgal biomass. Int J Hydrogen Energy 42(45):27600–27606

Sivasankari S (2010) Ethanol production from biomass: isolation, characterization and evaluation of cellulolytic and ethanologenic microbes on starch and lignocellulosic substrates, Ph.D Thesis. University of Madras, Tamil Nadu, India

Slocombe SP, Zhang Q, Black KD et al (2013) Comparison of screening methods for high-throughput determination of oil yields in micro-algal biofuel strains. J Appl Phycol 25:961–972

Solis CA, Mayol AP, San Juan JG, Ubando AT, Culaba AB (2020) Multi-objective optimal synthesis of algal biorefineries toward a sustainable circular bioeconomy. IOP Conf Ser Earth Environ Sci 463:012051. https://doi.org/10.1088/1755-1315/463/1/012051

Sotoudehniakarani F, Alayat A, McDonald AG (2019) Characterization and comparison of pyrolysis products from fast pyrolysis of commercial *Chlorella vulgaris* and cultivated microalgae. J Anal Appl Pyrolysis 139:258–273

Spruit CJP (1958) Simultaneous photoproduction of hydrogen and oxygen by *Chlorella*. Mededel Land bouwhoge school (Wageningen/Nederland) 58:1–17

Suganya T, Varman M, Masjuki HH, Renganathan S (2016) Macroalgae and microalgae as a potential source for commercial applications along with biofuels production: a biorefinery approach. Renew Sust Energ Rev 55:909–941

Sulfahri MS, Husain DR, Langford A, Tassakka ACMAR (2020) Fungal pretreatment as a sustainable and low cost option for bioethanol production from marine algae. J Clean Prod 265:121763. https://doi.org/10.1016/j.jclepro.2020.121763

Sydney EB, Sydney ACN, Carvalho JC, Soccol CR (2019) Microalgal strain selection for biofuel production. In: Biofuels from algae, pp 51–66. https://doi.org/10.1016/B978-0-444-64192-2.00003-2

Voith M (2009) Dow plans algae biofuels pilot. Chem Eng News 87(27):10

Waite TD, Spielman LA, Mitchell R (1972) Growth rate determination of the macrophyte Ulva in continuous culture. Environ Sci Technol 6:1096–1100

Wang X, Nordlander E, Thorin E, Yan J (2013) Microalgal biomethane production integrated with an existing biogas plant: a case study in Sweden. Appl Energy 112:478–484. https://doi.org/10.1016/j.apenergy.2013.04.087

Wang W, Xu Y, Xiaoxiao X, Zhang B, Tian W, Zhang J (2018) Hydrothermal liquefaction of microalgae over transition metal supported TiO_2 catalyst. Bioresour Technol 250:474–480

Wargacki AJ, Leonard E, Win MN, Regitsky DD, Santos CNS, Kim PB, Cooper SR, Raisner RM, Herman A, Sivitz AB, Lakshmanaswamy A, Kashiyama Y, Baker D, Yoshikuni Y (2012) An engineered microbial platform for direct biofuel production from brown macroalgae. Science 335:308–313

Webb J, Webb J, Sørensen P, Velthof G, Amon B, Pinto M, Rodhe L, Salomon E, Hutchings N, Burczyk P, Reid J (2013) An assessment of the variation of manure nitrogen efficiency throughout Europe and an appraisal of means to increase manure-N efficiency. Adv Agron 119:371–442. https://doi.org/10.1016/B978-0-12-407247-3.00007-X

Wi SG, Kim HJ, Mahadevan SA, Yang D-J, Bae H-J (2009) The potential value of the seaweed Ceylon moss (*Gelidium amansii*) as an alternative bioenergy resource. Bioresour Technol 100: 6658–6660

Wigmosta MS, Coleman AM, Skaggs RJ, Huesemann MH, Lane LJ (2011) National microalgae biofuel production potential and resource demand. Water Resour Res 47(3):1–13

Wooldridge SA (2013) Breakdown of the coral-algae symbiosis: towards formalizing a linkage between warm-water, bleaching thresholds and the growth rate of the intracellular zooxanthellae. Biogeosciences 10:1647–1658

Xiao C, Liao Q, Fu Q, Huang Y, Chen H, Zhang H, Xia A, Zhu X, Reungsang A, Liu Z (2019) A solar-driven continuous hydrothermal pretreatment system for biomethane production from microalgae biomass. Appl Energy 236:1011–1018. https://doi.org/10.1016/j.apenergy.2018.12.014

Yahmed B, Jmel N, Ben Alaya MA, Bouallagui MH, Marzouki MN, Smaali I (2016) A biorefinery concept using the green macroalgae *Chaetomorpha linum* for the coproduction of bioethanol and biogas. Energy Convers Manag 119:257–265

Yoon JJ, Kim YJ, Kim SH, Ryu HJ, Choi JY, Kim GS et al (2010) Production of polysaccharides and corresponding sugars from red seaweed. Adv Mater Res 93:463–466

Zaidi AA, Zhe FR, Malik A, Khan SZ, Bhutta AJ, Shi Y, Mushtaq K (2019) Conjoint effect of microwave irradiation and metal nanoparticles on biogas augmentation from anaerobic digestion of green algae. Int J Hydrogen Energy 44(29):14661–14670

Zainan NH, Srivatsa SC, Li F, Bhattacharya S (2018) Quality of bio-oil from catalytic pyrolysis of microalgae *Chlorella vulgaris*. Fuel 223:12–19

Zou S, Wu Y, Yang M, Li C, Tong J (2010) Bio-oil production from sub- and supercritical water liquefaction of microalgae *Dunaliella tertiolecta* and related properties. Energy Environ Sci 3: 1073–1078

Chapter 10
Algae–Bacterial Mixed Culture for Waste to Wealth Conversation: A Case Study

Somok Banerjee, Swatilekha Pati, and Shaon Ray Chaudhuri

Abstract The need for alternative sources of renewable fuel was felt in 1970–1980 with research being directed toward exploring algae as the starter material for biofuel production. Algal varieties are enormous and its potential for biofuel production is immense and much higher than any land plant-based product. However, there are limitations in terms of making them commercially viable. Extensive research has been conducted to address each step to make the application economically viable, such as the energy requirement mainly in the harvesting step, water requirement for cultivation during the time of fresh water scarcity, and algal growth medium requirement for cultivation. This chapter reports a case study of selective treatment of ammonia-rich dairy wastewater using a consortium of bacteria and microalgae, revealing the potential of the technology in dairy wastewater treatment with lipid-rich algal biomass production for biofuel extraction. Such approach cuts down on the use of fresh water and algal growth medium for algae cultivation and saves energy by alleviating the need for harvesting the biofilm-based algal biomass. The reclaimed water could be reused for secondary (non-potable) applications. In this way the waste could be substituted for feed and water for algal growth, while biofilm-based growth ensured energy savings and a rapid treatment within 48 h ensured more efficient treatment and biomass production compared to conventionally reported algal consortium.

Keywords Algae · Bacteria · Biofuel · Dairy wastewater · Metagenomics · Ammonia · Phosphate · Nitrate

S. Banerjee · S. Pati · S. Ray Chaudhuri (✉)
Department of Microbiology, Tripura University, Suryamaninagar, Tripura, India
e-mail: shaonraychaudhuri@tripurauniv.in

N. Srivastava, P. K. Mishra (eds.), *Technological Advancement in Algal Biofuels Production*, Clean Energy Production Technologies,
https://doi.org/10.1007/978-981-19-6806-8_10

10.1 Introduction

The primary fuel for transportation has been petroleum for ages. Its reserves are getting depleted with time in most of the oil-producing regions of the globe despite ever increasing demand (National Research Council 2012). This prompted the search for alternative renewable fuel for transportation and initiated research programs to gain better insight into algal varieties, biofuel production, and understanding the mechanism involved in stimulating oil production (Sheehan et al. 1998). The eukaryotic photosynthetic algae have been receiving increasing attention from the scientific community since 1978 (National Research Council 2012) due to its inherent potential of carbon dioxide sequestration along with nutrient (nitrate and phosphate) removal (Gonçalves et al. 2017).

Algae are autotrophs that use basic inorganic molecules in the presence of sunlight for metabolism through heterotrophic fermentation (Li et al. 2007). Their existence can be traced as early as three billion years ago (Menetrez 2012) and resides in each corner of the biosphere, contributing to oxygen generation. The relevance of exploring its potential was strengthened with the search of sources of renewable energy to sustain the ever increasing demand of the growing economy. While growing under adverse conditions, it can produce lipid, protein, and carbohydrates, among other by products (Hu et al. 2008). The prediction of the quantity of oil that can be generated from the photosynthetic algae per unit land is about 10^2 folds higher than the regular feedstock (soyabeans). Though such efficiency (as predicted) is yet to be achieved, the production from algal origin is much higher than from other land-based crops (Menetrez 2012; Hu et al. 2008). The varieties of algae known with potential for oil production are numerous with shorter doubling time than land plants. They can grow under very diverse conditions and have unlimited applications starting from cleaning of wastewater, production of biofuel from the accumulated lipid, and production of ethanol from the synthesized carbohydrate and protein-rich feed (both for human and animal consumption) (Adams et al. 2009). The by-product synthesis can be regulated through manipulation of the consortium composition, growth condition, duration of growth, and nutrient content in the growth medium (Adams et al. 2009; Becker 2007; Gouveia et al. 2009). However, when growth is unchecked, this leads to pollution of water bodies and health hazard for humans.

Algal growth with biomass enhancement is associated with chemical oxygen demand (COD), biological oxygen demand (BOD), total organic carbon (TOC), total nitrogen (ammonia, nitrate, nitrite), total phosphate (including ortho-phosphate) removal from the water (Bhandari et al. 2020; Mahapatra et al. 2014) during wastewater treatment. They are efficient in removal of toxins from the water (Suresh Kumar et al. 2015). Their nutrient requirement in minimal and can be adjusted as per the available situation. The removal is often better from consortium than pure culture (Zhu et al. 2019). The trick of developing an efficient system is selecting the right candidates and combining them in optimum proportion to produce the desired response. Microbial consortium is popularly used for eco-friendly bioremediation. The biomass for treatment is selected based on the wastewater composition and the pollutants to be removed. In natural habitat, perfect symbiotic association is observed among microalgae and bacteria. This symbiosis results in enhanced efficiency of nutrient removal (both organic and inorganic) with survivability under adverse conditions (Mamta et al. 2020). The growth of algae–bacterial consortia is interdependent. While algae use the carbon dioxide released and the plant growth hormones produced by the bacteria for its biomass enhancement, the bacteria use the oxygen released by the algae. The process continues in the presence of nutrients in an energy-efficient (passive mode) manner. This interaction has been efficiently used for wastewater treatment (Qi et al. 2019; Zhu et al. 2019; Su et al. 2011) with associated lipid-rich biomass development over extended period of time. However, a major limitation of such system is the prolonged growth period with nutrient removal (3 days and above) (Mujtaba and Lee 2017; Jagadevan et al. 2012). There is evidence of antagonism among the members of such consortium (Gonçalves et al. 2017), and hence the need for appropriate selection of the members and the cultivation condition. Some of the major limitations of large-scale algal cultivation with high lipid content are large land for cultivation due to prolonged doubling time; large volume of water for cultivation; energy for light, aeration, and harvesting of the biomass after growth as well as expensive medium for algal growth. Scientific intervention is currently focused on solving these issues along with making algal biofuel production sustainable at commercial scale.

Rapid progress in the industrial sector is deteriorating the water bodies day by day. Dairy industry is considered as the highest source of wastewater generator from the food sector because of the large volume of effluent with high organic content (Sandaruwani et al. 2018). Fresh water is one of the prime components required in the dairy industries, and the wastewater generated is substantially high whose safe disposal remains a challenge. The dairy industry itself produces about 3–11 million m^3 of wastewater annually, i.e., 1–3 times the volume of the milk that gets processed (Monroy et al. 1995). It mainly comprises huge quantities of milk constituents such as casein, lactose, inorganic salts, fats, dissolved sugars, traces of additives, and additionally sanitizers and detergents that are basically used in the cleaning process (Kolhe et al. 2009; Kumar and Desai 2011). The effluent that gets discharged is rich in nutrients (nitrogen, phosphorus) along with organic and inorganic contents and contains high biological oxygen demand (BOD) and chemical oxygen demand (COD). As a result, this nutrient-rich wastewater, if not treated properly, can cause eutrophication of the receiving water bodies and can harm the aquatic environment, pollute the ground water table, and thereby affect the ecosystem as a whole (Shinkar 2013; Gulhane and Shome 2019).

Thus, in order to cope up, this problem is drawing more and more attention from environmentalists and researchers lately. The conventional methods used are not that effective in the degradation of wastes despite undergoing a lengthy process (105–120 h), being costlier and consuming high energy along with bulk amount of sludge production post treatment. Moreover, the activated sludge produced is generally dumped inside the wastewater treatment facility itself or is landfilled, which is not feasible since more new land filling sites are required and it also increases the contamination risk of the environment with pollutants and pathogens (Tricolici et al. 2014). Lately, a study by Halder et al. (2020) showed an effective biofilm-based method of treating the dairy wastewater where it could be converted to ammonia-rich liquid biofertilizer within 16 h without generating sludge. The developed process could be scaled up to industrial level with sustained efficiency and the developed by-product was found to enhance the yield of a wide variety of economic crops, ensuring little use of freshwater and chemical fertilizer for agriculture (Gogoi et al. 2021a, b, c). However, this method has its limitations too, especially in urban areas with limited land for using the large volume of liquid biofertilizer around the effluent treatment plant. The ammonia-rich liquid fertilizer cannot be released into the aquatic environment as it would affect the same. In such situations, emphasis has to be given on developing alternative microbial processes for bioremediation of dairy wastewater (Biswas et al. 2019). The single unit operation reported in the previous study could bring the dairy wastewater close to dischargeable standards as per Central Pollution Control Board. However, such processes often need to be coupled with follow-up treatment to achieve dischargeable quality water (Biswas et al. 2021).

Phyco-remediation is one such effective approach which can be used to remove the residual pollutants from the partially treated wastewater. It is the process in which microalgae is used for biotransformation or bioremediation of environmental pollutants that includes nutrients and other xenobiotic compounds from wastewater and also sequesters CO_2 from the wastewater and air (Olguin et al. 2004; Mulbry et al. 2008). According to Asmare et al. (2013), microalgae increase its biomass by using waste as a nutrient source and hence can be used to treat wastewater, which can be directly discharged into the ecosystem.

Treatment of wastewater by phyco-remediation can be obtained by the addition of alga to the wastewater either in the form of monoculture, consortia, or combined systems. Monoculture or consortia system refers to the use of only a single or mixed algal species, respectively, which are used for the removal of nutrients and organic materials from the wastewater unlike that of the later one which requires both algae and bacteria for the treatment (Jia and Yuan 2017; Fito and Alemu 2019). The combined systems efficiently remove nitrogen, phosphorus, heavy metals, inorganic and organic carbon, and other pollutants (Mhedhbi et al. 2020).

In fact, the combined systems, comprising bacteria and microalgae in wastewater treatment, are more likely to achieve even higher efficacy of removing organic pollutants and nutrients than that of bacterial or microalgal stand-alone systems because of their synergistic cooperation with one another in the combined systems (Renuka et al. 2013; Foladori et al. 2018). In this symbiotic interaction, the microalgal photosynthesis produces oxygen, which is utilized by the aerobic bacteria for the mineralization of organic matter and this in turn provides the inorganic carbon required by the microalgae (Kwon et al. 2019; Gonzalez-Fernandez et al. 2015). Some other mutual interactions involve the production of various microalgal organic compounds composed of nucleic acids, proteins, and lipids, which are then used by bacteria as a substrate for their growth, while the bacterial growth results in the reduction of the oxygen concentration that leads to the enhancement of the microalgal metabolism (Gonzalez 2000; Liang et al. 2013). Additionally, in these systems, the bacterial degradation of the nitrogen compounds is more efficient because of the successive assimilation of ammonia by the microalgae present (Pires et al. 2013).

Thus, Biswas et al. (2021) conducted a study to investigate the phycoremediation of ammonia-rich wastewater obtained from three different dairy effluents by a microalgae–bacteria consortium enriched from Bheri water (wastewater fed aquaculture pond) at East Kolkata Wetland. The sources of dairy effluent include dairy wastewater from three different sources with different ammonia concentration (Biswas et al. 2019; Halder et al. 2020). Together, the microalgae–bacteria consortium was capable of reducing COD and ammonia by 93% and 87.2%, respectively, with the complete removal of nitrate and phosphorus from the wastewater samples as well. The study reported an enhanced algal biomass (67%), carbohydrate (55%), and lipid productivity (42%) while treating the dairy wastewater. Generally, the total carbohydrate content in microalgae is about 20% dry weight and this content may fluctuate depending on different cultivation conditions and time (Zhao et al. 2013; Laurens et al. 2012). Several other studies have also reported microalgal biomass and lipid enhancement when grown in different selected or synthetic media using either mixed or pure culture. However, for dairy wastewater, the study reported by Biswas et al. (2021) showed the highest biomass and lipid production till date with substantial increase in carbohydrate content. The present study primarily focuses on finding out the microbial community, their putative sources, and their functions to understand the exact mechanism of enhancement of algal biomass and lipid.

Some of the major limitations of the commercially available wastewater treatment process is the long retention time, hence requiring more area for Effluent Treatment Plant (ETP) setup with energy intense operation. Algae–bacterial consortium eliminates the need for mechanical aeration and hence could cut down on the energy requirement (Yang et al. 2020a, b). Numerous studies have simultaneously used microalgae–bacterial consortium for microalgal biomass production for different by-product generation with wastewater treatment (Yong et al. 2021; Khoo et al. 2021; Pérez-Nava et al. 2021; Cantera et al. 2021). The nutrient in wastewater and the water itself makes algal biomass production economical and self-sustainable while providing an option of reclaiming the treated water for secondary application. The proper selection of the members of the community is a must to make the operation viable (Jiang et al. 2021). Surface-attached growth of such consortium ensures minimized energy requirement for biomass harvesting (Yang et al. 2020a, b). There are several reports of enhanced performance efficiency of microbial systems in wastewater treatment using biofilm reactors (Gogoi et al. 2021a, b, c; Halder et al. 2020). Random development of algae–bacterial biofilms for wastewater treatment could require 2–10 days for treatment depending on the initial pollutant load (Yang et al. 2020a, b; Biswas et al. 2021). This treatment efficiency can be improved (within 12 h) by using aerobic granular technology (Yang et al. 2020a, b) and moving bed photobioreactor (Kang and Kim 2021). To overcome this limitations of conventional wastewater treatment, proper selection of the community members of a consortium and the right arrangement for biomass inside the reactor for optimum interaction of the pollutants with the microbes is essential (Kang and Kim 2021). Some studies have shown iron to be essential for promoting biomass development in algae–bacterial consortium (Rana and Prajapati 2021). A better understanding of the community profile and their interaction through omics can help in the development of an optimized system (Mu et al. 2020). In continuation with this approach, the current study reports a detailed interpretation of the whole metagenomics analysis of an microalgae–bacterial consortium that efficiently reduced the nutrient content of dairy wastewater generating enhanced biomass rich in lipid, carbohydrate, and protein (Biswas et al. 2021).

10.2 Combining Dairy Wastewater Treatment with Lipid-Rich Microalgae Cultivation: A Case Study

The algae–bacterial consortium in the laboratory was made to treat the dairy wastewater with different initial ammonia, nitrate, and phosphate concentration for 4 months with a hydraulic retention time of 48 h in batch mode. The biomass was collected aseptically and used for microscopic observation as well as sent for whole metagenomics analysis using Illumina HiSeqX platform. The light microscopic images of the different components of the consortium as observed using Leica-DM750 at a 40× magnification are presented in Fig. 10.1.

For whole metagenomics, the biomass was used for isolation of DNA. For DNA library preparation, the DNA was sheared using ultrasonic vibration and 350 base pair fragments were used for further analysis. The detailed procedure described earlier (Gogoi et al. 2021a, b, c; Biswas et al. 2021) was used to generate the data. The contigs over 200 base pairs from the merged assembly were selected for gene prediction using Prodigal software. The overall data obtained from the whole metagenomics analysis is given below as the Krona plot (Fig. 10.2), representing the diversity of the different domains.

The taxonomic diversity further reveals major population of *Desulfovibrio vulgaris*, followed by *Azoarcus* sp., *Chloroflexus aurantiacus*, *Azospirillum bracsilense*, *Curvibacter*, *Pseudomonas*, and *Cyanobactria* in the community (Fig. 10.3).

The gene ontology levels in terms of metabolic function, cellular components, and biological function (Fig. 10.4) reveal major genes belonging to metabolic pathways (Fig. 10.4a). Among them, the genes responsible for the molecular functions were highest, while those for cellular components were least (Fig. 10.4b). Among the molecular functions, genes pertaining to binding and the catalytic activity were the most prominent (Fig. 10.4c). Among the cellular components, the majority of the genes were associated with the cytoplasm (Fig. 10.4d). Genes pertaining to biological functions were maximum for cellular process followed by metabolic process (Fig. 10.4e), indicating a metabolically active community.

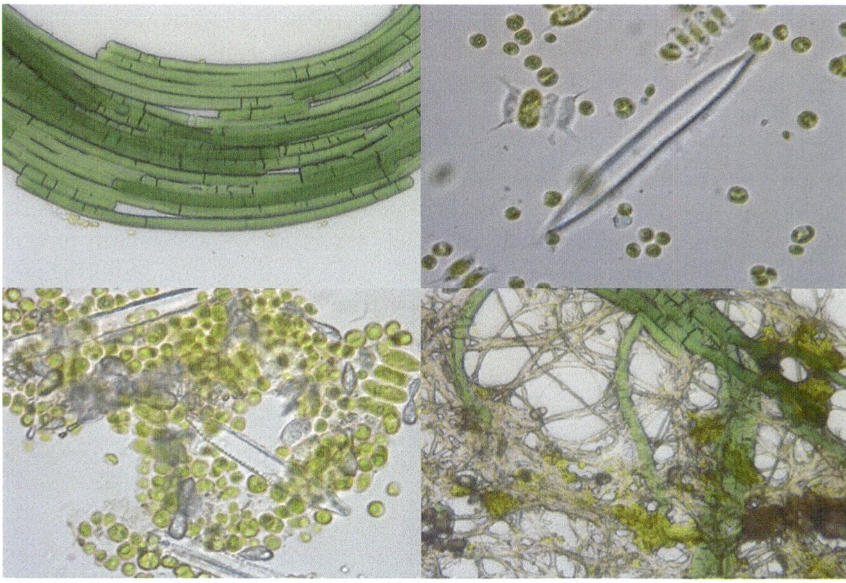

On the basis of their morphologies, the following algal genus were identified: *Chlorella, Chlamydomonas, Spirogyra, Oedogonium, Synedra, Oscillatoria, Desmodesmus* (*Scenedesmus* when spines are absent), *Pinnularia, Phaeodactylum tricornutum, Navicula, Nostocales, Achnanthes.*

Fig. 10.1 Light micrographs of the different algal components present in the algae–bacterial consortium involved in dairy wastewater treatment with lipid-rich algal biomass production

10.2.1 Understanding the Function of the Members in the Consortium

To understand the community profile, their origin, and function from the metagenomic data, a detailed screening of the literature was conducted for the phylum reports in the data. It was found that the community comprised 93% bacteria and the rest being the eukarya (58% of which was the algal population). In addition, to understand the exact mechanism of the microbial consortia and to find their putative function in the milk processing wastewater treatment, it was further classified up to its species level where the molecular mechanisms of each of these reported species were studied. The gross diversity obtained was plotted as a pie cart as represented in Fig. 10.5.

The bacterial population was divided into phylum Proteobacteria, Cyanobacteria, Chloroflexi, Bacteroidetes, Planctomycetes, actinobacter, and firmicutes. Among them Proteobacteria was the dominating bacterial community with 71% of abundance followed by Cyanobacteria (10%). The abundance of Proteobacteria in other wastewater treatment facilities has also been reported. For example, Jiang et al. (2008) reported Proteobacteria as the dominant phylum in the activated sludge at Gaobeidian Wastewater Treatment Plant, Beijing. Another study by Nascimento et al. (2018) on the microbial diversity in sewage sludge of 19 wastewater treatment plants also showed Proteobacteria as the most dominant community. Similarly, in the phylum Proteobacteria, Deltaproteobacteria was the most abundant class (34%) followed by Alphaproteobacteria (24%), Betaproteobacteria (22%), and Gammaproteobacteria (20%). Among these, *Desulfovibrio* belonging to the class Deltaproteobacteria was the most abundant genus followed by *Pseudomonas* and *Azospirillum*. *Desulfovibrio* is a genus of gram-negative sulphate reducing bacteria (SRB), members of which are mostly obligately anaerobic. These are mostly present in aquatic environments having high organic contents, in environments contaminated by metalloids, heavy metals, or some other pollutants or in water-logged soils. They are capable of detoxifying contaminated environments by an indirect chemical reduction of heavy metal via H_2S production (Goulhen et al. 2005). SRB are

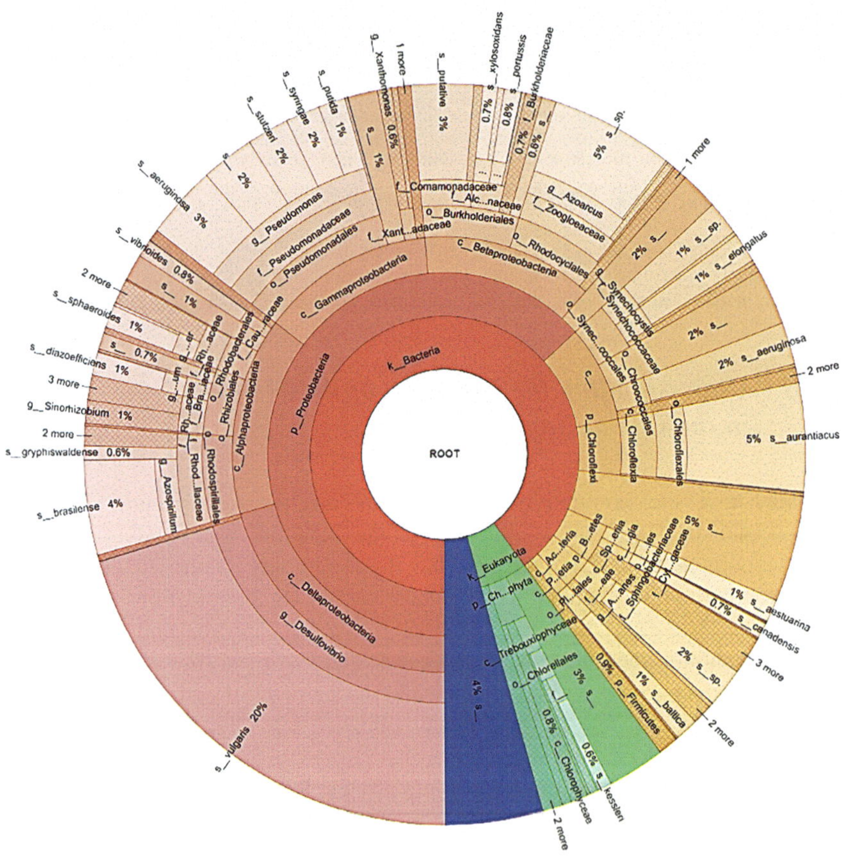

The Krona plot reflecting the diversity of the microbial community

Fig. 10.2 The whole metagenomics data revealing the microbial diversity as Krona plot for the total population and the domain Eukaryota

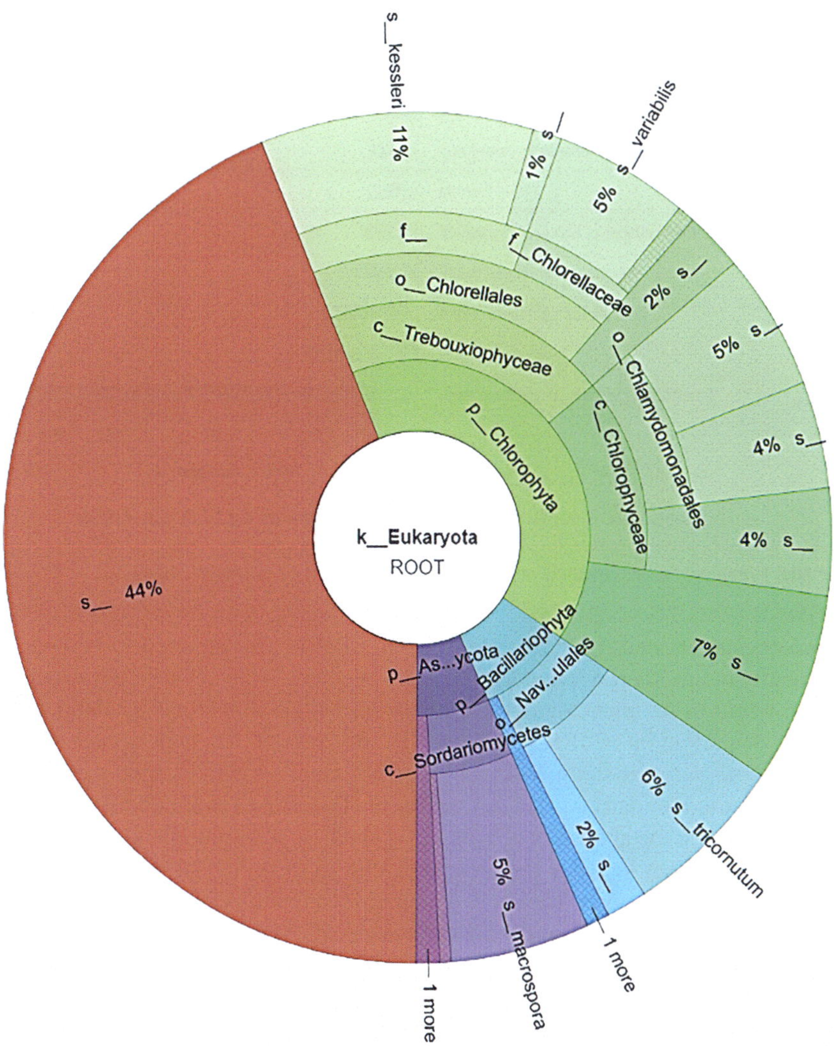

Krona Plot representing the diversity of the Eukaryotic Population

Fig. 10.2 (continued)

ubiquitous in sewer and in wastewater treatment plants (Hao 2003). Although they require a strictly anaerobic environment, yet their presence has been found in a number of aerobic regions (Gibson 1990) such as activated sludge and in aerobic fixed films. They also play a crucial role in the bioremediation of toxic metal ions. The increase in pH due to their metabolism causes toxic metal ions such as nickel, copper, and cadmium to precipitate as metal sulphides in acidic aquatic

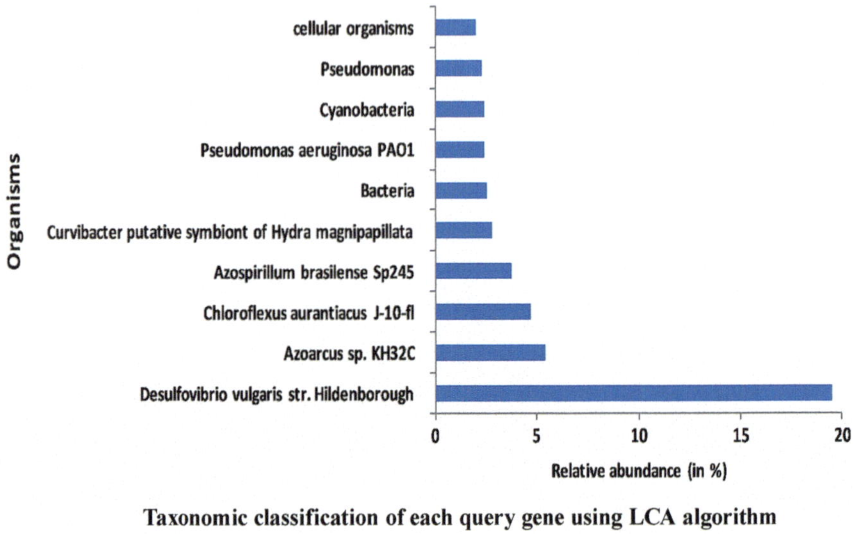

Fig. 10.3 Graphical representation of the community in terms of taxonomic diversity

environments (Heidelberg et al. 2004). Now, SRB is also reported to be present abundantly in the Bheri water at East Kolkata Wetland, which may be the source of *Desulfovibrio* in the microalgal–bacteria community as well (RayChaudhuri et al. 2008; Nasipuri et al. 2010). This may be the reason for high abundance of *Desulfovibrio vulgaris* strain *Hildenborough* (Clark et al. 2007) within the biofilm community used for dairy wastewater treatment. This is used as a model organism to study the energy metabolism of SRBs along with its economic importance in the environment, which includes the bioremediation of heavy metal ions. This genus particularly has been observed within the biofilm communities that are found in the interior surfaces of pipelines and metals (Zhu et al. 2003; Liu et al. 2007). Moreover, dairy wastewater is also reported to contain sulfate and heavy metals, which comes from the washing and packaging processes. It uses hydrogen, organic acids, or alcohols as electron donors for reducing sulfate and also plays a major role in the remediation of heavy metals. For example, chromium, a heavy metal, which is often found in dairy effluent (from the cleaning activity), exists in three oxidation states, among which chromate (Cr (VI)) is very toxic and is a carcinogenic agent that can cause health risks for humans. Chromate is soluble and readily absorbed by cells, while chromium in its reduced form, Cr (III), is insoluble. Trivalent chromium is more stable and sometimes is essential for mammals in trace amounts. *Desulfovibrio* sp. effectively reduces Cr (VI), thus remediates the toxic heavy metal (Goulhen et al. 2005).

In the phylum Alphaproteobacteria, Rhodospirillales and Rhizobiales are the most abundant order with 42% and 32% abundance, respectively, in the present consortium. Rhizobiales are known to have a symbiotic relationship with algae (lichen) and are most dominant in the algal phycosphere (Ramanan et al. 2016).

Since the algal sample was collected from Bheri water, it may be the source of Rhizobiales in the consortium as well. Soil may be another source of Rhizobiales into the system since most of the bacterial species belonging to Rhizobiales are known to be abundant in soil. *Bradyrhizobium diazoefficiens* USDA 110 is the most abundant species present in the community belonging to order Rhizobiales. Now, *Bradyrhizobium diazoefficiens* is a nitrogen fixing bacterium present in soil. Besides nitrogen fixation, it can also perform dissimilatory and assimilatory nitrate reduction and denitrification, which may be the reason for its abundance in the community

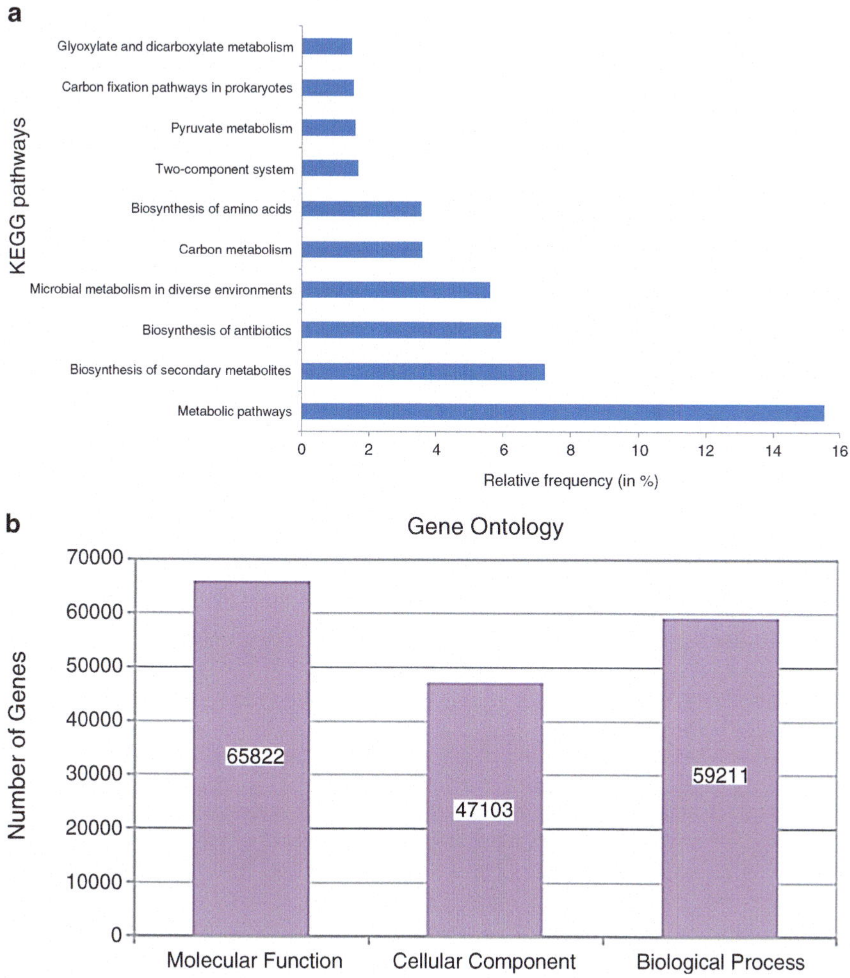

Fig. 10.4 Representation of gene ontology (molecular function, biological function, cellular components) pointing toward a metabolically active consortium. (**a**) KEGG orthology annotations for assessing their involvement in the metabolic pathways. (**b**) Gene ontology levels. (**c**) Gene ontology: molecular function. (**d**) Gene ontology: cellular components. (**e**) Gene ontology: biological function

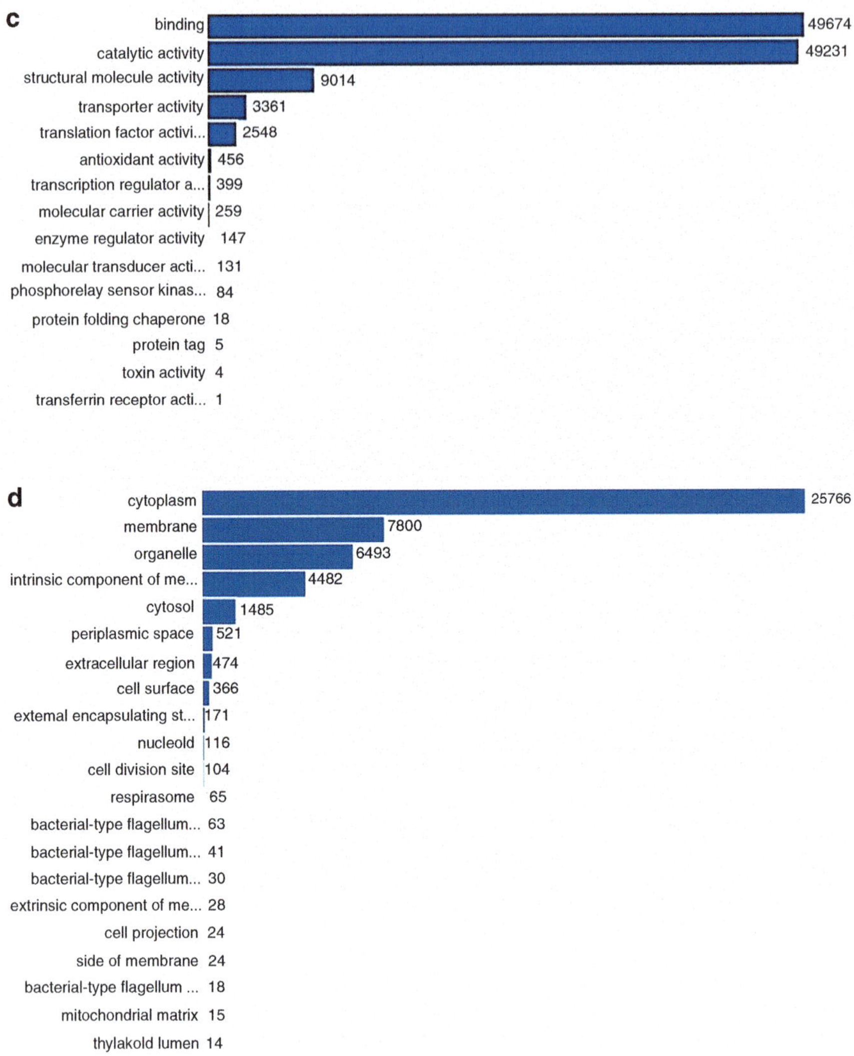

Fig. 10.4 (continued)

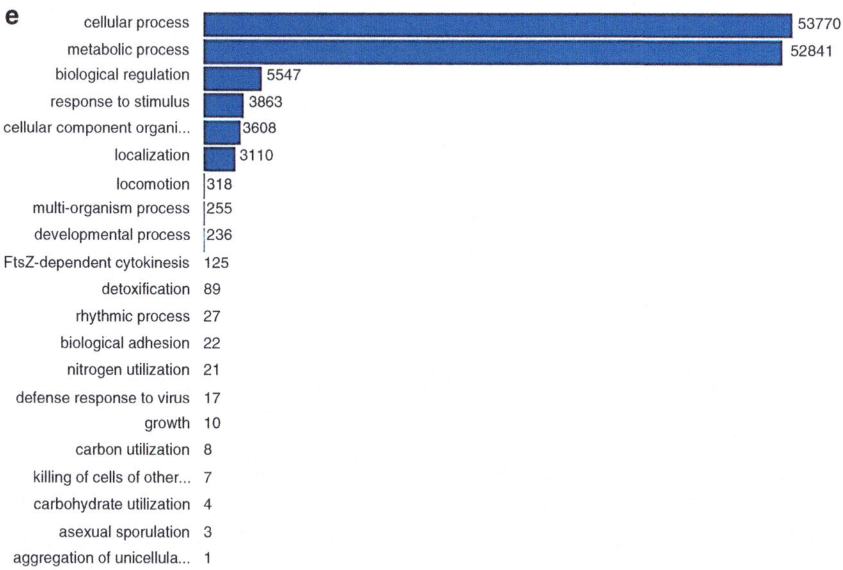

Fig. 10.4 (continued)

since the dairy effluents are rich in nitrate and ammonia. It also takes part in glyphosate and furfural degradation. Rhodospirillales such as *Azospirillum* sp. are capable of removing ammonium and phosphorus ions from synthetic wastewater more efficiently when co-immobilized with microalgae while enhancing microalgal growth (De-Bashan et al. 2002). This could be the reason for the abundance of *Azospirillum* sp. in the present microbial community.

Now, from the phylum Betaproteobacteria, strains belonging to order Rhodocyclales and Burkholderiales were found to be present in the consortium. According to Wang et al. (2020), Rhodocyclalaes is one of the most abundant orders present in different wastewater treatment plants and plays an important role in denitrification. Similarly, bacteria species belonging to Burkholderiales are also reported to be present in different wastewater treatment plants treating industrial and municipal waste (Cydzik-Kwiatkowska and Zielińska 2016). Since the microalgal culture was enriched from Bheri water, it could be the source of bacteria species belonging to Rhodocyclales and Burkholderiales. In the order Rhodocyclales, *Azoarcus* sp. KH32C is the most abundant bacteria present in the community, whereas *Curvibacter* putative symbiont of *Hydra mangipapillata* is the most abundant in order Burkholderiales. *Azoarcus* sp. is mainly reported to participate in degradation of soil contaminants that may be present in the dairy wastewater along with denitrification and nitrogen fixation. Now, *Curvibacter* sp. is known for its phosphate metabolism (Jechalke et al. 2013) and is present in the bioremedial sites and environmental aquatic samples. Dairy wastewater contains phosphate and these organisms may contribute to the phosphate reduction during the treatment process. This finding describes the reason behind the survival of these organisms in the bioreactor environment.

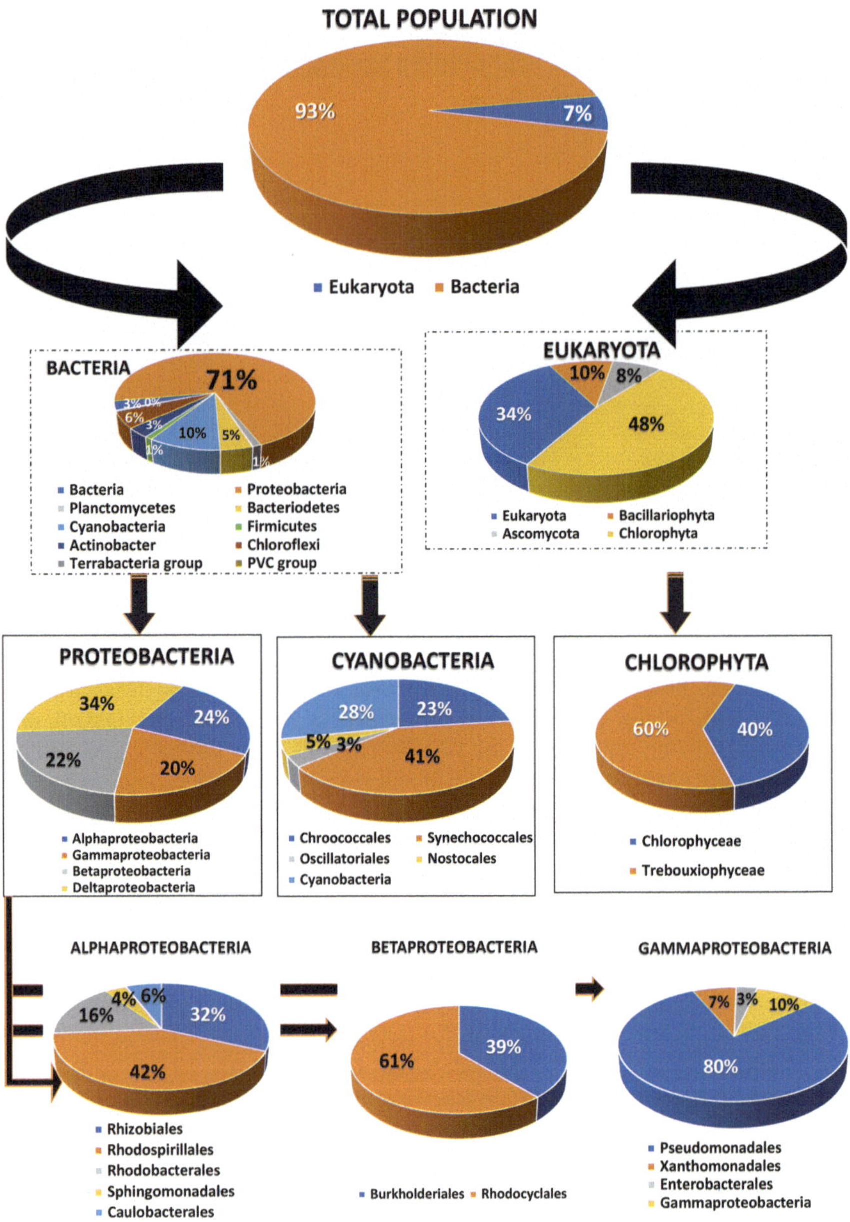

Fig. 10.5 Diversity of the microbial community as revealed through whole metagenomics analysis

In the phylum Gammaproteobacteria, the order Pseudomonadales is the most abundant population with 80% abundance. Al-Wasify et al. (2017) have previously isolated *Pseudomonas aeruginosa* from dairy wastewater and reported its role in degrading organic pollutants present in the effluent. Another study reported the high lipid and COD-degrading capability of *Pseudomonas* sp. isolated from dairy effluent (Sandaruwani et al. 2018). Similarly, the dairy effluent used by Biswas et al. (2021) could be the source of Pseudomonadales in the community. The above studies also justify the role of *Pseudomonas aeruginosa* PAO1 in the community. Another species present in the community was *Pseudomonas putida* strain KT2440, which is generally found in soil and water. It can convert nitrite to ammonia and can accumulate extracellular nitrate/nitrite. It is also capable of degrading naphthalene and can convert styrene oil into biodegradable plastic Polyhydroxyalkanoates (PHA).

Cyanobacteria belong to a group of photosynthetic bacteria and are the most diverse among all. Thus, because of its diversified nature, it is also found in soil and hence it may be added to the dairy effluent through leaching of the soil, which justifies its presence within the microbial community. It is capable of fixing nitrogen, decomposes organic wastes and residues, detoxifies heavy metals and other xenobiotic compounds, suppresses the growth of harmful microorganisms in soil and water (Martins et al. 2011). In dairy wastewater, they play an important role as a bioremediator, which shows a great potential for treating various contaminants like heavy metals either through accumulation or by degradation. For instance, toxic heavy metals (e.g., Mn, Pb, Cu, Zn) can be removed by certain species of cyanobacteria such as *Nostoc muscorum* and *N. rivularis* (Al-Amin et al. 2021). Recently, cyanobacteria are also used efficiently as economical bioremediating agents for treating nutrient-rich dairy wastewaters and then convert these nutrients into microalgae biomass. For instance, cyanobacterial species such as *Oscillatoria* sp. and *Formidium* sp. are capable of reducing nitrogen by 75.22% and 81.69% and phosphate by 86% and 94%, respectively, from the dairy effluent. The *Oscillatoria* sp. also showed high lipid content of about 175.02 mg L^{-1} (Kabariya and Ramani 2018). Moreover, it also helps in the reduction of pollution load while supporting the growth of other microbial populations, which can also reduce the BOD and COD levels of the wastewater.

The microalgae–bacteria consortia also contain a percentage of fungi that includes Ascomycota and Dikarya. Fungal strains such as *Aspergillus* sp. and *Cladosporium* sp. (both belong to division Ascomycota) had previously been isolated from dairy wastewater effluent (Hassan et al. 2020). They were also capable of degrading 72.5% of organic pollutants present in the effluent. Another study also reported the presence of *Altarnaria* sp., *Fusarium* sp., and *Aspergillus* sp. in dairy effluent and their capability of degrading 74.7% of the organics (chlorides, nitrate, calcium, carbonate) present in the dairy wastewater (Al-Wasify et al. 2017).

The community also contains an algal population (belongs to the order Chlorophyta and Bacillariophyta), which has been introduced into the consortium from the Bheri water to promote phyco-remediation of dairy wastewater.

10.2.2 *Understanding Biomass and Lipid Enhancement*

This study reported that the biomass of the microalgae was enhanced by 67% (from 1.38 to 2.3 g L^{-1}) when treated with the ammonia-rich dairy effluent besides showing lipid content enhancement by 42% (from 352.5 to 501 mg g^{-1}) within a hydraulic retention time of 48 h (Biswas et al. 2021) in each batch of water treatment. This approach of using algae is already reported for the production of lipid along with sustained biomass where the alga is grown in media. For instance, *Desmodesmus* sp. showed highest biomass (206–284 mg^{-1} L^{-1} day^{-1}) and lipid (13.85–52.86 mg^{-1} L^{-1} day^{-1}) production when grown in BBM 3N culture medium (Ho et al. 2013). Li et al. (2013) studied the high biomass (6.47 g L^{-1}) and lipid (5.78 g L^{-1}) production by *Chlorella protothecoides* under copper stress when grown in a defined media. Instead of defined media, there are other studies also reporting biomass and lipid enhancement in wastewater. One such example includes the maximum amount of biomass and lipid production by *Chlorella vulgaris* UTEX 265, i.e., 0.165 and 0.058 g^{-1} L^{-1} day^{-1}, respectively, when grown in swine wastewater (Nam et al. 2016). However, for dairy wastewater, the study reported by Biswas et al. (2021) showed the highest biomass and lipid enhancement till date. Hence, we are trying to understand the reason behind this. There are a number of possible reasons explaining the enhancement in biomass, carbohydrate (Cheng et al. 2017), and lipid content, namely nutrient starvation, effect of CO_2 level, cocultivation with bacteria, impact of light intensity, and impact of phytohormones, which are described below.

10.2.2.1 Nutrient Starvation

According to Zhu et al. (2016), nitrogen- and phosphorus-limiting conditions have a positive effect on lipid accumulation in microalgae. For example, under nitrogen starved conditions, the lipid content of *Chlorella* sp. increased remarkably from 15.48% in the initial 40 h to 50.34% at 120 h (Rai et al. 2017). An experiment by Meng et al. (2017) on *N. oceanica* showed that nitrogen deprivation in microalgae causes the breakdown of membrane lipids, which are mainly made up of phospholipids and also leads to the accumulation of triacylglycerol. Another study reported the increase in lipid content of *Tetradesmus obliquus* by 30% when placed under phosphate-limiting conditions (Mandal and Mallick 2009). However, the biomass enhancement was reported to be the result of two-stage cultivation in which microalgae is grown initially under optimum conditions (nutrient rich) for high biomass production and then grown under nutrient-limiting condition for inducing lipid accumulation (Mujtaba et al. 2012; Alvarez-Diaz et al. 2014; Mandal and Mallick 2009). Few studies showed a 41% and 51% enhancement of carbohydrate content in *Chlorella vulgaris* when grown under nitrogen (Dragone et al. 2011) and phosphorus (Brányiková et al. 2011) starved conditions, respectively. Another study reported a 55–65% carbohydrate enhancement in *Spirulina platens* if kept under nitrogen starved conditions (Sassano et al. 2010). Similar to the above studies, the dairy wastewater used by Biswas et al. (2021) was initially rich in ammonia, nitrates, and phosphates, which may have led to an increase in algal biomass, and as the treatment progressed, the reduction or removal of the above-mentioned nutrients from the effluent by the bacterial counterparts (which divide much faster) may have provided the nutrient-limiting condition that enhanced the lipid synthesis or accumulation in the algae. Similar explanations have been reported by other groups too (Yang et al. 2020a, b).

10.2.2.2 CO_2 Levels

Since more than 50% of algal biomass is constituted of carbon, the supply of carbon in the form of CO_2 is crucial for growth of microalgae. Many studies have shown enhanced biomass and lipid production under high CO_2 levels. Tang et al. (2011) reported increased lipid and PUFA accumulation in *T. obliquus* SJTU-3 and *C. pyrenoidosa* SJTU-2 in the presence of high levels of CO_2 (30–50%). However, optimum CO_2 level may vary among microalgal species since several studies have also reported increased biomass and lipid production at low CO_2 concentrations (1–5%). For instance, Fan et al. (2015) reported the maximum biomass (4.3 g L^{-1}) and lipid production (107 mg^{-1} L^{-1} day^{-1}) in *C. pyrenoidosa* at 5% CO_2 concentration. Another study carried out by Ho et al. (2012) showed highest lipid production (22.4%) in *S. armatus* at 2% CO_2. The rapid division of bacteria within the consortium generates higher concentration of CO_2, which stimulates lipid accumulation inside the algal biomass. On the other hand, Cheng et al. (2017) showed 77.6% increase in carbohydrate (DW) in *Chlorella* sp. AE10 when the CO_2 concentration was increased from 1% to 10%. An increment of 2.5-fold of carbohydrate was observed in *Chlorella* sp. when the CO_2 concentration was reduced from 3% to 0.04% (Izumo et al. 2007).

10.2.2.3 Cocultivation with Bacteria

The third reason could be the cocultivation of the microalgae with certain bacterial species that may be present in the consortium. Do Nascimento et al. (2013) showed 30% increase in chlorophyll, lipid content, and biomass in *Ankistrodesmus* sp. when cocultured with *Rhizobium* strain 10II. Another study reported 26% enhancement in the biomass of *Chlamydomonas reinhardtii* when cultivated with *Bradyrhizobium japonicum* in comparison to the pure microalgal strain (Wu et al. 2012). *Bradyrhizobium* sp. is also present in a substantial amount in the present microbial consortia used by Biswas et al. (2021) which could be triggering the biomass and lipid enhancement in its algal counterpart. Grover et al. (2020) showed a two- to three-fold increase in carbohydrates in coculture of microalgae, *C. vulgaris*, with nitrogen fixing bacteria, *Nitrobacter*, than that of monoculture. Similarly, another study by Choix et al. (2012) mentioned the enhancement in carbohydrate accumulation in *Chlorella* sp. when cocultured with *A. brasilense*. Now, *A. brasilense* is also present in substantial amount in the microbial consortia that may be triggering the carbohydrate enhancement in the algal counterpart.

10.2.2.4 Light Intensity

Few studies have also reported that the intensity of light used has some significant influence on biomass and lipid production in microalgae. Optimum light intensity results in the increased production of microalgal lipids (Hallenbeck et al. 2015) since adequate intensity of light is favorable for storing excess photoassimilates, which are then further transformed to chemical energy (Solovchenko et al. 2008). Mandotra et al. (2016) also reported an increased lipid production in *S. abundans* (from 21.20% to 32.77%) as the light intensity was raised from 3000 to 6000 lux. A similar study by Yeesang and Cheirsilp (2011) indicated that the microalga *Botryococcus* sp. showed highest lipid production of 35.9% at 6000 lux. Similarly, the light intensity of 6309 lux used by Biswas et al. (2021) may have also led to the enhancement of microalgal biomass and lipid content. De Philippis et al. (1992) reported an increase in carbohydrate content from 7% to 34% in *Spirulina maxima* when light intensity was increased.

10.2.2.5 Phytohormones

Another factor that plays a significant role in enhancing algal growth and lipid production is the addition of phytohormones (Chu 2017). Li et al. (2017) showed that the lipid production in *Monoraphidium* sp. QLY-1 was enhanced with the addition of phytohormone melatonin. Another phytohormone, indole acetic acid (IAA), also showed to enhance growth and lipid production in *T. obliquus* and other algal species (Salama et al. 2014; Fuentes et al. 2016). Now, phytohormones like IAA are also produced by some bacteria including *Bradyrhizobium* sp. (Wahyudi 2013; Seneviratne et al. 2016) and *Pseudomonas* sp. (Malik and Sindhu 2011), which are also present in the microalgae–bacteria consortium used in the dairy wastewater treatment by Biswas et al. (2021), indicating a possible involvement of the phytohormones in the current study.

A graphical representation of the finding is presented in Fig. 10.6.

10.3 Conclusion

The need to suffice the urge of the ever increasing population is increasing day by day. Thus, to manage the pollution that is being generated, several eco-friendly strategies are emerging. This study was conducted so as to solve the bottlenecks of the mechanism behind microbe-based wastewater treatment methods where this

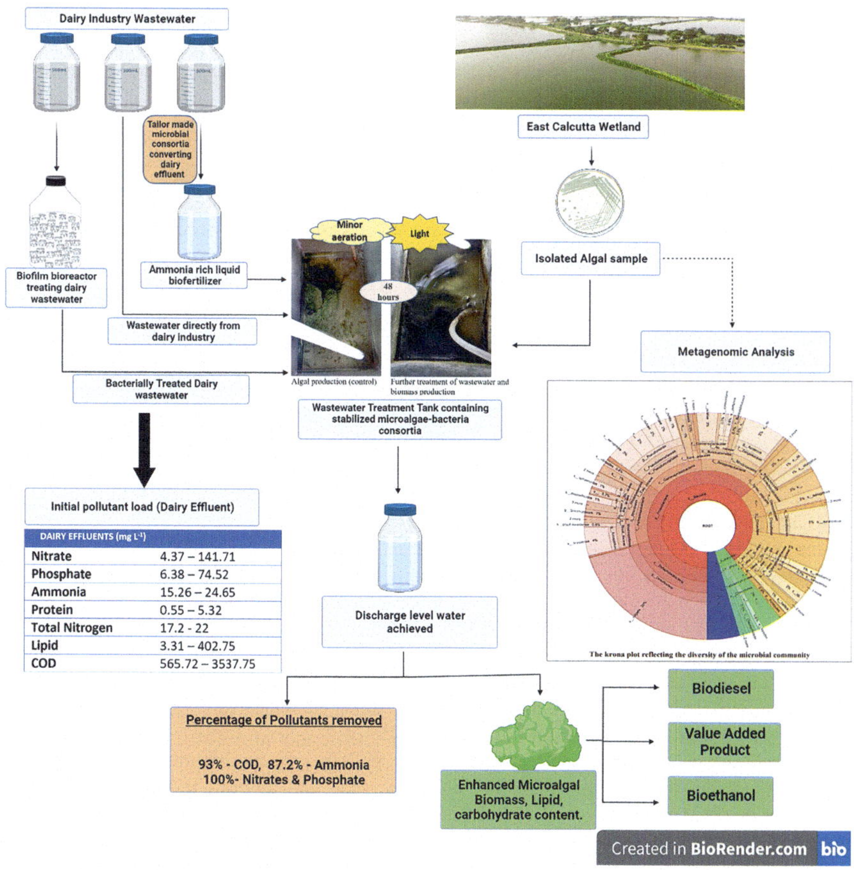

Fig. 10.6 The graphical representation of the proposed process at a glance. The nutrient-rich dairy wastewater gets treated with well-defined microalgae–bacterial consortium, generating enhanced lipid and carbohydrate-rich algal biomass

microalgae–bacteria consortium is capable of phyco-remediation of dairy wastewater with efficient reduction of COD and ammonia and complete removal of nitrates and phosphates from the effluent. The study also reported the enhancement in biomass, carbohydrate, and lipid content of the microalgae, which indicates that this consortium could also be served as a potential feedstock for sustainable biofuel production. In addition to this, it may also be utilized as the source for different value-added products such as biofertilizers as well as animal and aquaculture feed.

Acknowledgments The authors thank Tripura University for the computational facility; Biotechnology Industry Research Assistance Council, Government of India under the Biotechnology Ignition Grant [BIRAC/KIIT0200/BIG-10/17] and Ministry of Education under the Frontier Area of Science and Technology scheme [F.No. 5-1/2014-TS.VII dt 7th Aug 2014] for the financial assistance for conducting the work; and Tripura University and Maulana Abul Kalam Azad University of Technology, West Bengal, for the laboratory facility. The authors are thankful to the students and scholars whose work led to the generation of the metagenomics data that is being interpreted in this chapter.

References

Adams JM, Gallagher JA, Donnison IS (2009) Fermentation study on Saccharina latissima for bioethanol production considering variable pre-treatments. J Appl Phycol 21:569–574

Al-Amin A, Parvin F, Chakraborty J, Kim YI (2021) Cyanobacteria mediated heavy metal removal: a review on mechanism, biosynthesis, and removal capability. Environ Technol 10:44–57

Alvarez-Diaz PD, Ruiz J, Arbib Z, Barragan J, GarridoPerez C, Perales JA (2014) Lipid production of microalga Ankistrodesmus falcatus increased by nutrient and light starvation in a two-stage cultivation process. Appl Biochem Biotechnol 174:471–1483

Al-Wasify RS, Ali MN, Hamed SR (2017) Biodegradation of dairy wastewater using bacterial and fungal local isolates. Water Sci Technol 76:3094–3100

Asmare AM, Demessie BA, Murthy GS (2013) Baseline study on the dairy wastewater treatment performance and microalgae biomass productivity of an open pond pilot plant: Ethiopian case. J Algal Biomass Util 4:88–109

Becker EW (2007) Micro-algae as a source of protein. Biotechnol Adv 25:207–210

Bhandari M, Bhushan S, Rana MS, Raychaudhuri S, Simsek H, Prajapati SK (2020) Algae and bacteria-driven technologies for pharmaceutical remediation in wastewater. In: Shah MP (ed) Removal of toxic pollutants through microbiological and tertiary treatment. Elsevier, pp 373–408. ISBN 9780128210147

Biswas T, Chatterjee D, Barman S, Chakraborty A, Halder N, Banerjee S, Ray Chaudhuri S (2019) Cultivable bacterial community analysis of dairy activated sludge for value addition to dairy wastewater. Microbiol Biotechnol Lett 47:585–595

Biswas T, Bhushan S, Prajapati SK, Ray Chaudhuri S (2021) An eco-friendly strategy for dairy wastewater remediation with high lipid microalgae-bacterial biomass production. J Environ Manag 286:112196

Brányiková I, Maršálková B, Doucha J, Brányik T, Bišová K, Zachleder V, Vítová M (2011) Microalgae—novel highly efficient starch producers. Biotechnol Bioeng 108:766–776

Cantera S, Fischer PQ, Sánchez-Andrea I, Marín D, Sousa DZ, Muñoz R (2021) Impact of the algal-bacterial community structure, physio-types and biological and environmental interactions on the performance of a high rate algal pond treating biogas and wastewater. Fuel 302:121148

Cheng D, Li D, Yuan Y (2017) Improving carbohydrate and starch accumulation in Chlorella sp. AE10 by a novel two-stage process with cell dilution. Biotechnol Biofuels 10(75):75

Choix FJ, Bashan LE, Bashan Y (2012) Enhanced accumulation of starch and total carbohydrates in alginate-immobilized Chlorella spp. induced by Azospirillum brasilense: I. Autotrophic conditions. Enzym Microb Technol 51:294–299

Chu WL (2017) Strategies to enhance production of microalgal biomass and lipids for biofuel feedstock. Eur J Phycol 52:419–437

Clark ME, Edelmann RE, Duley ML, Wall JD, Fields MW (2007) Biofilm formation in Desulfovibrio vulgaris Hildenborough is dependent upon protein filaments. Environ Microbiol 9:2844–2854

Cydzik-Kwiatkowska A, Zielińska M (2016) Bacterial communities in full-scale wastewater treatment systems. World J Microbiol Biotechnol 32:66

De Philippis R, Sili C, Vincenzini M (1992) Glycogen and poly-{beta}-hydroxybutyrate synthesis in Spirulina maxima. J Gen Microbiol 138:1623–1628

De-Bashan L, Legorreta MM, Hernandez J, Bashan Y (2002) Removal of ammonium and phosphorus ions from synthetic wastewater by the microalgae Chlorella vulgaris coimmobilized in alginate beads with the microalgae growth-promoting bacterium Azospirillum brasilense. Water Res 36:2941–2948

Do Nascimento M, Dublan ML, Ortiz-Marquez JC, Curatti L (2013) High lipid productivity of an Ankistrodesmus–Rhizobium artificial consortium. Bioresour Technol 146:400–407

Dragone G, Fernandes BD, Abreu AP, Vicente AA, Teixeira JA (2011) Nutrient limitation as a strategy for increasing starch accumulation in microalgae. Appl Energy 88:3331–3335

Fan J, Xu H, Luo Y, Wan M, Huang J, Wang W, Li Y (2015) Impacts of CO_2 concentration on growth, lipid accumulation, and carbon-concentrating-mechanism-related gene expression in oleaginous Chlorella. Appl Microbiol Biotechnol 99:2451–2462

Fito J, Alemu K (2019) Microalgae–bacteria consortium treatment technology for municipal wastewater management. Nanotechnol Environ Eng 4:4

Foladori P, Petrini S, Nessenzia M, Andreottola G (2018) Enhanced nitrogen removal and energy saving in a microalgal–bacterial consortium treating real municipal wastewater. Water Sci Technol 78:174–182

Fuentes JL, Garbayo I, Cuaresma M, Montero Z, Valle MGD, Vílchez C (2016) Impact of microalgae-bacteria interactions on the production of algal biomass and associated compounds. Mar Drugs 14:100

Gibson GR (1990) A review: physiology and ecology of the sulphate-reducing bacteria. J Appl Bacteriol 69:769–797

Gogoi M, Biswas T, Biswal P, Saha T, Modak A, Gantayet LM, Nath R, Mukherjee I, Thakur AR, Sudarshan M, Ray Chaudhuri S (2021a) A novel strategy for microbial conversion of dairy wastewater into biofertilizer. J Clean Prod 293:126051

Gogoi M, Bhattacharya P, Bhushan S, Sen SK, Mukherjee I, Ray Chaudhuri S (2021b) Aquaculture effluent treatment with ammonia remover *Bacillus albus* (ASSF01). J Environ Chem Eng 9: 105697

Gogoi M, Mukherjee I, Ray Chaudhuri S (2021c) Characterization of ammonia remover *Bacillus albus* (ASSF01) in terms of biofilm formation ability with application in aquaculture effluent treatment. Environ Sci Pollut Res Int 29(41):61838–61855

Gonçalves AL, Pires JCM, Simões M (2017) A review on the use of microalgal consortia for wastewater treatment. Algal Res 24:403–415

Gonzalez L (2000) Increased growth of the microalga Chlorella vulgaris when co-immobilized and co-cultured in alginate beads with the plant-growth-promoting bacterium Azospirillum brasilense. Appl Environ Microbiol 66:1527–1531

Gonzalez-Fernandez C, Sialve B, Molinuevo-Salces B (2015) Anaerobic digestion of microalgal biomass: challenges, opportunities and research needs. Bioresour Technol 198:896–906

Goulhen F, Gloter A, Guyot F, Bruschi M (2005) Cr(VI) detoxification by Desulfovibrio vulgaris strain Hildenborough: microbe–metal interactions studies. Appl Microbiol Biotechnol 71:892–897

Gouveia L, Marques AE, da Silva TL, Reis A (2009) Neochloris oleabundans UTEX #1185: a suitable renewable lipid source for biofuel production. J Ind Microbiol Biotechnol 36:821–826

Grover S, Tirkey SR, Veeramallegowda, Yadav S, Sibi S (2020) Improved biomass through mutualistic co-culturing of *Chlorella vulgaris* with *Nitrobacter* in sewage water. Biotechnology 19:1–9

Gulhane V, Shome SD (2019) Treatment efficiency enhancement of dairy effluent by bioaugmentation using bacterial species. In: Proceedings of sustainable infrastructure development & management (SIDM)

Halder N, Gogoi M, Sharmin J, Gupta M, Banerjee S, Biswas T, Agarwala B, Gantayet L, Sudarshan M, Mukherjee I, Roy A, Ray Chaudhuri S (2020) Microbial consortium–based conversion of dairy effluent into biofertilizer. J Hazard Toxic Radioact Waste 24:04019039

Hallenbeck PC, Grogger M, Mraz M, Veverka D (2015) The use of design of experiments and response surface methodology to optimize biomass and lipid production by the oleaginous marine green alga, Nannochloropsis gaditana in response to light intensity, inoculum size and CO_2. Bioresour Technol 184:161–168

Hao O (2003) Sulfate reducing bacteria. In: Mara D (ed) Handbook of water and wastewater microbiology, pp 459–469. ISBN 0-12-470100-0
Hassan RG, El-Said MA, Mohamed LA (2020) Assessment of some bacterial and fungal strains for dairy wastewater treatment. Egypt J Appl Sci 35:272–283
Heidelberg JF, Seshadri R, Haveman SA, Hemme CL, Paulsen IT, Kolonay JF (2004) The genome sequence of the anaerobic, sulfate-reducing bacterium Desulfovibrio vulgaris Hildenborough. Nat Biotechnol 22:554–559
Ho SH, Chiang CY, Chen CN, Chang JS (2012) Effect of light intensity and nitrogen starvation on CO_2 fixation and lipid/carbohydrate production of an indigenous microalga Scenedesmus obliquus CNW-N. Bioresour Technol 113:244–252
Ho SH, Lai YY, Chiang CY, Chen CN, Chang JS (2013) Selection of elite microalgae for biodiesel production in tropical conditions using a standardized platform. Bioresour Technol 147:135–142
Hu Q, Sommerfeld M, Jarvis E, Ghirardi M, Posewitz M, Seibert M, Darzins A (2008) Microalgal triacylglycerols as feedstocks for biofuel production: perspectives and advances. Plant J 54: 621–639
Izumo A, Fujiwara S, Oyama Y, Satoh A, Fujita N, Nakamura Y, Tsuzuki M (2007) Physicochemical properties of starch in Chlorella change depending on the CO_2 concentration during growth: comparison of structure and properties of pyrenoid and stroma starch. Plant Sci 172:1138–1147
Jagadevan S, Jayamurthy M, Dobson P, Thompson IP (2012) A novel hybrid nano zerovalent iron initiated oxidation—biological degradation approach for remediation of recalcitrant waste metalworking fluids. Water Res 46:23952404
Jechalke S, Franchini AG, Bastida F, Bombach P, Rosell M, Seifert J, Bergen von M, Vogt C, Richnow HH (2013) Analysis of structure, function, and activity of a benzene-degrading microbial community. FEMS Microbiol Ecol 85:14–26
Jia H, Yuan Q (2017) Removal of nitrogen from wastewater using microalgae and microalgae—bacteria consortia. Cogent Environ Sci 31:1–15
Jiang X, Mingchao M, Li J, Lu A, Zhong Z (2008) Bacterial diversity of active sludge in wastewater treatment plant. Earth Sci Front 15:163–168
Jiang L, Li Y, Pei H (2021) Algal–bacterial consortia for bioproduct generation and wastewater treatment. Renew Sust Energ Rev 149:111395
Kabariya JH, Ramani VM (2018) Dairy wastewater treatment by cyanobacteria for removal of nutrients with extraction of high value compounds from biomass. Int J Curr Microbiol App Sci 7:1527–1538
Kang D, Kim K (2021) Real wastewater treatment using a moving bed and wastewater-borne algal–bacterial consortia with a short hydraulic retention time. Processes 9:116
Khoo KS, Chia WY, Chew KW, Shaw PL (2021) Microalgal-bacterial consortia as future Prospect in wastewater bioremediation, environmental management and bioenergy production. Indian J Microbiol 61:262–269
Kolhe AS, Ingale SR, Bhole RV (2009) Effluents of dairy technology. Int Indexed Refereed Res J 5: 459–461
Kumar D, Desai K (2011) Pollution abatement in milk dairy industry. Curr Pharma Res 1:145–152
Kwon G, Kim H, Song C, Jahng D (2019) Co-culture of microalgae and enriched nitrifying bacteria for energy-efficient nitrification. Biochem Eng J 152:107385
Laurens LM, Dempster TA, Jones HD, Wolfrum EJ, Van Wychen S, McAllister JS, Rencenberger M, Parchert KJ, Gloe LM (2012) Algal biomass constituent analysis: method uncertainties and investigation of the underlying measuring chemistries. Anal Chem 84:1879–1887
Li X, Xu H, Wu Q (2007) Large-scale biodiesel production from microalga Chlorella protothecoides through heterotrophic cultivation in bioreactors. Biotechnol Bioeng 98:764–771
Li Y, Mu J, Chen D (2013) Production of biomass and lipid by the microalgae Chlorella protothecoides with heterotrophic Cu(II) stressed (HCuS) coupling cultivation. Bioresour Technol 148:283–292

Li D, Zhao Y, Ding W, Zhao P, Xu JW, Li T, Ma H, Yu X (2017) A strategy for promoting lipid production in green microalgae Monoraphidium sp. QLY-1 by combined melatonin and photoinduction. Bioresour Technol 235:104–112

Liang Z, Liu Y, Ge F (2013) Efficiency assessment and Ph effect in removing nitrogen and phosphorus by algae-bacteria combined system of Chlorella vulgaris and Bacillus licheniformis. Chemosphere 92:1383–1389

Liu H, Huang L, Huang Z, Zheng J (2007) Specification of sulfate reducing bacteria biofilms accumulation effects on corrosion initiation. Mater Corros 58:44–48

Mahapatra DM, Chanakya HN, Ramachandra TV (2014) Bioremediation and lipid synthesis through mixotrophic algal consortia in municipal wastewater. Bioresour Technol 168:142–150

Malik DK, Sindhu SS (2011) Production of indole acetic acid by Pseudomonas sp.: effect of coinoculation with Mesorhizobium sp. Cicer on nodulation and plant growth of chickpea (*Cicer arietinum*). Physiol Mol Biol Plants 17:25–32

Mandal S, Mallick N (2009) Microalga Scenedesmus obliquus as a potential source for biodiesel production. Appl Microbiol Biotechnol 84:281–291

Mandotra SK, Kumar P, Suseela MR, Nayaka S, Ramteke PW (2016) Evaluation of fatty acid profile and biodiesel properties of microalga Scenedesmus abundans under the influence of phosphorus, pH and light intensities. Bioresour Technol 201:222–229

Martins J, Peixe L, Vasconcelos VM (2011) Unraveling cyanobacteria ecology in wastewater treatment plants (WWTP). Microb Ecol 62:241–256

Mamta, Nandan N, Bhushan S, Chaudhuri SR, Simsek H, Prajapati SK (2020) Chapter 15. Algae – and bacteria – driven technologies for pharmaceuticals remediation in wastewater. In: Shah MP (ed) Removal of toxic pollutants through microbiological and tertiary treatment, 1st edn. Elsevier, ISBN 9780128210147, pp 373–408

Menetrez MY (2012) An overview of algae biofuel production and potential environmental impact. Environ Sci Technol 46:7073–7085

Meng Y, Cao X, Yao C, Xue S, Yang Q (2017) Identification of the role of polar glycerolipids in lipid metabolism and their acyl attribution for TAG accumulation in Nannochloropsis oceanic. Algal Res 24:122–129

Mhedhbi E, Khelifi N, Foladori P, Smaali I (2020) Real-time behavior of a microalgae–bacteria consortium treating wastewater in a sequencing batch reactor in response to feeding time and agitation mode. Water 12:1893

Monroy HO, Vazquezz M, Derramadero JC, Guyot JP (1995) Anerobic-aerobic treatment of dairy wastewater with national technology in Mexico: the case of "El Sanz". In: 3rd international symposium on waste management problems in agro-industries, Mexico city, 4–6 October 1995, pp 202–209

Mu R, Jia Y, Ma G, Liu L, Hao K, Qi F, Shao Y (2020) Advances in the use of microalgal-bacterial consortia for wastewater treatment: community structures, interactions, economic resource reclamation, and study techniques. Water Environ Res 11:1496

Mujtaba G, Lee K (2017) Treatment of real wastewater using co-culture of immobilized Chlorella vulgaris and suspended activated sludge. Water Res 120:174184

Mujtaba G, Choi W, Lee CG, Lee K (2012) Lipid production by Chlorella vulgaris after a shift from nutrient-rich to nitrogen starvation conditions. Bioresour Technol 123:279–283

Mulbry W, Kondrad S, Pizarro C, Kebede-Westhead E (2008) Treatment of dairy manure effluent using freshwater algae: algal productivity and recovery of manure nutrients using pilot-scale algal turf scrubbers. Bioresour Technol 99:8137–8142

Nam K, Lee H, Heo SW, Chang YK, Han JI (2016) Cultivation of Chlorella vulgaris with swine wastewater and potential for algal biodiesel production. J Appl Phycol 29:1171–1178

Nascimento AL, Souza AJ, Andrade PAM, Andreote FD, Coscione AR, Oliveira FC, Regitano JB (2018) Sewage sludge microbial structures and relations to their sources, treatments, and chemical attributes. Front Microbiol 9:1462

Nasipuri P, Pandit GG, Thakur AR, Chaudhuri SR (2010) Comparative study of soluble sulfate reduction by bacterial consortia from varied regions of India. Am J Environ Sci 6:152–158

National Research Council (2012) Sustainable development of algal biofuels in the United States. The National Academies Press, Washington, DC
Olguin EJ, Sanchez G, Mercado G (2004) Cleaner production and environmentally sound biotechnology for the prevention of upstream nutrient pollution in the Mexican coast of the Gulf of Mexico. Ocean Coast Manag 47:641–670
Pérez-Nava J, Hernández-Aldana F, Martínez-Valenzuela C, Rivera A (2021) Pseudomonas sp. isolated from wastewater and their interaction with microalgae. J Biochem Technol 12(2): 1–5
Pires JCM, Martins FG, Simões M (2013) Wastewater treatment to enhance the economic viability of microalgae culture. Environ Sci Pollut Res 20:5096–5105
Qi Y, Chen X, Hu Z, Song C, Cu Y (2019) Bibliometric analysis of algal-bacterial symbiosis in wastewater treatment. Int J Environ Res Public Health 16:1077
Rai V, Muthuraj M, Gandhi MN, Das D, Srivastava S (2017) Real-time iTRAQ-based proteome profiling revealed the central metabolism involved in nitrogen starvation induced lipid accumulation in microalgae. Sci Rep 7:45732
Ramanan R, Kim BH, Cho DH, Oh HM, Kim HS (2016) Algae-bacteria interactions: evolution, ecology and emerging applications. Biotechnol Adv 34:14–29
Rana MS, Prajapati SK (2021) Resolving the dilemma of iron bioavailability to microalgae for commercial sustenance. Algal Res 59:102458
RayChaudhuri S, Salodkar S, Sudarshan M, Mukherjee I, Thakur AR (2008) Role of water hyacinth-mediated phytoremediation in waste water purification at East Calcutta wetland. Environ Sci 5:53–62
Renuka N, Sood A, Ratha SK (2013) Evaluation of microalgal consortia for treatment of primary treated sewage effluent and biomass production. J Appl Phycol 25:1529–1537
Salama ES, Kabra AN, Ji MK, Kim JR, Min B, Jeon BH (2014) Enhancement of microalgae growth and fatty acid content under the influence of phytohormones. Bioresour Technol 172:97–103
Sandaruwani A, Kumarasinghe C, Samarakoon D, Ariyadasa TU, Gunawardena SHP (2018) Investigation of the efficacy of dairy wastewater treatment using lipid-degrading bacterial strains. In: Moratuwa engineering research conference (MERCon), pp 362–366
Sassano CEN, Gioielli LA, Ferreira LS, Rodrigues MS, Sato S, Converti A, Carvalho JCM (2010) Evaluation of the composition of continuously-cultivated Arthrospira (Spirulina) platensis using ammonium chloride as nitrogen source. Biomass Bioenergy 34:1732–1738
Seneviratne M, Gunaratne S, Bandara T, Weerasundara L, Rajakaruna N, Seneviratne G, Vithanage M (2016) Plant growth promotion by *Bradyrhizobium japonicum* under heavy metal stress. S Afr J Bot 105:19–24
Sheehan J, Dunahay T, Benemann J, Roessler P (1998) A look back at the U.S. Department of Energy's Aquatic Species Program: biodiesel from algae. National Renewable Energy Laboratory, Golden, CO
Shinkar SB (2013) Comparative study of various treatments for dairy industry wastewater. Int J Curr Eng Technol 3:42–47
Solovchenko AE, Khozin-Goldberg I, Didi-Cohen S, Cohen Z, Merzlyak MN (2008) Effects of light intensity and nitrogen starvation on growth, total fatty acids and arachidonic acid in the green microalga Parietochloris incisa. J Appl Phycol 20:245–251
Su Y, Mennerich A, Urban B (2011) Municipal wastewater treatment and biomass accumulation with a wastewater-born and settleable algal-bacterial culture. Water Res 45:3351–3358
Suresh Kumar K, Dahms H-U, Won E-J, Lee J-S, Shin K-H (2015) Microalgae—a promising tool for heavy metal remediation. Ecotoxicol Environ Saf 113:329–352
Tang D, Han W, Li P, Miao X, Zhong J (2011) CO_2 biofixation and fatty acid composition of Scenedesmus obliquus and Chlorella pyrenoidosa in response to different CO_2 levels. Bioresour Technol 102:3071–3076
Tricolici O, Bumbac C, Postolache C (2014) Microalgae-bacteria system for biological wastewater treatment. J Environ Prot Ecol 15:268–276
Wahyudi A (2013) Production of IAA by Bradyrhizobium sp. WASET 74:152–155

Wang Z, Li W, Li H, Zheng W, Guo F (2020) Phylogenomics of Rhodocyclales and its distribution in wastewater treatment systems. Sci Rep 10:3883
Wu S, Li X, Yu J, Wang Q (2012) Increased hydrogen production in co-culture of Chlamydomonas reinhardtii and Bradyrhizobium japonicum. Bioresour Technol 123:184–188
Yang J, Li Z, Lu L, Fang F, Guo J, Ma H (2020a) Model-based evaluation of algal-bacterial systems for sewage treatment. J Water Process Eng 38:101568
Yang J, Shi W, Fang F, Guo J, Lu L, Xiao Y, Jiang X (2020b) Exploring the feasibility of sewage treatment by algal–bacterial consortia. Crit Rev Biotechnol 40(2):169–179
Yeesang C, Cheirsilp B (2011) Effect of nitrogen, salt, and iron content in the growth medium and light intensity on lipid production by microalgae isolated from freshwater sources in Thailand. Bioresour Technol 102:3034–3040
Yong JJJY, Chew KW, Khoo KS, Show PL, Chang J-S (2021) Prospects and development of algal-bacterial biotechnology in environmental management and protection. Biotechnol Adv 47:107684. ISSN 0734-9750
Zhao C, Bruck T, Lercher JA (2013) Catalytic deoxygenation of microalgae oil to green hydrocarbons. Green Chem 15:1720–1739
Zhu XY, Lubeck J, Kilbane JJ (2003) Characterization of microbial communities in gas industry pipelines. Appl Environ Microbiol 69:5354–5363
Zhu LD, Li ZH, Hiltunen E (2016) Strategies for lipid production improvement in microalgae as a biodiesel feedstock. Biomed Res Int 1-8:1
Zhu S, Huo S, Feng P (2019) Developing designer microalgal consortia: a suitable approach to sustainable wastewater treatment. In: Microalgae biotechnology for development of biofuel and wastewater treatment, pp 569–598

The manufacturer's authorised representative in the EU is Springer Nature Customer Service Centre GmbH, Europaplatz 3, 69115 Heidelberg, Germany. If you have any concerns regarding our products, please contact ProductSafety@springernature.com

Printed and bound by CPI Group (UK) Ltd, Croydon, CR0 4YY
30/09/2026
02225317-0009